Slavic Linguistics in a Cognitive Framework

Marcin Grygiel / Laura A. Janda (eds.)

Slavic Linguistics in a Cognitive Framework

PETER LANG
Frankfurt am Main · Berlin · Bern · Bruxelles · New York · Oxford · Wien

Bibliographic Information published by the Deutsche Nationalbibliothek
The Deutsche Nationalbibliothek lists this publication in the Deutsche Nationalbibliografie; detailed bibliographic data is available in the internet at http://dnb.d-nb.de.

Cover Design:
Olaf Glöckler, Atelier Platen, Friedberg

ISBN 978-3-631-61239-2

www.peterlang.de

Table of contents

Introduction: New perspectives in the study of Slavic languages

Over the past several years, enthusiasm for Cognitive Linguistics has grown rapidly among Slavic linguists and Slavists who previously did not have easy access to linguistic literature written in English. A big impetus for the propagation of cognitive concepts and development of general interest in cognitive frameworks among linguists in Central and Eastern Europe was given by the 10th International Linguistics Conference organized in Poland in 2007 by Prof. Elżbieta Tabakowska. There are also regular conferences held under the auspices of Russian, Polish and Slavic Cognitive Linguistics Associations and Central European Cognitive Linguistics conferences taking place in Croatia which contribute to the popularization of cognitive approaches to Slavic linguistics and cognitive linguistic analysis in general.

The present volume is also proof that a progressive reorientation is taking place in the universities of Central and Eastern Europe, with cognitive linguistics gaining the upper hand. Hence there is a growing demand for cognitive linguistics publications, especially those focusing on Slavic data and preferably written in English. In this respect English offers much wider possibilities for dissemination of ideas and scientific debate, helping to bridge the gap between the Western and Eastern linguistic traditions.

The Slavic group of languages, with 315 million speakers and 13 main languages, is the fourth largest Indo-European sub-branch and one of the major language families of the world. It is also one of the most often studied and most written-about groups. Based on genetic relatedness, the Slavic language group is classified into three branches which also indicate their geographical distribution: the South Slavic branch, with two subgroups – Slovene-Croatian-Bosnian-Serbian-Montenegrin and Bulgarian-Macedonian; the West Slavic branch, with three subgroups – Czech-Slovak, Sorbian, and Lekhitic (Polish and related tongues); and the East Slavic branch, comprised of Russian, Ukrainian and Belarusian. Despite being relatively numerous, the Slavic languages are a closely-knit group due to shared lexicon and grammatical structure.

Within the Indo-European family of languages, Slavic has shown the strongest tendencies to preserve most of the inherited complicated nominal and verbal morphology: person, number, gender, case, tense, mood and aspect distinctions, while undergoing numerous phonological and phonetic changes. This is in contrast to languages such as English which lost most of their morphological categories.

On the whole, Slavic languages are synthetic and rely on inflection as the main method of differentiating grammatical meanings through the use of affixes as well as vowel and consonant alterations. Analytic markers are also present, but visible to a greater degree in nouns than in verbs because many functions of the noun case endings may be taken over by prepositions. This method of analytic expression has led to the almost complete disappearance of the case system in Bulgarian and Macedonian and a slight reduction in the number of cases or their restricted use in other Slavic languages. In Slavic linguistics the grammatical category of case is a widely studied phenomenon and its analysis has yielded a large body of literature focusing not only on the evolution and distinctiveness of the various case forms but also on the meanings associated with individual cases.

Apart from cases and prepositions, another area attracting the attention of linguists studying Slavic languages is aspect. Unlike most other European languages with aspectual distinctions, Slavic marks aspect overtly on all verb forms. The most fundamental aspectual distinction is between perfective and imperfective. Essentially, the perfective aspect schematizes an event as a complete action, while the imperfective aspect presents it as an unfolding, repeated or habitual process. This distinction is widespread among languages of the world, but whereas other languages mark grammatical aspect within inflection (Romance) or by means of periphrastic constructions (English), Slavic languages mark aspect derivationally. As a result, each Slavic verb is inherently marked as either perfective or imperfective and functions as a separate lexical entry.

Moving beyond verbal and nominal categories, another long-standing problem in Slavic linguistics has been linearization and analysis of clitic systems. Clitics are lexical items that are characterized by the lack of prosodic accent or stress and tend to be incorporated into a single phonetic group or phrase with an autonomous stressed word. Slavic has a relatively free word order, but clitics exhibit very rigid positioning patterns. Many Slavists have tried to describe the distribution of clitics in various Slavic languages establishing differences in clitic placement, their types and possible inventory. A major point of theoretical contention concerning the analysis of Slavic clitics has been what role syntactic factors play and what role prosodic ones play, the general assumption being that semantic factors are irrelevant since clitics are devoid of lexical meaning.

It is generally assumed that the similarities among the various languages of the Slavic group indicate that as late as the seventh or eighth century AD they were practically identical. As a consequence, the Slavic languages constitute a more homogeneous group than the languages of some of the other branches and they are still to a large degree mutually comprehensible. This also makes Slavic quite unique and invites linguists to the treasure house of Slavic lexicology. Slavic is much less overlaid by foreign lexical borrowings than other languages and its rich vocabulary has preserved many archaic features. Lexicology is well developed as a named field of descriptive Slavic linguistics and the lexicons of the Slavic languages have been intensively studied for over the last two centuries.

Cognitive Linguistics offers an enticing possibility to approach these time-honored problems from a fresh standpoint and view them in a new dimension. First of all, Cognitive Linguistics refutes the validity of purely formal, logical and algorithmic approaches to language, believing that language is primarily about meaning and studying form for its own sake misses the main point. The task of Cognitive Linguistics is to describe the structure of language in terms of its function and in this respect Cognitive Linguistics is not a break with, but a continuation of a long linguistic tradition focusing on the correlation between linguistic form and its usage.

Students taking an introductory course in linguistics are often taught that language is a means of communication. However, Cognitive Linguistics sees language in a much broader perspective. Language is more than a coding mechanism for sending and receiving messages. Language facilitates interaction with the world. In order to do so, language schematizes meaning. Linguistic meaning is never purely objective; it is instead processed by general cognitive mechanisms. As such, meaning very often turns out to be vague, stereotypical, changeable – in other words, very human.

Cognitive Linguistics is not a unified theory, but rather a large movement constituting an open forum for wide-ranging discussion. This is evident from the articles that appear in the present volume. The research area addressed herein covers a vast spectrum of different topics and reveals a range of theoretical models they follow. All the contributions in this volume, however, are Slavic-oriented and thoroughly grounded in the framework of Cognitive Linguistics.

We present the articles in three thematic sections representing fields of interest and simultaneously corresponding to levels of linguistic organization. The first section, Part I: *Cognitive approaches to grammar*, opens with Laura Janda's paper offering a comprehensive definition of completability as an important parameter in the description of the Russian aspect system, backed up by morphological, syntactic and semantic evidence. Russian aspect is also the subject of Stephen M. Dickey's article in which he applies Langacker's theory of subjectification to the analysis of Russian perfective aspect. Finally, a wider view on the interaction between aspect and modality in Slavic is offered by Dagmar Divjak who uses corpora and statistical techniques to find a variable which best predicts aspect choice in Polish modal constructions.

A number of articles in the present volume analyze Slavic case systems. Starting from the hypothesis that case forms represent symbolic units, Christoph Rosenbaum and Wolfgang Schulze analyze the 'semantic space' of the morphological case architecture in Slovak, paying particular attention to the category of animateness. Mirjam Fried focuses on the fine-grained affectedness distinctions expressed by Czech dative-marked nominals and presents the role of pragmatic factors in motivating the dative of empathy. The Russian instrumental of comparison is discussed in Ekaterina Rakhilina and Elena Tribushinina's article, where they who propose a new integrated approach to instrumentals inspired by construction grammar. Construction grammar is also the inspiration for Marcin Grygiel's account of Serbian and other Slavic realizations of affirmation. Liljana Mitkovska's analysis of possessive-locative constructions in Macedonian expressed implicitly, without any special possessive marker, are the topic of the article that concludes this section.

Part II of the volume, entitled *Cognitive semantics*, includes contributions focused mainly on analysis of the Slavic lexical inventory. Nevertheless, the boundaries between semantics and syntax are to a large extent a matter of convention and the articles reflect this fact. Mario Brdar and Rita Brdar-Szabó explore cases of polysemy triggered by metonymy in their analysis of Slavic and Germanic vocabulary associated with the concepts WOOD and TREE. Diana Prodanović-Stankić presents the results of a corpus-based study of conceptual metaphors and metonymies in Serbian proverbs containing names of animals. The last article in this section, authored by Agnieszka Będkowska-Kopczyk, investigates emotions as causes of actions, illustrating this problem with Slovene and Polish prepositional expressions and cases.

The last section shows wider applications of Cognitive Linguistics in areas which still remain largely unexplored. Part III: *Cognitive discourse analysis and applied linguistics* includes a sample of studies exploring text, spoken discourse, communication and presenting applications of Cognitive Linguistics in translation studies and lexicography. The section starts with Andrej A. Kibrik's analysis of local discourse structure in a Russian corpus of narratives in which children and adolescents retell their night dreams. Metaphorical expressions are a theme in the article authored by Piotr Twardzisz, where Polish legal language is analyzed in the context of Polish-English translation. The volume ends with Danko Šipka's report on metaphor validation in highly polysemous structures in nonstandard Serbo-Croatian vocabulary from the author's *SerboCroatian – English Colloquial Dictionary*.

The volume was realized with the help of a great many people. First of all we would like to thank all the authors for their contributions and cooperation. Thanks are also due to Ute Winkelkötter and Kathrin Kummer of Peter Lang International Academic Publishers for their professional advice and patience. Last but not least, we would like to thank Wojciech Pączek for his technical support.

Marcin Grygiel
Laura A. Janda

I.
Cognitive approaches to grammar

Completability and Russian aspect

Laura A. Janda, University of Tromsø

1. Introduction

Janda (2007a) introduced the term "Completability" as a parameter relevant to the Russian aspect system. Though the term has been used in numerous subsequent publications (Janda 2007b, 2008a, 2008c, 2009, 2010, Janda and Korba 2008), it has not been adequately defined or investigated. The purpose of this article is to revisit the concept of Completability, explore the range of linguistic phenomena associated with Completability, and address the larger theoretical implications of this parameter. This article presents several types of evidence for Completability that have not been previously identified, in addition to summarizing findings from a number of sources.[1]

2. What is Completability?

We commence with a brief definition (1), illustrated with an example (2).

(1) A Completable situation prototypically leads to a result.
(2) *Professor pišet stat'ju.* 'The professor is writing an article.'

Since Completability contrasts with Non-Completability, it is best to define (3) and illustrate (4) Non-Completability as well.

(3) A Non-Completable situation prototypically does not lead to a result.
(4) *Professor rabotaet v kabinete.* 'The professor is working in the office.'

In the Completable situation, the professor begins writing an article on page 1, every bit of engagement brings the article further along on its trajectory, and if the professor continues working, s/he will get to the end of the article and have a result. The Completable situation in (2) is heterogeneous, with a beginning, middle, and end. In the Non-Completable situation (4), working in one's office is interpreted as a homogeneous activity not entailing any trajectory or progress toward a result. Examples (2) and (4) use imperfective verbs, and both verbs can be perfectivized without changing the Completability of the situations, as in (5) and (6):

---

1 The author wishes to thank her employer, the University of Tromsø. Thanks are also due to Anna Baydimirova, who checked all the Russian examples for accuracy. All remaining errors can be attributed to the author.

(5) *Professor napisal stat'ju.* 'The professor wrote an article.'
(6) *Professor porabotal v kabinete nekotoroe vremja, a zatem vyšel na progulku.* 'The professor worked in the office for a while and then went for a walk.'

Although the tense and aspect are different, (5) describes the same Completable situation as (2), where the professor has worked and worked and then come to a conclusion and stopped. (6) is likewise parallel to (4) in terms of Non-Completability: the professor worked for a while and then simply stopped, perhaps in order to do something else, but not because something was completed.

Strictly speaking, Completability is a property of situations, not specific lexical items (verbs).

(7) *Professora pišut stat'i.* 'Professors write articles.'

Note that the lexical items used in (2) and (7) are the same, but they describe different situations. (2) is Completable, whereas (7) is Non-Completable, yielding a gnomic statement of a job description: writing articles is just something that professors do, akin to working in their offices. For convenience we will sometimes refer to verbs as being Completable or Non-Completable, but this is shorthand for saying that the given verbs are associated with Completable or Non-Completable situations.

All verbs have a profile of situations they are associated with. For some verbs, like *pisat'* 'write', *čitat'* 'read', this profile is rather broad, encompassing both Completable and Non-Completable situations. For other verbs this profile is very narrow, restricted only to one type of situation with regard to Completability. Verbs like *krepnut'* 'get stronger' and *gibnut'* 'perish' are strongly associated with Completable situations: it is not possible to engage in the activities of getting stronger or perishing without achieving any result. Verbs like *stonat'* 'moan' and *prygat'* 'hop' are strongly associated with Non-Completable situations: no amount of moaning or hopping in and of itself will lead to a result. Some verbs have a primary association, but can be focused toward the other end of the Completability spectrum. *Rabotat'* 'work', for example, is primarily Non-Completable, but the addition of a prefix can yield a verb associated with Completable situations, as in *pererabotat'* 'revise, edit'.

Russian aspectual morphology aligns with Completability (additional evidence of this alignment is presented in 3.1). Base imperfective verbs may have a range of Completability associations (cf. verbs cited in the previous paragraph), whereas all other types of verbs are much more restricted in the situations they describe. Base perfective verbs like *dat'* 'give' are strongly Completable. Prefixed perfective verbs are either strongly Completable (like *napisat'* 'write' and *perepisat'* 'revise, edit') or strongly Non-Completable (like *postonat'* 'moan for a while' or *zaprygat'* 'start hopping', which quantify some engagement in an action without implying any result). Suffixed perfectives are strongly Non-Completable (like *prygnut'* 'hop once', which again quantizes engagement without result, since the agent winds up back in the same place at the end). Secondary Imperfectives tend to be Completable (cf. *perepisyvat'* 'revise, edit'), but can yield Non-Completable construals (licensing forms like *poperepisyvat'* 'spend some time revising', which do not entail progress toward a result, just quantized engagement, cf. Mehlig 1994, 1996, 1997).

Completability is not a feature. Completability is conceptually related to some features that have been proposed for Russian aspect, such as "totality" and "boundedness", but differs from these features in several ways. For one, features are usually used as synonyms for "perfective", such that perfective verbs "have" the feature, whereas imperfective verbs do not. Thus perfective verbs are labeled "+bounded/+totality", while imperfective verbs are labeled "-bounded/-totality". However, Completability does not designate one aspect or the other; both perfective and imperfective verbs can be either Completable or Non-Completable, and imperfective verbs can in addition be ambiguous for Completability. Secondly, Completability is not reducible to "plus" vs. "minus" values. It is a scalar phenomenon, and many verbs are ambiguous or variable in their Completability, while others show strong associations with one end of the continuum or the other. Thirdly, Completability is conceptually more complex than a feature, for it combines a number of factors, among them an expected result that usually entails a change of state and a heterogeneous structure, with a beginning, middle (where progress is made) and end. All of these factors are present in prototypical examples of Completability, but some may be attenuated. For example, verbs that have connative ("tyring") interpretations, like *ugovarivat'* '(try to) convince', can describe situations that have a beginning, middle and end, but there is no guarantee of the outcome, and there may be no progress in the middle, since the agent may have little or no control over the patient. Still, this verb is primarily Completable, and behaves as such according to many measures (cf. section 3). Bundling of factors is typical of grammatical parameters that are metaphorically motivated (cf. the many properties of aspect motivated by properties of matter in Janda 2004). I argue in Section 4.2 (and elsewhere, cf. Janda 2008a, 2008c, 2009, 2010) that Completability is metaphorically motivated by the physical experience of journeys, which have a beginning (departure point), middle (when progress is made along a trajectory) and an expected conclusion, namely arrival at the destination. Fourthly, Completability is subject to construal, which means that the "same" objective situation can be construed differently, depending upon what factors are most salient for the speaker. A speaker confronted with a professor hard at work could utter (2), (4), or even (7), depending upon whether the speaker wanted to emphasize progress toward a result or not. Finally, there are important theoretical differences between a parameter such as Completability and a +/- feature, and these are addressed in 4.2.

To summarize, Completability is a parameter of situations described by verbs. Completable situations are the metaphorical analogs of journeys in that progress is made toward an expected conclusion. Completability is a scalar phenomenon and can be influenced by construal. Base imperfective verbs can associate with either end of the scale or be ambiguous. Perfective verbs tend to be associated with one end of the scale.

This description of Completability is primarily conceptual. Just because a concept can be identified does not mean that it is grammatically relevant (cf. Talmy 2005 for examples of concepts that are not encoded in grammars). Section 3 presents a variety of evidence for the grammatical relevance of Completability in Russian and Section 4 presents practical and theoretical advantages of recognizing the role of Completability in Russian.

3. Evidence for Completability

If Completability is a valid parameter in Russian, it should align with observable phenomena in the language, such as morphological, syntactic, and semantic patterns. Indeed we find all three types of evidence for Completability, indicating that Completability is a systemic factor in the Russian verbal lexicon. Most of this evidence has surfaced since Completability was first proposed (Janda 2007a), indicating that Completability is significant beyond the domain where it was initially recognized. This section provides the first attempt at a comprehensive list of Completability phenomena in Russian.

3.1. *Morphological evidence*

As mentioned in Section 2, Completability is tangible in certain patterns of morphological marking. Indeed, it was the differentiation of various types of perfective verbs in Russian that led to identification of Completability in the first place (3.1.1.). The observations of Completability in relation to morphological marking here are more complete than those found in previous works on the topic.

3.1.1. *Various types of perfectives*

Traditional analyses of Russian aspect yielded the "pair" model, which presumes that Russian verbs exist in aspectual pairs, each containing a single imperfective verb and a single perfective verb (cf. Vinogradov 1938, Šaxmatov 1941, Bondarko 1983, Čertkova 1996, Zaliznjak & Šmelev 2000). Janda 2007a presents the alternative "cluster" model, according to which Russian verbs exist in clusters that may contain from zero to four distinct types of perfectives. The cluster model elaborates on the pair model in that pairs, where they exist, are either a type of cluster structure, or, more frequently, pairs are embedded in larger cluster structures containing multiple types of perfectives. In addition, the cluster model allows for structures that do not contain pairs at all. Overall, the cluster model is more accurate than the pair model in that it provides a comprehensive account of the real complexity of the Russian aspect system.

Crucial to the cluster model is the recognition of four major types of perfectives, based on their semantics and morphological behavior. The main division between the types of perfectives is motivated by Completability, since two types of perfectives are Completable and two are Non-Completable. The four types of perfectives are presented here, along with their Completability status and descriptions of the morphological behavior associated with each type:

Natural Perfectives like *napisat'* 'write' and *svjazat'* 'tie' generally correspond to the perfective partners of imperfective verbs according to the pair model and share the same lexical meaning as their imperfective correlates *pisat'* 'write' and *vjazat'* 'tie'. Natural Perfectives are strongly Completable, since they refer to the natural conclusion of a goal-oriented situation. Most Natural Perfectives are formed via prefixation of base imperfective verbs, and such Natural Perfectives often resist the formation of secondary imperfectives. Natural Perfectives can be formed from imperfectives that are either strongly Completable themselves (cf. perfective *okrepnut'* from *krepnut'* 'get stronger') or are ambiguous as to Completability (cf. perfective *napisat'* from *pisat'* 'write'). There are a few base perfective verbs, usually referring to punctual events such as *dat'* 'give' that serve as Natural Perfectives and derive the corresponding imperfective via suffixation, as in *davat'* 'give'. In addition, there are non-prototypical morphological relationships between Natural Perfectives and corresponding imperfectives, including homonymy (with bi-aspectual verbs like *likvidirovat'* 'liquidate', where one form is both perfective and imperfective), suppletion (e.g., *skazat'* 'say' perfective and *govorit'* 'say' imperfective), and semi-suppletion (e.g., *brosit'* 'throw' perfective and *brosat'* 'throw' imperfective).

Specialized Perfectives like *perepisat'* 'revise' and *razvjazat'* 'untie' are lexically distinct from their corresponding base verbs because the prefixes direct the action in a way not inherent in the base verb. Specialized Perfectives are strongly Completable, for they refer to situations that are goal-oriented and conclude with a result (a revised manuscript or an untied knot). Specialized Perfectives can be formed from all three types of imperfectives: Completable imperfectives (such as *blokirovat'* 'block' which forms *razblokirovat'* 'unblock'), ambiguous imperfectives (such as *pisat'* 'write', which forms *perepisat'* 'revise'), and Non-Completable imperfectives (such as *rabotat'*, which forms *pererabotat'* 'revise'). There are, however, some Non-Completable imperfectives that cannot form Specialized Perfectives (such as *stonat'* 'moan'). Unlike Natural Perfectives, Specialized Perfectives nearly always form secondary imperfectives (such as *perepisyvat'* 'revise' and *razvjazyvat'* 'untie'; but note that some verb stems resist the formation of any secondary imperfective, like *kipjatit'* 'boil', which forms no secondary imperfective from Specialized Perfectives like *perekipjatit'* 'overboil').

Complex Act Perfectives like *postonat'* 'moan for a while' or *zaprygat'* 'start hopping' quantize engagement in an action that lacks goal-directedness. Complex Act Perfectives focus on starting (ingressives) or stopping (terminatives) an activity or engaging in an activity for a while (delimitatives and perduratives). Since Complex Act Perfectives entail neither progress nor a result, they are strongly Non-Completable. Only imperfective base verbs that have a Non-Completable construal can form Complex Act Perfectives. Thus base verbs that are either Non-Completable (like *stonat'* 'moan') or ambiguous for Completability (like *čitat'* 'read') can form Complex Act Perfectives (*postonat'* 'moan for a while' and *počitat'* 'read for a while'), whereas strongly Completable base imperfectives tend not to form Complex Act Perfectives (such perfectives are not formed from verbs like *krepnut'* 'get stronger' and *likvidirovat'* 'liquidate'). Complex Act Perfectives are formed by means of prefixation (primarily *po-*, *pro-*, and *za-*, though other prefixes such as *ot-*, *raz-* and *na-* also appear), and strongly resist the formation of secondary imperfectives.

Table 1. The four types of perfectives in Russian

Type of perfective	Completability of perfective	Completability of base imperfective	Most typical morphological marking	Presence of secondary imperfective
Natural Perfective	Completable	Must have Completable construal	Prefix	Some
Specialized Perfective	Completable	No restriction	Prefix	Yes
Complex Act Perfective	Non-Completable	Must have Non-Completable construal	Prefix	No
Single Act Perfective	Non-Completable	Must have Non-Completable construal	Suffix	No

Single Act Perfectives like *prygnut'* 'hop once' describe a single cycle of repeated events that take place without focus on a result. Like Complex Act Perfectives, Single Act Perfectives are formed from base imperfectives that have a Non-Completable construal and resist the formation of secondary imperfectives. Single Act Perfectives are formed primarily via *-nu* suffixation, though the prefix *s-* can also form such perfectives, particularly from indeterminate motion verbs (such as *xodit'* 'walk', which forms the Single Act Perfective *sxodit'* 'walk someplace and come back once').

Table 1 summarizes the relationship between Completability and morphological markings in Russian aspect. Note that the table describes the most typical patterns (deviations are noted above), and that restrictions on the Completability of the base imperfective verb do not exclude ambiguity (situations where a verb has both Completable and Non-Completable construals, as in *čitat'* 'read').

3.1.2. *Motion verbs*

The Russian verbs of motion give the most concrete evidence for Completability, since this parameter is marked directly in their stems. Determinate stems, such as *idti* 'walk', describe Completable situations with goal-directed progress, namely proceeding along a path or toward a destination. Indeterminate stems, such as *xodit'* 'walk', describe Non-Completable situations such as non-directed motion and ability to move. When indeterminate stems are used with destinations, as in *oni xodili v kino* 'they went to the movies', they refer to round-trips: achievement of the destination is annulled since the subject returns to the starting point. Often indeterminate stems refer to a series of round-trips, as in *deti xodjat v školu* 'the children go to school'. The marking of morphologically distinct determinate and indeterminate stems was considerably more widespread in Old Russian and Old Church Slavonic, and it is likely that this distinction played a crucial role in the development of Russian aspect (cf. Van Wijk 1929, Vaillant 1948, Mayo 1985, Bermel 1997, Andersen 2006, Dickey 2007). The presence of determinate (Completable) vs. indeterminate (Non-Completable) stems for a portion of the Old Russian verbal lexicon including non-motion verbs, such as *glasiti* vs. *glašati* 'call' and *mesiti* vs. *měšati* 'mix' points to a general cognitive strategy according to which many activities could be construed as engagement either in a goal-oriented or in a non-goal oriented (and often repetitive) fashion. Determined (a.k.a. Completable) stems refer to specific acts of delivering a message or getting a certain set of ingredients combined, whereas indeterminate (Non-Completable) stems refer to repeated acts with no focus on progress or result. In Janda 2008c I suggest that the grammaticalization of aspect in Russian led to the extension of determinacy to the entire verbal lexicon. In its extended use, determinacy developed into Completability, and developed a new means of expression, namely in terms of the types of perfectives that could be formed (Natural and Specialized Perfectives being Completable, and Complex and Single Act Perfectives being Non-Completable). When prefixation took over as the primary means for expressing Completability, the former means of expression, via distinct determinate and indeterminate stems, could contract, and now it is restricted only to a handful of motion verbs.

Completability is especially tangible in the clusters of the modern Russian verbs of motion (Janda 2009, 2010). Compare this pattern with that found in Table 1. Natural Perfectives such as *pojti* 'walk' are formed from the determinate (Completable) stem *idti* 'walk'. Specialized Perfectives, such as *prijti* 'arrive' and *iznosit'* 'wear out' are formed from both determinate (Completable) stems like *idti* 'walk' and indeterminate (Non-Completable) stems like *nosit'* 'carry'. Complex Act Perfectives such as *poxodit'* 'walk for a while' and Single Act Perfectives such as *sxodit'* 'walk someplace and come back once' are formed only from indeterminate (Non-Completable) stems like *xodit'* 'walk'. This pattern of formations is entirely parallel with that found in the rest of the verbal lexicon, with the difference that Completability is unambiguous for motion verbs, since each verb is marked for only one end of the continuum or the other.

3.1.3. *Bi-aspectual verbs*

Completability yields a prediction for bi-aspectual verbs that is confirmed by empirical study. Bi-aspectual verbs express both the Natural Perfective and the corresponding imperfective with the same form, as we see in *likvidirovat'* 'liquidate'. In order for a verb to function as a Natural Perfective without the support of a prefix, it must be strongly associated with Completable situations. Note that although many borrowed verbs are bi-aspectual in Russian, about 40% of foreign verbs enter the language as ordinary imperfective base verbs and use prefixes to form their Natural Perfectives, as we see in *fil'trovat'* 'filter', which has *profil'trovat'* as its Natural Perfective (contra claims by some scholars that all foreign verbs are bi-aspectual; see Janda 2007b). In other words, the criteria for bi-aspectuality are semantic: only verbs that are strongly Completable are bi-aspectual. Since we know that bi-aspectual verbs are strongly Completable, we should expect them to strongly resist the formation of Complex Act Perfectives, since these perfectives require the presence of a Non-Completable construal. A statistical analysis of 555 foreign verbs in Russian shows that there is indeed a strong correlation between bi-aspectuality and resistance to formation of Complex Act Perfectives (see Janda 2007b for data and full discussion).

3.1.4. *Summary of morphological evidence*

Completability yields a major division among types of perfective verbs that can be formed, with Natural and Specialized Perfectives associated with Completable situations on the one hand, and Complex Act and Single Act

Perfectives associated with Non-Completability on the other hand. Completability is directly marked in the determinate vs. indeterminate stems of motion verbs and accurately predicts what types of perfectives are formed from which motion verb stems. Finally, we see that Completability leads to a correct prediction concerning bi-aspectual verbs and the formation of Complex Act Perfectives.

3.2. *Syntactic evidence*

Whereas the morphological evidence for Completability presented in 3.1 is motivated by the framework of cognitive linguistics, one can also find syntactic evidence for Completability motivated by the work of other scholars, mainly in the framework of generative linguistics. These parallels have not been drawn before. The scholars cited in this section have identified patterns in relation to limited subsets of verbs, concentrating only on the behaviors of Specialized Perfectives and Complex Act Perfectives. I add to these observations a) the claim that the syntactic patterns are part of the overall phenomenon of Completability, b) corpus evidence that confirms their observations (which were based on speaker intuitions), and c) additional corpus evidence that extends these observations to the remaining types of verbs, namely Natural Perfectives and Single Act Perfectives. This research is based on data from the Russian National Corpus (www.ruscorpora.ru; RNC; all examples from this section are culled from this source).

3.2.1. *Passivization*

Baydimirova (2009) and Bondarko and Bulanin (1967: 158-163) point out that certain types of Russian verbs do not passivize despite the fact that they are transitive, among them delimitatives prefixed in *po-* and ingressives prefixed in *za-*.[2] Both types of verbs are Complex Act Perfectives, representing Non-Completable situations. This observation is confirmed by data in the RNC. Examples (8) and (9) illustrate the use of both types of verbs in transitive active constructions:

(8) *Smex pomog mne, ja razogrel sup, vyučil uroki,* ***počital knigu****, starajas' otgonjat' mysli o Mirone.*

2 There are in addition some isolated examples of verbs that do not passivize, such as *minovat'* 'pass' (which is bi-aspectual, and thus a Natural Perfective when used perfectively), however, these are exceptions rather than broad classes like the delimitatives and ingressives.

'Laughter helped me, I warmed up the soup, studied my lessons, and **read the book for a while**, trying to ward off thoughts of Miron.' [Al'bert Lixanov. Kikimora (1983)]

(9) *A potom, oborvav sebja na poluslove,* ***zapel pesnju*** *tex samyx pampasov ili, možet byt', prerij i prinjalsja izobražat' kovboja verxom na lošadi.*
'And then, interrupting himself in the middle of a word, **he started singing a song** of those very pampases, or perhaps prairies, and began to portray a cowboy mounted on a horse.' [Irina Medvedeva, Tat'jana Šiškova. Diktatura bezumija (2004) // "Naš sovremennik", 2004.02.15]

However, there are no examples of passivization of such phrases in the RNC: one cannot say **pesnja byla zapeta* 'the song was started singing' or **kniga byla počitana* 'the book was read for a while'. Futhermore, it seems that this generalization holds for all types of Non-Completable perfectives, though it is harder to find good examples of transitive uses for most of them. Perduratives prefixed in *pro-* can have accusative objects, although these are limited to time expressions and thus not prototypical direct objects, as in (10). Single Act Perfectives are largely intransitive (*prygnut'* 'hop once') or govern objects in the instrumental (*kivnut'* golovoj 'nod one's head once') or genitive case (*xlebnut' supa* 'take one gulp of soup'), but there are a few examples of transitives with accusative objects, as in (11).

(10) *Odnaždy muž ne vzjal pevicu na rybalku (nemudreno: ona dolžna byla vot-vot rodit'), i ta* ***ves' den' proplakala****.*
'Once the husband did not take the songstress with him to go fishing (it wasn't a good idea: she was just about to give birth), and she **cried the whole day**. [Natal'ja Skljarova. Esli by u medvedja bylo ruž'e (2002) // "Večernjaja Moskva", 2002.02.07]

(11) *Nu, ja* ***kusnul jabloko****.*
'Well, I **bit the apple once**.' [Vasilij Šukšin. Pečki-lavočki (1970-1972)]

These transitives do not passivize either: there are no examples of the type **ves' den' byl proplakan* 'the whole day was cried through' or **jabloko bylo kusnuto* 'the apple was bitten once'. We can conclude that Completability plays a role in Russian passivization, namely that Non-Completable perfectives do not passivize, even if they are transitive.

3.2.2. *VP internal vs. VP external*

Svenonius (2004a, 2004b, forthcoming) and Ramchand (2004) describe a series of differences between the syntactic behavior of "lexical" prefixes (cf. also Spencer & Zaretskaya 1998) and "superlexical" prefixes. Lexical prefixes are present in the formation of perfective verbs that are semantically distinct from their base imperfectives, as for example the use of *za-* and *o-* to form *zapit'* 'wash down' from *pit'* 'drink' or *zapisat'* 'record, enroll' and *opisat'* 'describe' from *pisat'* 'write'. Superlexical prefixes are present in the formation of perfective verbs that quantize an action, as for example the use of *po-* to form *popit'* 'drink for a while' from *pit'* 'drink' or *popisat'* 'write for a while' from *pisat'* 'write'. Mutatis mutandis, lexical prefixes are present in Specialized Perfectives, whereas superlexical prefixes characterize Complex Act Perfectives. Since Specialized Perfectives are Completable and Complex Act Perfectives are Non-Completable, I argue that the pattern of syntactic differences described by Svenonius and Ramchand is also indicative of differences in Completability. In addition, I present further evidence that the same syntactic differences are also relevant for the other types of perfectives in Russian, namely Natural Perfectives and Single Act Perfectives.

Svenonius and Ramchand demonstrate that lexical prefixes are low in the derivational tree, VP internal, and can be idiomatic (i.e., the meaning of the perfective verb is not strictly predictable from the prefix, as we see in *zapit'* 'wash down', *zapisat'* 'record, enroll' and *opisat'* 'describe'). I concur with these claims based on the Specialized Perfectives and suggest that the same observations hold also for the other group of Completable perfectives, namely Natural Perfectives such as *vypit'* 'drink' and *napisat'* 'write'. Indeed, according to some accounts, the relationship between a Natural Perfective and its imperfective correlate is so close that they can be considered inflectional forms of a single paradigm (cf. discussion in Janda 2007c). Furthermore, it is the case that many imperfectives have more than one Natural Perfective, and generally there is no straightforward way to predict which of the meanings of the imperfective is focused in each Natural Perfective. The imperfective *est'* has three meanings, with three corresponding Natural Perfectives, *poest'* 'get something to eat', *s"est'* 'eat up', and *raz"est'* 'corrode', but there is no apriori way to know which prefix should go with which meaning.

Superlexical prefixes are located high in the derivational tree, VP external (serving like adverbs) and highly predictable in their meanings (i.e., a verb prefixed with superlexical *po-* will always mean 'do X for a while'). Again I concur with these observations for Complex Act Perfectives and suggest that they are also valid for the other group of Non-Completable perfectives, namely the Single Act Perfectives, such as *prygnut'* 'hop once' and *kusnut'* 'bite once', since these perfectives always quantize an action by selecting a single cycle of a repeatable behavior.

Svenonius (2004b) futher suggests two symptoms that correlate with the different positions of lexical and superlexical prefixes, involving nominalization and ordering of prefixes. Verbs with lexical prefixes can nominalize, whereas verbs with superlexical prefixes as a rule resist nominalization. I concur with this observation, since a Specialized Perfective like *opisat'* 'describe' can have a nominalization such as *opisanie* 'description', whereas Complex Act Perfectives lack such nominalizations: **popisanie* 'writing for a while' is not attested in the RNC. Furthermore, this observation can be extended to the remaining groups of perfectives. Natural Perfectives, which are Completable (like Specialized Perfectives), can be nominalized, as we see in *napisanie* 'writing', but Single Act Perfectives, which are Non-Completable (like Complex Act Perfectives) resist nominalization: **kusnutie* 'biting once' and **prygnutie* 'hopping once' are not attested.

When prefixes are stacked, the derivational position of the superlexical prefix dictates that it appears outside the lexical prefix, as we see in *pozapisyvat'* 'record for a while', where the superlexical *po-* is outside the lexical *za-*. This observation is valid, but cannot be extended to the other types of perfectives due to morphological restrictions.[3]

3.2.3. *Predicational structure*

Both Svenonius and Ramchand identify a set of differences in argument structure associated with lexical vs. superlexical prefixes. These differences include transitivity and the type of objects allowed as well as added elements of argument structure. Lexical prefixes tend to add transitivity: base verbs that are either intransitive or of variable transitivity can become transitive. Thus imperfective intransitive *dumat'* 'think' is transitivized in a Specialized Perfective like *pridumat'* 'think up, invent'. This observation also holds for Natural Perfectives: imperfective *pisat'* 'write' can be either transitive or intransitive, but perfective *napisat'* 'write' is transitive. Superlexical prefixes do not increase transitivity: *popisat'* 'write for a while' and *popit'* 'drink for a while' are not more transitive than their corresponding imperfectives. Again, this observation can be extended to Single Act Perfectives, which are strongly intransitive.

In terms of argument structure, lexical prefixes can change the type of objects that appear and can add elements of structure, whereas superlexical prefixes cannot effect such changes. As a base imperfective, *pit'* 'drink' can take an accusative (12) direct object that refers to a beverage, but cannot govern this object in the instrumental case, so it is not possible to say **on pil vodoj* 'he drank by means of water'. But the Specialized Perfective *zapit'* 'wash down' takes an accusative direct object that refers to a non-beverage (food or medicine) and has an additional object in the instrumental case that does refer to a beverage (13). The Complex Act Perfective *popit'* 'drink for a while' has the same argument structure (14) as the imperfective (12).

3 Natural Perfectives often resist both secondary imperfectivization and further prefixation, and since Single Act Perfectives are typically marked with a suffix, it would not be possible to say whether the suffix or the prefix is more "external" in examples like *zaxlopnut'* 'slam shut once'.

(12) *Ničego ne el, tol'ko* ***pil vodu*** *iz grafina.*
'He didn't eat anything, just **drank water** from the carafe.' [Fridrix Gorenštejn Kuča (1982) // "Oktjabr'", № 1, 1996]

(13) *— My umoljaem tebja: uspokojsja! — vtalkivaja mne v rot piljuju i zastavljaja* ***zapit' ee vodoj****, pričitala mama. — Vyxod, bessporno, est'.*
'— We beg you: Calm down! — stuffing a pill in my mouth and forcing me **to wash it down with water**, mom wailed. —Surely there is a solution.' [Anatolij Aleksin. Razdel imuščestva (1979)]

(14) *A my by prosto* ***popili*** *s toboj* ***čaj*** *i legli spat'.*
'We would just spend some time **drinking tea** with you and then go to bed.' [Sergej Kozlov. Pravda, my budem vsegda? (1969-1981)]

Similarly, imperfective *pisat'* 'write' has an optional accusative direct object that refers to a document (15), whereas the Specialized Perfective *zapisat'* 'record, enroll' can have an accusative direct object that refers to a person plus an additional argument consisting of a directional phrase, such as *v školu*, literally 'into the school' (16), but this structure is not possible for the base imperfective. The Complex Act Perfective reflects the same argument structure (17) as the imperfective (15).

(15) *Ja soznatel'no* ***pisal roman****, no roman osobogo roda — sociologičeskij.*
'I **wrote a novel** intentionally, but it was special kind of novel — a sociological one.' [Aleksandr Zinov'ev. Russkaja sud'ba, ispoved' otščepenca (1988-1998)]

(16) *God nazad* ***zapisala ego v školu*** *iskusstv.*
'A year ago **I enrolled him in** art **school**.' [Pomogite, požalujsta! (2002) // "Domovoj", 2002.09.04]

(17) *Nu, čast' vremeni ja potraču na uničtoženie nedopisannyx stixov; nemnožko ešče* ***popišu*** *prozu.*
'Well, I will spend some of the time destroying unfinished poems; in addition I **will spend some time writing** prose.' [Julij Danièl'. Pis'ma iz zaključenija (1966-1970)]

Data from the RNC support Svenonius' and Ramchand's division between lexical and superlexical prefixes based on syntactic behavior. Whereas Natural Perfectives do not change the argument structure (18-19) of the imperfective base, they do tend to narrow down the options (largely in the direction of transitivization), but this identity with the imperfective is likely motivated by the special status of the Natural Perfective (approximating an inflectional form, see above). Single Act Perfectives, like the similarly Non-Completable Complext Act Perfectives, do not augment the argument structure: (20) shows the same structure for the imperfective base *kusat'* 'bite' as we see for the Single Act Perfective *kusnut'* 'bite once' in (11).

(18) *Kogda bokal kosnulsja gub Vojcika, on žadno pril'nul gubami k krasnovatomu steklu i* ***vypil vodu****.*

'When the chalice touched Vojcik's lips, he greedily clutched the reddish glass with his lips and **drank up the water**.' [Ju. O. Dombrovskij. Obez'jana prixodit za svoim čerepom, čast' 2 (1943-1958)]

(19) *Podobno bol'šinstvu žurnalistov, ja mečtal* ***napisat' roman***.
'Like most journalists, I dreamed of **writing a novel**.' [Sergej Dovlatov. Čemodan (1986)]

(20) *Namin melanxolično ogljadyvalsja, tumanno ulybalsja, kuril kal'jan, sonno lil na šal'vary čto nalili v bokal,* ***kusal jabloko*** *i v diskussijax učastija ne prinimal, ne mešaja devuškam obščat'sja.*
'Namin looked around sadly, smiled in a hazy way, smoked a hookah, sleepily poured onto the shalwars what had been poured in the chalice, **bit an apple** and did not participate in the discussions, not interfering with the girls' socializing.' [Nikolaj Klimintovič. La vie de jour (1996) // "Kommersant"-Daily", 1996.01.27]

3.2.4. *Summary of syntactic evidence*

Table 2 gathers the evidence presented in this section. "Pass." refers to capacity for passivization, and "Nomin." refers to capacity for nominalization. "Int. order" represents internal order in the stacking of prefixes, according to which lexical prefixes are internal (+), while superlexical prefixes are external (-). "⇑ trans." compares perfectives that increase transitivity (with respect to the base imperfectives) with those that do not. "Δ arg." compares perfectives that change the argument structure of the base imperfective with those that do not. "NA" signals "not applicable" where relevant observations are not available.

Table 2: Summary of syntactic evidence for Completability

Status	Type	Pass.	Nomin.	Int. order	⇑ trans.	Δ arg.
Completable	Nat Perf	+	+	NA	+	NA
	Spec Perf (lexical)	+	+	+	+	+
Non-Completable	Comp Act Perf (superlexical)	-	-	-	-	-
	Sing Act Perf	-	-	NA	-	-

Table 2 is presented as a means of visualizing the pervasive divide between Completable and Non-Completable status of types of perfectives, but it suppresses some of the detail presented in this section. It should be taken as a generalization over typical examples, where "+" and "-" represent trends rather than absolute feature values.

Overall we see that the observations suggested by Svenonius and Ramchand for Specialized Perfectives vs. Complex Act Perfectives (here additionally supported by usage-based corpus data) hold also for Natural Perfectives and Single Act Perfectives, confirming Completability as an endemic parameter in Russian grammar.

3.3. Semantic evidence

Here I offer two kinds of semantic evidence for Completability, one involving the metonymical relationships between imperfective and perfective verbs, and the other involving the structure of semantic networks of verbs and prefixes. Though both phenomena have been described in earlier works, in neither case has the connection with Completability been explored before. For the first time here I demonstrate the parallels between these two phenomena and Completability.

3.3.1. Metonymic relationships between imperfective and perfective verbs

Peirsman and Geeraerts (2006) describe metonymy in terms of a radial category based on a prototypical core of spatial part-whole relationships such as *Where's the redhead?* where a part (red hair) refers to an entire individual. The primary dimension structuring the category of metonymy is contiguity, which is a scale of strength of contact. The contiguity scale has four points, ranging from the strongest and most prototypical part-whole relationship to the weakest and most peripheral relationship of adjacency. Here are brief descriptions of the four points with examples cited by Peirsman and Geeraerts (2006):

- Part-whole (cf. *grading papers* where an entire process is named only by one part of the process, the actual recording of the grades)
- Containment (cf. German *Frauenzimmer* 'woman', where a word that once meant 'womanhood' now denotes a single individual from that collection)
- Contact (cf. cause & effect metonymy of French *lumiére* where 'light' stands for 'lamp', and the lamp as a cause is in contact with the light it produces)
- Adjacency (cf. German *Tafelrunde* 'roundtable' where the table stands for the people at it).

Whereas Peirsman and Geeraerts (2006), like nearly all work on metonymy (cf. Kövecses and Radden 1998, Panther and Radden 1999, Dirven and Pörings 2002), focuses on lexical metonymy, recent work (Janda forthcoming, Nesset 2009) has begun to explore the role of metonymy in grammar. Also in this vein, Janda 2008b analyzes the four types of Russian perfectives in terms of the contiguity scale suggested by Peirsman and Geeraerts (2006). Metonymy is an appropriate model for understanding the types of Russian perfectives because perfectivization always involves a narrowing of the range of meanings available in the imperfective. Here is a brief inventory that demonstrates the parallels between the points on the contiguity scale and the types of perfectives, illustrated with diagrams derived from Peirsman and Geeraerts (2006). Note that perfectives are symbolized by a solid line (since they are bounded), and imperfectives are symbolized by a dashed line (since they are unbounded):

Complex Act Perfectives

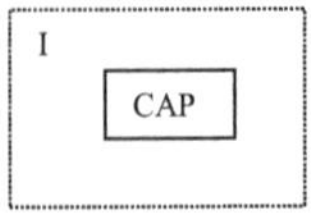

Part-whole: A Complex Act Perfective (CAP) such as *postonat'* 'moan for a while' or *zaprygat'* 'start hopping' quantifies a portion of an unbounded imperfective activity (I) such as *stonat'* 'moan' or *prygat'* 'hop', usually the beginning, ending, or another period of time.

Single Act Perfectives

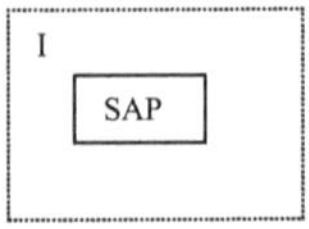

Containment: The relationship here is between an imperfective (I) like *prygat'* 'hop' that describes an unbounded series of repetitions and a Single Act Perfective (SAP) that describes only one item contained the series *prygnut'* 'hop once'.

Natural Perfectives

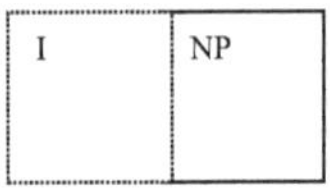

Contact: A Natural Perfective (NP) like *napisat'* 'write' describes the natural culmination of the Imperfective (I) activity *pisat'* 'write'. The contiguity is clear in that the two verbs have the same meaning, but differ only in aspect. There is contact in the temporal domain because the activity shares a temporal boundary with the result it produces. The Natural Perfective denotes an immediate consequence of the corresponding imperfective situation.

Specialized Perfectives

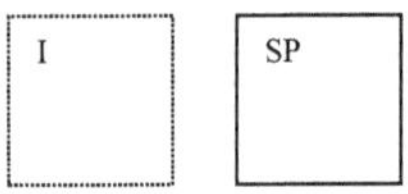

Adjacency: In relation to an unprefixed imperfective (I), a Specialized Perfective (SP) like *perepisat'* 'revise' and *razvjazat'* 'untie' represents a related action that is more specific in its focus. The actions are not the same; there is some semantic difference here that corresponds to distance, and this distance is captured by the adjacency relationship.

The results of this analysis are summarized in Table 3, which additionally indicates the prototypicality of the metonymy and the Completability status.

Table 3. Distribution of types of perfectives along contiguity scale

Status	Type	Contiguity point	Prototypicality
Non-Completable	Complex Act Perf	part-whole	most prototypical
	Single Act Perf	containment	prototypical
Completable	Natural Perf	contact	peripheral
	Specialized Perf	adjacency	most peripheral

In Table 3, as in Table 2, "Status" shows whether the type of perfective is Completable or Non-Completable, and "Type" names the four types of perfectives. "Contiguity point" shows the alignment of the types of perfectives with the four points on the contiguity scale of metonymy, and "Prototypicality" identifies where each point on the contiguity scale lies in the radial network of metonymy. We see a clear pattern here: Non-Completable perfectives reflect prototypical instances of metonymy, whereas Completable perfectives reflect weaker, more peripheral instances of metonymy. We thus observe a parallel between Completability and the prototypicality of the metonymy that motivates the perfective.

3.3.2. *Semantic networks*

Janda and Nesset (forthcoming) present a semantic network for the Russian prefix *raz-* (Figure 1). When analyzed from the perspective of Completability, we find yet another parallel.

Figure 1 presents in condensed form the results of a study of approximately 1,000 perfectives prefixed in *raz-*. The prefix *raz-* forms three types of perfectives: Natural Perfectives (NP), Specialized Perfectives (SP), and Complex Act Perfectives (CAP). Figure 1 shows that the two Completable types of perfectives, the Natural Perfectives and Specialized Perfectives, occupy overlapping portions of the network, namely in submeanings 1-8. In these submeanings, the meanings that the prefix *raz-* adds to the base verb in the Specialized Perfective parallel the meanings of the base verbs that form their Natural Perfectives with *raz-*. The examples cited in Figure 1 illustrate this (here we take only three sets of examples, but for a full discussion see Janda and Nesset forthcoming): *pilit'* 'saw', *toptat'* 'stamp one's feet', and *katat'* 'roll' receive additional meanings corresponding to *apart* in *raspilit'* 'saw apart', *crush* in *rastoptat'* 'trample', and *spread* in *raskatat'* 'roll out (dough)'. The corresponding meanings are already present in the base verbs that form Natural Perfectives: *bit'* 'break' presumes *apart*, *davit'* means 'crush', and *vetvit'sja* 'branch out' presumes *spread*. There is one place in the network where Natural Perfectives and Specialized Perfectives do not overlap, but this involves meanings of "undoing" (10-11 in Figure 1) that are incompatible with the formation of Natural Perfectives (if the base verb and the perfective have the same lexical meaning, the relationship between them cannot be "undo"), and does not detract from the otherwise complete overlap. The Complex Act Perfectives (cf. 9 in Figure 1) like *razvolnovat'sja* 'become upset, start fussing' (an ingressive), on the contrary, do not overlap with either the Natural Perfectives or the Specialized Perfectives; they occupy their own part of the network. We thus see a division that parallels Completability: perfectives of similar Completability (Natural Perfectives and Specialized Perfectives) can have overlapping meanings, whereas perfectives of dissimilar Completability (Complex Act Perfectives in opposition to the other two) do not have overlapping meanings.

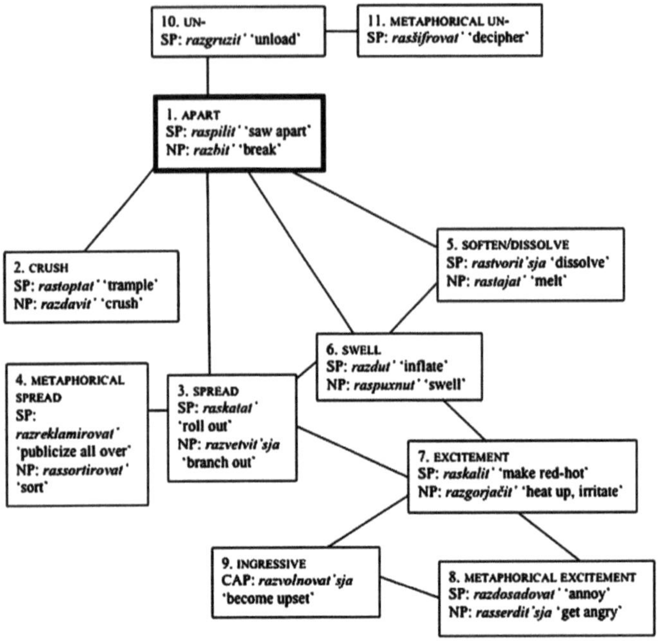

Figure 1. Semantic network of the Russian prefix raz-

3.3.3. *Summary of semantic evidence*

Both the semantic mechanism of metonymy and the semantic networks of prefixes give evidence for parallels to Completability. Though all types of perfective formations involve a narrowing of meaning and thus a metonymical relationship, the types of metonymies invoked parallel the Completability status of the resulting perfective verbs: prototypical metonymies produce Non-Completable perfectives, whereas non-prototypical metonymies produce Completable perfectives. In the semantic network of a prefix, Completable perfectives can have overlapping meanings, but the meanings of Completable perfectives do not overlap with the meanings of Non-Completable perfectives. In other words, for both phenomena we see that perfectives of like Completability pattern together, but perfectives of unlike Completability pattern separately. These patterns are not random and further confirm the prominence of Completability in the Russian aspect system.

4. Advantages of Completability

The evidence for Completability presented in Section 3 suffices to demonstrate the pervasiveness of Completability in Russian aspect. But aside from uniting these various parallels under one conceptual umbrella, what advantage is there to recognizing Completability? I argue that Completability presents practical advantages for linguistic analysis and pedagogical application, as well as theoretical advantages in terms of the elegance and sophistication.

4.1. *Practical advantages*

Completability forms the backbone for the cluster model of Russian aspect, which provides a more comprehensive and accurate description of the Russian aspect system than the traditional pair model. In addition to recognizing the four types of perfectives, the cluster model accommodates additional formations, such as Specialized Single Act Perfectives like *zaxlopnut'* 'slam shut once', which is a Specialized Perfective formed via prefixation of the Single Act Perfective *xlopnut'* 'slam once', itself formed via suffixation of the imperfective *xlopat'* 'slam' (for a discussion of this type of perfective, see Makarova and Janda 2009). Also accommodated in the cluster model are instances of "prefix stacking": such forms occur when a Complex Act Perfective such as *pozapisyvat'* 'record for a while' is formed from the secondary imperfective (in this case *zapisyvat'* 'record') of a Specialized Perfective (*zapisat'* 'record'), which is in turn derived from a base imperfective (*pisat'* 'write'; for more discussion see Janda 2007a). As argued by Janda and Korba (2008), the descriptive gains achieved by this model can be applied to language instruction, helping learners to achieve a more native-like grasp of Russian aspect.

4.2. *Theoretical advantages*

Theoretically Completability represents an advance over feature-based analysis. Completability also integrates both Aktionsart and the verbs of motion into a comprehensive account of aspect instead of setting them aside as "exceptions".

Traditional "pair model" descriptions of Russian aspect are typically built on the structuralist assumption of semantic features such as "boundedness" (cf. Bondarko 1971, Jakobson 1971[1957], Avilova 1976, Padučeva 1996, Talmy

2000) or "totality" (cf. Maslov 1965, Bondarko 1971, Vinogradov 1972, Comrie 1976, Smith 1991, Durst-Andersen 1992, Dickey 2000,) with absolute +/- values. Despite the fact that most linguists have moved on from structuralism, it has been harder to shed the habit of feature assignment. A scalar parameter such as Completability that admits manipulation via construal is more compatible with the contemporary usage-based approach of cognitive linguistics. Corpus data reveals that few, if any, linguistic relationships involve absolute binary values; most phenomena are more complex and scalar (Newman 2008).

Aktionsart (a.k.a. actionality) phenomena such as delimitatives, ingressives and semelfactives fall outside the pair model because such verbs are "unpaired" for aspect and thus exceptional. The pair approach ignores the important role such verbs play in shaping the Russian aspect system. The problem is that the pair model focuses exclusively on relationships that are Completable, overlooking relationships that describe Non-Completable situations. When we recognize that many base imperfective verbs admit a variety of Completability construals, it is possible to connect the both types of relationships in a single model. More concretely, the pair model limits our view of aspectual relationships to pairs with a base verb and a Natural Perfective such as *pisat'/napisat'* 'write', plus pairs with a Specialized Perfective and a secondary imperfective such as *razdut'/razduvat'* 'inflate' (cf. imperfective *dut'* 'blow'). However, in reality there are more aspectual relationships involved, such as the relationship of base verbs to Complex Act Perfectives like *popisat'* 'write for a while' and *zadut'* 'begin to blow', and the relationship of base verbs to Single Act Perfectives such as *dunut'* 'blow once'.

Another type of "exception" from the point of view of the pair model is the class of motion verbs. However, as argued above (Section 3.1.2), motion verbs are not only accommodated, but indeed central to a model based on Completability. Motion verbs serve as the metaphorical source domain for the concept of Completability, with movement through space toward a destination mapping onto progress through an event toward a result. Motion verbs also display the maximal structure of an aspectual cluster since they have all four types of perfectives (cf. examples in 3.1.2).

5. Conclusions

I bring together a lot of loose ends in this article. I offer a comprehensive definition of Completability. I gather up morphological evidence for Completability from scattered sources, and add syntactic and semantic evidence that has not been previously connected to Completability. I argue that Completability yields a more thorough and precise account of the Russian aspect system than has been possible before, and that this account comports well with contemporary theory and practice in cognitive linguistics.

References

Andersen, Henning. 2006. Grammation, regrammation, and degrammation: Tense loss in Russian. *Diachronica* 23, 231–258.

Avilova, Natal'ja S. 1976. *Vid glagola i semantika glagol'nogo slova* [Verbal aspect and the semantics of the lexical verb]. Moscow: Akademija nauk SSSR.

Baydimirova, Anna. 2009. Russian passive morphology: a fuzzy category. Paper presented at the Symposium on linguistic categorization and the nature of linguistic categories, University of Tromsø, 23-24 April.

Bermel, Neil. 1997. *Context and the Lexicon in the Development of Russian Aspect* (Linguistics 129). Berkeley: University of California Press.

Bondarko, Aleksandr V. 1971. *Vid i vremja russkogo glagola* [Aspect and tense of the Russian verb]. Moscow: Prosveščenie.

Bondarko, Aleksandr V. 1983. *Principy funkcional'noj grammatiki i voprosy aspektologii* [Principles of functional grammar and issues of aspectology]. Leningrad: Nauka.

Bondarko, A. V. & L. L. Bulanin. *Russkij glagol* [The Russian verb]. Leningrad: Prosveščenie.

Čertkova, Marina Ju. 1996. *Grammatičeskaja kategorija vida v sovremennom russkom jazyke* [The grammatical category of aspect in the contemporary Russian language]. Moscow: Moscow State University.

Comrie, Bernard. 1976. *Aspect*. Cambridge: Cambridge U. Press.

Dickey, Stephen M. 2000. *Parameters of Slavic Aspect. A Cognitive Approach.* Stanford: CSLI Publications.

Dickey, Stephen M. 2007. A Prototype Account of the Development of Delimitative PO- in Russian. In Dagmar Divjak and Agata Kochanska (eds.), *Cognitive Paths into the Slavic Domain. Cognitive Linguistics Research*, 326–371. Berlin & New York: Mouton de Gruyter.

Dirven, R. & R. Pörings (eds.). 2002. *Metaphor and Metonymy in Comparison and Contrast* (Cognitive Linguistics Research 20). Berlin: Mouton de Gruyter.

Durst-Andersen, Per. 1992. *Mental Grammar: Russian Aspect and Related Issues*. Columbus, OH: Slavica.

Jakobson, Roman O. 1971 [1957]. Shifters, verbal categories, and the Russian verb. In *Selected Writings II*, 130-147. The Hague: Mouton.

Janda, Laura A. 2004. A metaphor in search of a source domain: the categories of Slavic aspect. *Cognitive Linguistics* 15(4), 471-527.

Janda, Laura A. 2007a. Aspectual clusters of Russian verbs. *Studies in Language* 31(3), 607-648.

Janda, Laura A. 2007b. What makes Russian Bi-aspectual verbs Special. In Dagmar Divjak and Agata Kochanska (eds.), *Cognitive Paths into the Slavic Domain. Cognitive Linguistics Research*, 83-109. Berlin & New York: Mouton de Gruyter.

Janda, Laura A. 2007c. Inflectional morphology. In Dirk Geeraerts and Hubert Cuyckens (eds.), *Handbook of Cognitive Linguistics*, 632-649. Oxford: Oxford U Press.

Janda, Laura A. 2008a. Semantic Motivations for Aspectual Clusters of Russian Verbs. In Christina Y. Bethin (ed.), *American Contributions to the 14th International Congress of Slavists, Ohrid, September 2008*, 181-196. Bloomington, IN: Slavica Publishers.

Janda, Laura A. 2008b. Metonymy via Perfectivization of Russian Verbs. *Studia Slavica Helsingiensia* 35, 77-85.

Janda, Laura A. 2008c. Motion Verbs and the Development of Aspect in Russian. *Scando-Slavica* 54, 179-197.

Janda, Laura A. 2009. Totally normal chaos: The aspectual behavior of Russian motion verbs. *Harvard Ukrainian Studies* 28, 183-193.

Janda, Laura A. 2010. Prefixed Perfectives from Non-Determined Motion Verbs in Russian. In Viktoria Driagina-Hasko and Renee Perelmutter (eds.), *New Approaches to Slavic verbs of motion* (= Studies in Language Companion Series 115). Amsterdam/Philadelphia: John Benjamins, 125-140.

Janda, Laura A. forthcoming. The role of metonymy in Czech word-formation. *Slovo a slovesnost.*

Janda, Laura A. & John J. Korba. 2008. Beyond the pair: Aspectual clusters for learners of Russian. *Slavic and East European Journal* 52(2), 254-270.

Janda, Laura A. & Tore Nesset. forthcoming. Taking Apart Russian RAZ-. MS.

Kövecses, Zoltan & Günter Radden. 1998. Metonymy: Developing a cognitive linguistic view. *Cognitive Linguistics* 9, 37-77.

Makarova, Anastasia & Laura A. Janda. 2009. Do It Once: A Case Study of the Russian *-nu-* Semelfactives. *Scando-Slavica* 55, 78-99.

Maslov, Jurij S. 1965. Sistema osnovnyx ponjatij i terminov slavjanskoj aspektologii. *Voprosy obščego jazykoznanija*, 53-80.

Mayo, Peter J. 1985. *The Morphology of Aspect in Seventeenth-Century Russian (Based on the Texts of the Smutnoe Vremja).* Columbus, OH: Slavica.

Mehlig, Hans Robert. 1994. Gomogennost' i geterogennost' v prostranstve i vremeni [Homogeneity and heterogeneity in space and time]. *Revue des etudes Slaves* 66, 595-606.

Mehlig, Hans Robert. 1996. Some analogies between the morphology of nouns and the morphology of aspect in Russian. *Folia Linguistica* 30, 87-109.

Mehlig, Hans Robert. 1997. Nekotorye zamečanija po povodu opisanija katergorii vida v russkom jazyke [Some notes on the description of the category of aspect in Russian]. *Russian Linguistics* 21, 177-193.

Nesset, Tore. 2009. Metonymy of aspect/aspects of metonymy. *Scando-Slavica* 55, 65-77.

Newman, John. 2008. Aiming low in linguistics: Low-level generalizations in corpus-based research. *Proceedings of the 11th International Symposium on Chinese Languages and Linguistics, National Chiao Tung University, Hsinchu, Taiwan. May 24 2008.* http://www.ualberta.ca/~johnnewm/Aiming%20Low.pdf. (Accessed 10 May 2009).

Padučeva, Elena V. 1996. *Semantičeskie issledovanija* [Semantic research]. Moscow: Jazyki russkoj kul'tury.

Panther, Klaus-Uwe & Günter Radden (eds.). 1999. *Metonymy in Language and Thought.* Amsterdam & Philadelphia: John Benjamins.

Peirsman, Yves & Dirk Geeraerts. 2006. Metonymy as a prototypical category. *Cognitive Linguistics* 17, 269-316.

Ramchand, Gillian. 2004. Time and the event: The semantics of Russian prefixes. *Nordlyd* 32(2), 323-361.

Šaxmatov, A. A. 1941. *Sintaksis russkogo jayzka* [Russian syntax]. Leningrad. Učpedgiz.

Smith, Carlota S. 1991. *The Parameter of Aspect*. Dordrecht: Kluwer.
Spencer, Andrew & Marina Zaretskaya. 1998. Verb prefixation in Rusian as lexical subordination. *Linguistics* 32, 913-968.
Svenonius, Peter. 2004a. Slavic prefixes and morphology: An introduction to the *Nordlyd* volume. *Nordlyd* 32(2), 177-204.
Svenonius, Peter. 2004b. Slavic prefixes inside and outside VP. *Nordlyd* 32(2), 205-253.
Svenonius, Peter. forthcoming. Russian prefixes are phrasal. *Proceedings of FDSL 5*.
Talmy, Leonard. 2000. *Toward a Cognitive Semantics. Vol. 2. Typology and Process in Concept Structuring*. Cambridge, MA: MIT Press.
Talmy, Leonard. 2005. In Beate Hampe (ed.), *From perception to meaning: Image schemas in cognitive linguistics*, 199-234. Berlin: Mouton de Gruyter.
Vaillant, André. 1948. *Manuel du vieux slave*. v. I : *Grammaire*. Paris: Institut d'études slaves.
Van Wijk, Nicolaas. 1929. Sur l'origine des aspect du verb slave. *Revue des études slaves* 9, 237-252.
Vinogradov, V. V. 1938. *Sovremennyj russkij jazyk. Grammatičeskoe učenie o slove* [Contemporary Russian. Grammatical study of the word]. Moscow: Učpedgiz.
Vinogradov, Viktor V. 1972. *Russkij jazyk* [Russian], 2nd ed. Moscow: Vysšaja škola.
Zaliznjak, Anna A. & Aleksej D. Šmelev. 2000. *Vvedenie v russkuju aspektologiju* [Introduction to Russian aspectology]. Moscow: Jazyki russkoj kul'tury.

Subjectification and the Russian perfective

Stephen M. Dickey, University of Kansas

1. Introduction

The very term for aspect in Russian, *vid* (literally 'view'), reflects a common impression that the aspect category in Russian (as well as the other Slavic languages) is a fairly subjective category. That is to say, the choice between the perfective and imperfective aspects is determined by the speaker's construal of an action in time, e.g., whether the action is viewed as having reached its endpoint or not, and not necessarily by the "objective" temporal constituency of the action. However, the degree to which one considers Russian aspectual usage to be subjectively motivated in fact depends partly on one's approach to the meanings of the perfective and imperfective as well as linguistic semantics in general, and various studies have downplayed the subjective nature of Russian aspect. Forsyth (1970:2), for example, suggests that although "there frequently is freedom for the individual's 'subjective choice' of aspectual form in Russian [...] the importance of this feature has been greatly exaggerated." Similarly, Sasse (2002:210, fn. 11) suggests that "aspect choice is no more subjective than, say, the choice between a nominalization and a finite verb form."

Today most aspectologists would probably agree with Forsyth and Sasse that aspectual choice in Russian (and other languages) is generally not subject to the conceptual vicissitudes of individual speakers. However, the synchronic and diachronic facts of Russian aspect strongly support Holden's (1990:153) view that Russian aspect has evolved "in the direction of increasingly abstract, subjective, and pragmatic meanings," i.e., it has developed from the expression of the objective, prototypical perception of events to the construal of reality "from alternative perspectives" (Holden [1990:151]). This article attempts to contribute to our understanding of the subjective nature of the Russian category of aspect in its current state, and also to shed some light on its development from a relatively "more objective" category to a relatively "more subjective category". In order to do so it takes a very specific approach to the notion of subjectivity in language, that of Langacker's theory of subjectification (cf., e.g., Langacker [1999]).

This approach to subjectivity is applied to the semantic profile of the Russian perfective aspect in general, and in particular to certain kinds of Russian procedural verbs, known as "intensive-resultative verbs". It is argued that processes of subjectification have contributed to the distinctive nature of aspect in Russian (and East Slavic) in comparison to other Slavic languages, and this difference is considered with regard to the theory of the *east-west Slavic aspect division* developed by Dickey (2000, 2003, 2005, to appear), Dickey and Hutcheson (2003) and Dickey and Kresin (2009). According to this theory, the Slavic languages break down in terms of important parameters of aspectual usage and aspectual morphology into an eastern group (Russian, Ukrainian, Belarusian, Bulgarian, Macedonian) a western group (Czech, Slovak, Slovene, Sorbian) and two transitional zones (Polish and Bosnian/Croatian/Serbian). Finally, the high degree of subjectification of the Russian (East Slavic) perfective is argued to be circumstantial evidence for the view that aspect in Russian (East Slavic) has been grammaticalized to a higher degree than in other Slavic languages.

2. Langacker's theory of subjectification

This section presents a brief outline of Langacker's theory of subjectification (for details, see Langacker [1990], [1999], [2006]). Section 2.1 presents the basics of the most recent version of the theory (taken from Langacker [1999]). Section 2.2 presents the theory as applied to the English perfect in Langacker (1990:337–341) and Carey (1995), which involves many of the issues addressed in this analysis.

2.1. *The basic theory of subjectification according to Langacker (1999)*

A linguistic unit conveys content within its *overall scope* (the full range of conceptual content that it evokes) as well as within its more limited *immediate scope*, i.e., its locus of attention (also known as the *onstage region*). The particular entity or relationship that a linguistic unit foregrounds is called its *profile*. For example, the word *elbow* has as its overall scope the entire human body, whereas its immediate scope is the human arm, and its profile is the joint between the upper and lower arm and their immediately adjacent regions.

This analysis is concerned with temporal relationships as the profiles of verbs. Profiles of relationships presuppose the existence of their participants, chiefly a trajector that has some relationship with one or more landmarks. The temporal relationships profiled by verbs are ultimately the product of sequential scanning, which tracks the path of the trajectory as it changes its location over time relative to the landmark, though it is by summary scanning that a composite profile of all stages is constructed. An example is the verb *fall*, schematized in Figure 1.

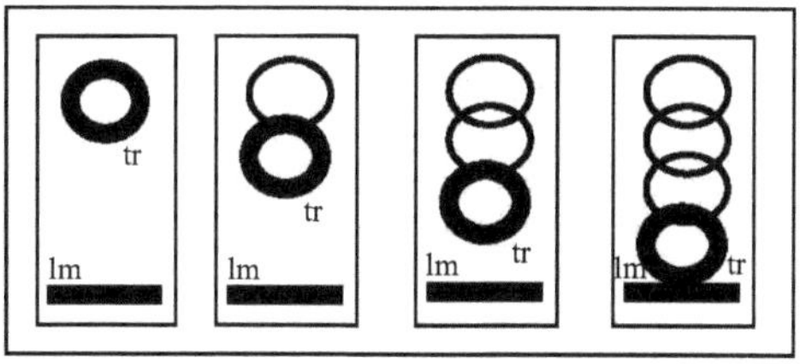

Figure 1. Summary scanning of *fall*

As described above, the profile of a linguistic unit consists of the foregrounded elements in its immediate scope, i.e., its onstage region. Such onstage elements, as they are the object of conception, are said to be *objectively construed*. The conceptualizer, on the other hand, remains in the simplest case completely off-stage, which means that it is only implicitly present in the conceptualization and thus *subjectively construed* to a maximum degree in the act of conceptualization. The conceptualizer, which we may equate with the speaker, is part of the *ground*, i.e., the immediate constituents and circumstances of the speech event, including the speaker and hearer as well as its time and place. Certain linguistic units, called *grounding predications*, e.g., definite articles and deictic elements, invoke the ground indirectly by including it in the overall scope of a construction. Such indirect apprehension of the ground is nevertheless a case of subjective construal, though to a lesser degree than when the ground remains completely off-stage.

The most recent version of the theory of subjectification accounts for the change from the objective construal to the subjective construal of a relation in terms of the *attenuation* of the objective basis for the construed relationship. In particular, the attenuation involves in one way or another the subject's control of the relationship. This is illustrated in Figure 2.

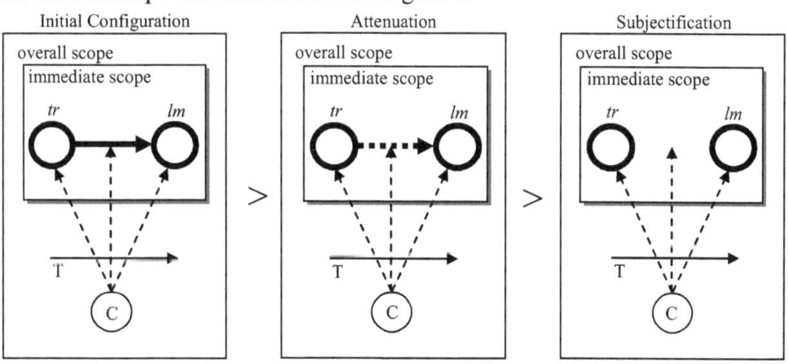

Figure 2. Langacker's (1999: 152) attenuation in degree of conceptual content

In Figure 2, the trajector is related to the landmark by an asymmetrical relationship (represented by the bold arrow), which is conceived by the conceptualizer (C) as s/he makes repeated scans (represented by the dashed lines) of the onstage region proceeding from the trajector to the landmark during the processing time (T). In the first stage, the initial configuration, there is an objectively construed relationship; in the second stage, attenuation, the objective (onstage) situation provides less motivation for the construal of the asymmetrical relationship in question; in the final stage, subjectification, there is no objective basis for the construal of the relationship, which is established solely on the basis of the conceptualizer's activity. As this established relationship "inheres in the process of conceptualization itself [and is not] an onstage object of conception", the relationship is subjectively construed. Langacker emphasizes that the subjective construal is in fact present in the initial configuration, and that the subjectification process really amounts to an attenuation of the objectively construed relationship, leaving the subjectively construed relationship as all that remains.

Langacker (1999:155) observes that attenuation of subject control occurs with respect to least four parameters: (1) status (from actual to potential, specific to generic), (2) focus (the extent to which elements of the conceptualization stand out as a focus of attention), (3) a shift in domain (from a physical interaction to a social or experiential one), and (4) a change in the locus of activity (such as a change in the mover from the onstage trajector to an offstage one in the ground). Without going into detail, we can exemplify this process on the basis of the examples of English *go* in (3), taken from Langacker (1999:158).

(3) a. *Sam was going to mail the letter—but he never made it to the post office.*
b. *Sam was going to mail the letter—but he never got around to it.*
c. *If Sam isn't careful he's going to fall off that ladder.*
d. *It's going to be summer before we know it.*

Example (3a) represents the initial configuration, i.e., the spatial sense of *go*, in which the construal conveyed by the verb is maximally objective; note that the subject (Sam) is not only the trajector, but also acting volitionally and as such is the locus of responsibility. Example (3b) represents attenuation as a shift in domain—the physical motion of *go* has disappeared, leaving only subjectively construed motion though time; however, the subject is still the locus of responsibility. In (3c) there is further attenuation in that the subject is no longer acting volitionally, i.e., there is little or no subject control, though the subject is still to some extent the locus of responsibility. In (3d) the attenuation has reached a maximum (the stage of full subjectification): not only is the motion subjectively construed, but the subject is no longer acting volitionally or the locus of responsibility in any way.

Although Langacker's analyses of subjectification processes tend to be made on the basis of synchronic data, subjectification and loss of subject control are clearly involved processes of change that result in grammaticalization. Such involvement is evident from the cases of grammaticalized categories that he discusses—the English *go*-future, "raising" constructions, modals, *be*-auxiliaries and the *get*-passive. The following section discusses the subjectification process involved with grammaticalization of the English perfect.

2.2. *Subjectification and the grammaticalization of the English perfect*

Langacker's (1990) subjectification analysis of the English perfect proceeds from a related earlier *have* + participle construction, which resembles the perfect but is a kind of resultative construction indicating "roughly, that the subject stands in a relationship of accomplishment vis-à-vis the finishing, or that the prior occurrence of finishing remains relevant to him" (Langacker [1990:339]). This construction is diagramed in Figure 3.

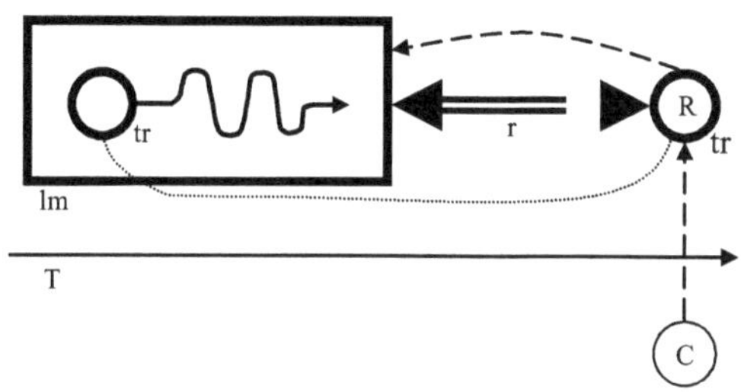

Figure 3. Langacker's (1990: 340) diagram of the resultative *have* + participle construction

According to this account, the trajector stands in a relationship of "potency or relevance (r) with respect to a prior event that constitutes a relational landmark"; the potency/relevance relationship is profiled by *have*. As this construction is a *reference-point construction*, it also profiles a reference-point function (the thin dashed line), which represents "the mental path by which the speech-act participants establish mental contact with the target" (Langacker 1990:338), which is in this case the prior event expressed by the participle. The reference point (R) is also the trajector of the event expressed by the participle; this identity is signaled by the thin dotted line.

The *have*-perfect, exemplified by *He has finished*, represents an attenuation of the potency/relevance relationship, which no longer exists between the landmark event (*finished*) and the subject, but between the landmark event and some other reference point, which often coincides with the ground (where the conceptualizer, C, is located). Further, as the potency/relevance relationship is attenuated, the reference-point function becomes the profile of the construction. This is diagramed in Figure 4.

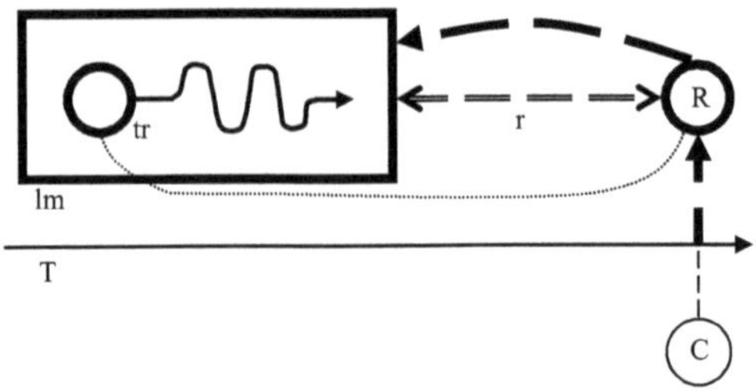

Figure 4. Langacker's (1990: 340) diagram of perfect *He has finished*

The removal of the subject as the reference-point in the perfect is a kind of loss of subject control in the diagram of subjectification given in Figure 2. This account of subjectification in the English perfect is in fact made in an earlier version of the theory (that of Langacker [1990], which is evident in the inclusion of the conceptualizer's scanning as part of the onstage profile of the construction). However, this treatment of the English perfect shares the core elements of the newer version of subjectification theory in that it posits an attenuation of the potency/relevance relationship with the trajector (which amounts to a loss of subject control), leaving the reference-point relationship as the profile.[1]

Carey (1995) applies Langacker's subjectification analysis to the historical development of the English perfect. She points out that "current relevance is an inherently subjective notion in that the link between the past event and the current situation is dependent on the attitude/judgment of the speaker" (Carey [1995:99–100]). She views current relevance as a kind of "wider result" of a verbal action, following Dahl's (1985:134–135) views on the connection between results and the perfect:

> "The term 'result' may be understood in a wider or narrower sense… For instance, if a person dies, the result in a narrow sense is that he is dead: the results in a wider sense include e.g. his relatives being sad. It is results in the narrower sense that

---

1 A more current account of subjectification in the English perfect would presumably rely solely on the attenuation of the potency/relevance relationship. Updating Langacker's (1990) version of the subjectification of the English perfect to his current (1999) version is beyond the scope of this paper; the aforementioned commonalities should be sufficient for the purpose at hand.

characterize resultative constructions: in the case of the PFCT [perfect], the delineation is much harder to draw."

Based on a quantitative examination of Old and Middle English data, Carey argues that the Old English *have* + participle construction developed into the Middle English perfect via a twofold process: (1) an increase in the salience of the anterior event; (2) a shift in the locus of relevance from the subject to the discourse itself. Carey argues that communication verbs played a crucial role in these shifts, as they "typically refer to a resultant knowledge state in the subject" (Carey [1995:99]) and are used to sum up the speakers' progress in the evolving discourse. Evidence for her view is the fact that verbs of communication appear to have increased in the *have* + participle construction early in the grammaticalization process. Later, the locus of relevance widens even further to "any aspect of the discourse context that the speaker chooses to construe as involved in a cause-effect relationship with the anterior event" (Carey [1995]: 88).

Carey points out that the path of grammaticalization of the perfect suggested by the historical data comports with Langacker's (1990) hypothesis of the subjectification of the English perfect: (1) the reference point of the construction shifts from the subject to the ground, and (2) the locus of relevance becomes increasingly diffuse. I would add that the increased salience of the anterior event may be interpreted as the result of the anterior event remaining as an onstage profile of the construction while the potency/relevance relationship is attenuated, leaving the relationship with the reference point as a subjectively construed entity. It would take us too far afield to examine the details of Carey's diachronic analysis; what is important for the purpose at hand is the overall analysis that the grammaticalization of the English perfect appears to be a clear case of subjectification, involving the attenuation of a profiled relationship and a loss of subject control. It is argued in section 4 that in the grammaticalization of the Russian perfective these processes have occurred in a very similar manner. However, before proceeding to the analysis, it is necessary to introduce a minimum of basic facts concerning Russian and Slavic aspect; these are presented in section 3.

3. A brief overview of Russian aspect and perfective procedural verbs

The following sections give a brief description of the aspect category in Russian (Slavic) as well as of the closely related category of perfective procedural (*Aktionsart*) verbs. These topics are complex to say the very least and many points remain in dispute; only the bare facts necessary for this discussion are mentioned (for detailed discussions in English of Slavic aspect and/or procedurals, see Forsyth [1970], Townsend [1975], Leinonen [1982], Schuyt [1990], Stunová [1998], Dickey [2000], Janda [2007], Dickey and Janda [2009]). Examples are taken from Russian. Aspectual pairs of verbs are indicated by a double slash: imperfective verbs precede the double slash and perfective verbs follow it. Individual imperfective and perfective verbs are marked with the subscripts $_{IMPF}$ and $_{PF}$ respectively. Note that the category of procedural verbs also includes various imperfective formations, but these are omitted from the discussion as they are not directly relevant to this analysis.

3.1. *The Slavic aspect category*

With various exceptions, in the Slavic languages verbal notions are expressed by derivationally related pairs of verbs, e.g., *pisat'//napisat'*, *vyxodit'//vyjti* 'go out', *zapisyvat'//zapisat'* 'note down'. From a cross-Slavic perspective, the prototypical aspect pair refers to a telic situation, either an accomplishment (e.g., *pisat'//napisat'* 'write', *pit'//vypit'* 'drink', *vyxodit'//vyjti* 'go out') or an achievement (e.g., *naxodit'//najti* 'find', *terjat'//poterjat'* 'lose', *vyskakivat'//vyskočit'* 'jump out'). Telic predicates are also historically the origin of the Slavic aspect category. In particular, the imperfective : perfective opposition existed regularly for accomplishment predicates first, subsequently spreading to achievement predicates and later activity predicates (cf. Bermel [1997]).

Much has been written about the meanings/conditions of use of the perfective and imperfective aspects. Here the perfective aspect is the object of primary concern. Let us begin with a recent and influential account of the perfective, that developed by Chung and Timberlake (1985). According to Chung and Timberlake (1985:224), in order for a predicate to be coded perfective it must reach closure with respect to the event frame "in that the frame includes the event and the event reaches its inherent limit." With regard to the resultative predicates listed above, reaching closure entails the existence of the result produced by the process in question: for *napisat'*$_{PF}$ 'write' it is the existence of a text on some surface; for *vypit'*$_{PF}$ 'drink' it is the empty state of the glass; for *vyjti*$_{PF}$ 'go out' it is the state of the trajector outside of the landmark container, etc. Thus, for such predicates the use of the perfective depends on whether the conceptualizer/speaker conceives and asserts the existence of the result in the relevant temporal context (event frame)—if /she does not, the imperfective is used. Though more rigorous in its formalization, Chung's and Timberlake's version of the semantics of the Russian aspect opposition recalls Isačenko's (1960:133–133) characterization of the aspect opposition as reflecting alternative perspectives: the imperfective presents the perspective of a viewer situated within an action, whereas the perfective presents the perspective of a viewer from without, which encompasses the action as a whole.[2] Inasmuch as this common element of their respective approaches is accurate, it seems beyond dispute that the Russian aspect opposition does involve a certain degree of subjectivity, even if the alternatives tend to be conventionalized in various contexts.

According to this account, perfective verbs, as they encode closure, tend to be used to profile single completed events. Imperfective verbs are used in cases in which closure is either not reached or not salient in the conceptualization, e.g., for states, events in process, habitually repeated events, and cases in which an event is construed as complete but the result is either irrelevant or absent. The latter are known by the term "imperfective general-factual" or "constative imperfective"; an example is given in (1).

(1) *Žitel'nica postradavšego doma Svetlana Bodrunova,*
resident.F suffered.PTCP.M house.M Svetlana Bodrunova.F
proživajuščaja na 6-m ètaže, utverždaet, čto strojka,
residing.PTCP.F on 6.M.LOC storey.M.LOC claim.3SG that construction
gde proizošlo ČP, velas' s narušeniem
where occurred.PV.3SG incident led.3SG.REFL with violation
pravil, a sam kran uže odnaždy ***padal****.*
rules.GEN and self.M.NOM craneM.NOM already once fell.IPFV.M.3SG
'A resident of the destroyed house, Svetlana Bodrunova, residing on the sixth floor, claims that the construction project where the incident occurred was conducted in violation of the rules, and the crane itself had already fallen once.' [Internet]

Such usage is of course a complex phenomenon in its totality, and deserves detailed commentary, which cannot be given here (for accounts of the Russian imperfective general-factual, cf. Israeli [1996, 1998, 2001] and Zel'dovič [2002]); I have chosen here an achievement predicate *padal*$_{\text{IMPF}}$ 'fell', which necessarily reaches closure and produces a result (even if in [1] that result had been annulled in the meantime—the crane had been return to its upright position), in combination with the explicit adverbial *odnaždy* 'once', in order to illustrate just how "complete" single events coded imperfective can be in Russian. Note in this connection that it is erroneous to analyze the imperfective *padal*$_{\text{IMPF}}$ 'fell' as signaling that closure was not reached in the frame. This is one of the reasons that some analyses employ different definitions of the perfective, which are discussed in the next section.

---

2 This account of the perfective is necessarily simplified, and is in fact not one that I would endorse for Russian (regarding Chung and Timberlake [1985], see the remarks in Dickey [2000:41–42]). In my view, it is more adequate as an account of the perfective in languages of the western group such as Czech.

3.2. *The Russian perfective: beyond closure*

Since Galton (1976) emphasized temporal succession as a key semantic element of the Slavic perfective aspect, several analyses of Russian aspect have departed from the idea that closure (or an equivalent notion) is sufficient to characterize its meaning. Dickey (2000), following Leinonen (1982), argues on the basis of cross-Slavic data that the Russian perfective expresses a category labeled temporal definiteness. A situation is temporally definite if it is unique in the temporal fact structure of a discourse, i.e., if it is viewed as qualitatively different from preceding and subsequent states of affairs (the uniqueness condition), whereby it is necessarily viewed as a complete whole (though not necessarily having reached closure). In other words, if it can be located in time inasmuch as it is differentiated from the situations in preceding and subsequent temporal intervals. In this respect, Dickey (2000) shares Barentsen's (1998) view that "sequential connection" is the constant inherent semantic feature of the Russian perfective aspect (as well as the perfective in the other eastern languages; note that the differences between Dickey 2000 and Barentsen 1998 are largely theory-internal).

The theory of Russian aspect offered by Zel'dovič (2002) is in its essentials compatible with that offered by Dickey (2000), and formalizes a version of the uniqueness condition in a very interesting way. According to Zel'dovič (2002: 29), a perfective verb expresses the meaning given in (2):

(2) "(a) the speaker conceives of one or more sets *M* of situations, and in every *M* there is a situation expressed as '*P*' or '*P* + *A*' (*A* is some [adverbial] comment about *P*) such that (b) within *M* (every individual *M*) the situation expressed by '*P*' or '*P* + *A*' occupies a single temporal interval; (c) *P* [i.e., the lexical content of the verb]."

Zel'dovič's formulation of the meaning of the Russian perfective aspect has specific consequences: Since the speaker conceives of a (contextual) set of situations *M* in which the situation in question *P* occupies a single (unique) temporal interval, when using a perfective verb the speaker must have in mind some situation(s) *other* than *P*, which Zel'dovič (2002:31) labels *P'*. According to Zel'dovič, it is "indeed the request to reconstruct *P'* [that] determines the main traits in the behavior of the pf aspect," and for telic verbs in conversational discourse, P' is usually a situation subsequent to the situation expressed by the verb. This subsequent situation can very easily be the direct result of the situation expressed by the verb, as mentioned above in section 3.1. But Zel'dovič (2002:34) points out that in addition to *direct* results, actions can produce a myriad of *oblique* results that may be the situation that concerns the speaker (cf. Dahl's [1985] observations on results cited in section 2.2). Thus, for example, *Ivan sel*$_{PF}$ 'Ivan sat down' and *Ivan svaril*$_{PF}$ *sup* 'Ivan cooked soup' have the direct results *Ivan sidit* 'Ivan is sitting' and *sup suščestvuet* 'the soup exists' (respectively); the oblique results with each of these clauses could be that Ivan carried out an order or promise to sit down or cook soup, or that he won a bet by performing these actions. It is important to point out that there is no defined set of oblique results—they depend on the concerns of the speaker in the larger discourse context. Zel'dovič (34) points out that the past-tense of perfective verbs are used "on account of some oblique result, and not the direct result of a predicate *P* (it may be clear in the meantime that the direct result of *P* is not still in force at the moment of speech)."

It is useful to illustrate the effects of communicating a direct versus an oblique result on the basis of Zel'dovič's analysis of the following exchange (taken from Durst-Andersen 1993:139):

(3) *A ty pozvonil emu v čas?*
And you called.PV.2SG him in hour
***Pozvonil**, no bylo zanjato.*
called.PV.1SG but was occupied
'And did you call him at one?'
'I called, but it was busy.'

The direct result (*P'*) presupposed by the speaker using *pozvonil*$_{PF}$ 'called/made a call' in the question is that the subject talked to the recipient. The response, however, asserts that the call took place, but that there was no conversation because the line was busy. The reason that the perfective verb *pozvonil*$_{PF}$ 'called' is nevertheless felicitous in the response is that it does not activate the direct result as *P'*, but rather the oblique result of the speaker having fulfilled his obligation to make the call. Evidence that this is the case is the fact that a response involving an affirmative answer to the question is questionable: $^{?}$*Da, pozvonil*$_{PF}$*, no bylo zanjato* 'Yes, I called, but it was busy.' The reason for the infelicity of the explicitly affirmative response is that answering *da* 'yes' signals that the speaker accepts the presuppositions of the question, i.e., that the result of interest (*P'*) is the subsequent conversation, and asserts that this result ensued, which contradicts the subsequent statement that the line was busy.
This introduction has presented the basic facts of the meaning of the Russian perfective aspect required for this analysis. Before going on, however, it is necessary to describe some particular classes of Russian perfective procedural verbs. These are discussed in the next section.

3.3. *Some types of Russian perfective procedural verbs*

As mentioned in section 3.1, the prototypical aspect pair refers to a telic situation. But not all pairs fit this characterization, and various kinds of perfective verbs are unpaired with their imperfective source verbs. Very often the prefixation of atelic verbs produces special perfective verbs that are commonly known as procedural (*Aktionsart*) verbs. The atelic predicates involved include stative verbs (e.g., *stojat'*$_{IMPF}$ 'stand', *ležat'*$_{IMPF}$ 'lie') and activities (e.g., *tancevat'*$_{IMPF}$ 'dance', *rabotat'*$_{IMPF}$ 'work'). Also included are imperfective verbs that enter into accomplishment aspectual pairs, as they also allow an activity construal. For example, *pisat'*$_{IMPF}$ 'write' is ambiguous between the prototypical, goal-oriented/heterogeneous (i.e., accomplishment) construal—*pisat'*$_1$—and a peripheral/homogeneous (i.e., activity) construal—*pisat'*$_2$ (cf. Dickey 2006). The latter, *pisat'*$_2$ 'be writing/a writer', is a source verb for the derivation of perfective procedural verbs.

With the exception of semelfactives suffixed in *-nu-*, the great bulk of Russian perfective procedural verbs are formed via prefixation. Prefixation forming procedural verbs is one of three kinds of perfectivizing prefixation in Russian (the other two being lexical prefixation, e.g., *s-pisat'*$_{PF}$ 'copy', and purely perfectivizing prefixation, e.g., *na-pisat'*$_{PF}$ 'write', the former being quite common and the latter being relatively infrequent). Procedural prefixation is distinct from lexical prefixation in that it does not alter the basic lexical meaning of the source verb, but "modifies it in some way, usually with respect to time or intensity" (Townsend 1975:118). This analysis proceeds from the categorization of Russian prefixed perfective procedural verbs given by Zaliznjak and Šmelev (2000:104–118), who distinguish eight basic types : *ingressive* verbs (e.g., *za-svistet'*$_{PF}$ 'start whistling'), *delimitative* verbs (e.g., *po-guljat'*$_{PF}$ 'stroll a while'), *perdurative* verbs (e.g., *pro-govorit'*$_{PF}$ 'spend some [long, specified] amount of time talking'), *finitive* verbs (e.g., *ot-cvesti*$_{PF}$ 'finish blooming'), *cumulative* verbs (e.g., *na-kupit'*$_{PF}$ 'buy a lot of'), *satiative* verbs (e.g., *na-est'-sja*$_{PF}$ 'eat to satiety'), and a number of varied formations known as *intensive-resultative* verbs (e.g., *do-zvonit'-sja*$_{PF}$ 'call and get through/reach by telephone').

Not all of these types are equally important for the analysis presented here. The delimitative *Aktionsart* is particularly important for the Russian aspectual system as a whole, and I have argued (cf. Dickey [2006]) that delimitative verbs in *po-* for all intents and purposes comprise aspectual pairs with their atelic source verbs, e.g., *ležat'//poležat'* 'lie' and *pisat'*$_2$*//popisat'* 'write [activity]' (as opposed to *pisat'*$_1$*//napisat'* 'write [accomplishment]). As I have also argued (Dickey and Hutcheson [2003], Dickey [2005, 2007]), the status of *po-* delimitatives as perfective partners of their imperfective source verbs has played a role in the grammaticalization of Russian aspect by systematically extending the imperfective : perfective opposition to the class of atelic activity predicates as a whole.

Let us turn next to the last class mentioned, the intensive-resultatives, as they include a number of distinct types of verbs that form one of the main foci of this analysis, and thus deserve a more detailed description. The intensive-resultatives are interesting from a formal standpoint in that they are formed with a prefix in conjunction with the reflexive/middle suffix *-sja* (henceforth referred to as a circumfix) as are the semantically closely related satiatives. These verbs may be characterized as profiling an intensive/prolonged episode of the action in question, and often with a result that is either beyond the control of the subject, unintended by the subject, or undesired (adverse) for the subject. In any case, the result is always oblique. One common type, which Forsyth (1970: 21) calls absorptives, is formed by the circumfix *za-...-sja* (lit. 'behind...REFL'), and profiles the continuation of the activity to the point where it occurs "independently of the will of the subject (usually because s/he has become too absorbed in the action) and possibly lead[ing] to some adverse consequences" (Zaliznjak and Šmelev [2000:117], c.f., e.g., *zagovorit'sja*$_{PF}$ 'become absorbed in conversation', *zagljadet'sja*$_{PF}$ 'become absorbed in looking at something'.

Another category are attainment verbs formed with the circumfix *do-...-sja* (lit. 'up to...REFL'). An older group of these verbs profiles the attainment of a desired result, e.g., *dokopat'sja*$_{PF}$ 'discover by digging', *dozvonit'sja*$_{PF}$ 'call and get through/reach someone by telephone'. Another, newer group, which is particularly characteristic of Russian and is sometimes known as *ironic-resultative* verbs, profiles the continuation of the activity "to the point of an undesired and previously unforeseen result" (Zaliznjak and Šmelev [2000:117]), e.g. *dokupat'sja*$_{PF}$ *(do vospalenija legkix)* 'bathe to the point of (pneumonia)', *dorabotat'sja*$_{PF}$ *(do nervnogo sloma)* 'work to the point of (a nervous breakdown)'. As Zaliznjak and Šmelev point out, ironic-resultatives are most often formed from activity source verbs that do not profile goal-oriented activities such as *igrat'* 'play', *guljat'* 'stroll', etc.

The absorptives and ironic-resultatives are the chief representatives of the intensive-resultatives with which we are concerned, and appear to be quite productive in colloquial Russian (cf. Isačenko [1960:250] concerning the latter, as well as the relatively recent formation *donarkomanit'sja*$_{PF}$ 'take narcotics to the point of an adverse result'). However, other types exist, such as what Isačenko (1960: 250) terms *resultative-qualificational* verbs, which are formed with the circumfix *iz-...-sja* (lit. 'out... REFL') and which profile the habitual performance of a negative activity to the point where it becomes a habit of the subject, e.g. *izolgat'sja*$_{PF}$ 'lie to the point of becoming an incorrigible liar', *izvorovat'sja*$_{PF}$ 'steal to the point of becoming an incorrigible thief'.

Other types have no labels and have not to my knowledge been the subject of a detailed treatment in the literature apart from Rutkowska (1981). She presents a description of several minor types of intensive-resultatives that exist in Russian but not Polish, which cannot be reproduced here. Some of the more interesting types with representative examples are given in (4); the meaning of the circumfix is glossed in brackets:

(4) a. *Ot(o)-...-sja* ['away...REFL']: 'avoid something unpleasant or dangerous by performing the activity of the source verb, e.g., *otgovorit'sja*$_{PF}$ 'avoid something by

talking/talk one's way out of something', *otsidet'sja*~PF~ 'avoid a danger by sitting in shelter/hiding';

b. *Ot(o)-...-sja* ['away...REFL'], *pro-...-sja* ['through...REFL']: 'return to equilibrium/a normal state via the activity of the source verb', e.g., *otležat'sja*~PF~ 'lie to the point where one returns to a normal state', *prospat'sja*~PF~ 'sleep to the point where one returns to a normal, sober state';

c. *Pro-...-sja* ['through...REFL']: 'ruin oneself, lose one's money and property as a result of the activity of the source verb', e.g., *proferšpilit'sja*~PF~ 'ruin oneself losing at cards or spending money', *prostroit'sja*~PF~ 'squander all one's money building a house';

d. *O-/Obo-...-sja* ['around...REFL']: 'perform the action of the source verb erroneously or unsuccessfully', e.g., *ostupit'sja*~PF~ 'misstep', *oslyšat'sja* 'hear incorrectly';

e. *O-/Obo-...-sja* ['around...REFL']: 'become accustomed to performing the action of the source verb', e.g., *obterpet'sja*~PF~ 'become accustomed to enduring physical or mental pain', *o(b)gljadet'sja*~PF~ 'become accustomed to looking in the dark';

f. *Pri-...-sja* ['to...REFL']: 'become accustomed to something via the activity of the source verb', e.g., *prigljadet'sja*~PF~ 'get used to looking at something', *pripit'sja*~PF~ 'get used to a beverage by drinking it';

g. *S-...-sja* ['together...REFL']: 'reach harmony/accord as a result of the activity of the source verb', e.g., *srabotat'sja*~PF~ 'reach harmony/accord in work', *spet'sja*~PF~ 'reach harmony/accord in singing';

h. *Pod(o)-...-sja* ['up to from below...REFL']: 'adapt to someone/gain someone's favor via the activity of the source verb', e.g., *podol'stit'sja*~PF~ 'gain someone's favor by flattery', *podslužit'sja*~PF~ 'gain someone's favor through service'.

According to Rutkowska's descriptions, all of these models are productive in the contemporary language, even if they are considered colloquial or substandard. From the general definitions of these derivational models as well as the examples it should be clear that these intensive-resultative verbs, like the absorptives and the ironic-resultatives, profile the attainment of various oblique results. In this respect, we can also connect the absorptives and intensive-resultatives to *po-* delimitatives, which, as I have argued (cf. Dickey [2006]), also often profile some oblique result of an action (as opposed its direct result). Consider the following examples:

(5) a. *Abxazskaja milicija slavno* ***porabotala***.
Abzhkazi police gloriously PO-worked
'The Abkhazi police did a terrific job.'

b. ***Popisal*** *i rešil vot vpečatlenija v ètot dnevnik*
PO-wrote.1S and decided.1S PRESENTATIVE impressions into this diary
pozapisyvat'.
to.write.down.DISTR
'I wrote a little while and decided thus to write down all my impressions in this diary.'

In (5a), the source verb *rabotat'*$_{IMPF}$ 'work' has no direct result as it is strictly an activity verb, so that any profiled result is an oblique result; delimitative *porabotal*$_{PF}$ 'PO-worked' profiles the situation that resulted from the work, i.e., that the wife of a business executive was freed from captivity. In (5b), the result of the delimitative *popisal*$_{PF}$ 'PO-wrote' is the satisfaction with writing that leads to the decision to keep a diary, which is an oblique result (the direct result is the text that was produced).

The number of types of perfective procedural verbs presented above demonstrates clearly the degree to which the category of the Russian perfective is a lexical as well as a grammatical category. And conversely, the lexical encoding of such non-canonical results in single derived verbs is a compelling piece of evidence that the schema of the Russian perfective aspect includes as a salient/foregrounded element some subsequent state of affairs (but not necessarily a direct result). The lexical coding of such results—which, it bears pointing out, tend to be expressed periphrastically in other languages (including Slavic languages outside of East Slavic)—also indicates that the schema of the Russian perfective enjoys considerable organizing force not only in the grammatical structure of Russian but in its linguistic structure as a whole.

The remainder of this paper applies Langacker's theory of subjectification to the oblique results profiled by the kinds of Russian perfective verbs discussed in sections 3.1–3.3, focusing on synchronic facts but considering some basic diachronic issues as well. Further, it considers the implications of the subjectification analysis of the Russian perfective for an account of the grammaticalization of aspect in Russian and other Slavic languages.

4. A subjectification analysis of the Russian perfective

The lengthy preliminary sections 1–3 have provided the background for a relatively simple subjectification analysis of the Russian perfective that is presented in the following sections. Section 4.1 analyzes the simple cases of the usage of telic perfective verbs in Russian. Section 4.2 analyzes the Russian perfective procedural verbs introduced in section 3.3. Section 4.3 presents some further discussion and considers the implications for a theory of the grammaticalization of the Russian perfective aspect.

4.1. *Subjectification and telic perfective verbs in Russian*

Let us begin with the original sense of the perfective members of the prototypical aspect pairs described in section 3.1, e.g., *napisat'*$_{PF}$ 'write', *vypit'*$_{PF}$ 'drink', *vyjti*$_{PF}$ 'go out'. As resultative perfective verbs, they profile the activity in its totality as well as the existence of the intended result. In terms of subjectification, this profile corresponds to the initial configuration in Figure 2. The initial configuration of the Russian perfective aspect is diagramed in Figure 5.

The trajector is the agent/subject of the process (a landmark) that stands in an asymmetrical relationship of *effecting* (E) the direct result (a secondary landmark); this relationship is henceforth termed the effecting-relationship.[3] I consider it likely that what I am calling the effecting-relationship is a variant of Langacker's (2006:28) conceptual archetype INTEND in a configuration involving a verbal process and its result as landmarks, but I cannot argue for this view in the necessary detail here.

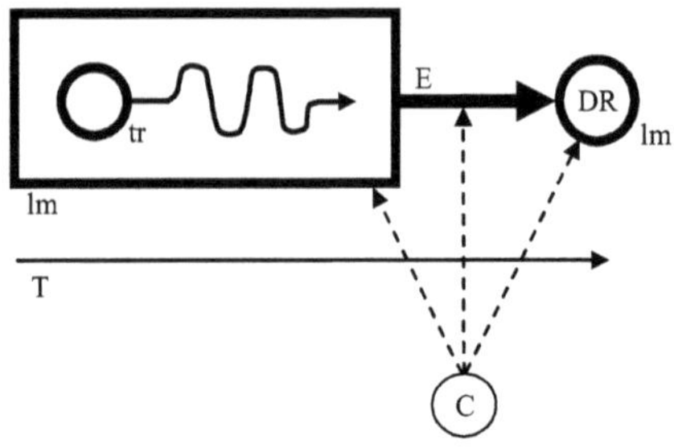

Figure 5. The initial configuration of the Russian perfective

The sense of the perfective diagramed in Figure 5 is activated when resultative verbs are used to foreground the existence of the direct result, as in the example *Ivan sel*$_{PF}$ 'Ivan sat down' (direct result = *Ivan sidit*$_{IMPF}$ 'Ivan is sitting'), or the following example taken from the Internet:

(6) *Ja **napisal** stixotvorenie, kotoroe mne očen' nravitsja.*
I wrote.PV poem which me very pleases.
'I wrote a poem that I like a lot.'

The focus of (6) , i.e., *P'* is the existence of a new poem.

With such telic resultative verbs, which are the prototypical Russian (Slavic) perfective verb, the possibility of significant subjectivization is rather limited. The reason for this lies in the lexical semantics of the verb, e.g., *napisat'*$_{PF}$ write, which in combination with its perfectivity necessarily asserts the ensuing result.

3 Note that in Figure 5 the direct result is not a reference point, unlike the corresponding entity in the diagram of the perfect (Figure 4); despite the well-known similarity between Russian past-tense perfective usage and the English perfect (cf. Zel'dovič [2002]), the content communicated by Russian perfective verbs cannot in my view be considered as involving a reference-point in Langacker's sense of the term, due to the difference in form.

If, as Zel'dovič (2002) points out, *P′* is an oblique result, as in the case of *Ivan sel*$_{PF}$ (where *P′* = *Ivan poslušalsja*$_{PF}$ *prikaza* 'Ivan complied with an order'), for many cognitively basic verbs the oblique result ensues only as a consequence of the direct result, i.e., *Ivan sel*$_{PF}$ 'Ivan sat down' > *Ivan sidit*$_{IMPF}$ 'Ivan is sitting' > *Ivan poslušalsja*$_{PF}$ *prikaza* 'Ivan has complied with an order'.

However, for telic verbs that are not prototypically resultative verbs, the possibility of subjectification is correspondingly greater. Consider again example (3), in which in the response *pozvonil*$_{PF}$ 'made the call' the situation *P′* is not the direct result (talking to the recipient), but rather the oblique result of the speaker having fulfilled an obligation to place the call. In this case, we may say that there has been attenuation of the effecting-relationship, as no direct result is conceptualized; the effecting-relationship links the process with an independent oblique result, some element of the "here and now of the speaker" (Carey [1995:99]). This attenuation of the effecting-relationship is diagramed in Figure 6, in which OR is the oblique result.

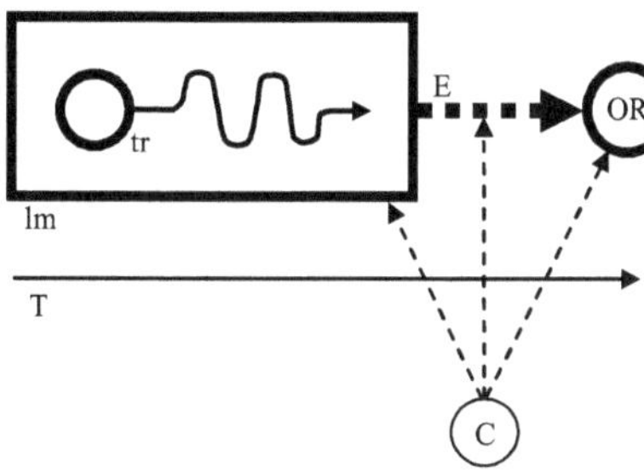

Figure 6. Attenuation of the perfective effecting-relationship

I suggest that the attenuation of the effecting-relationship is also present with certain uses of perfective communication verbs. Consider the examples in (7), taken from Israeli (2001:75).

(7) a. *Mne **govorili**, čto ty zdes'.*
me.DAT told.IPFV.3PL that you here
'I was told that you are here.'

a. *Mne **skazali**, čto ty zdes'.*
me.DAT told.PV.3PL that you here
'I was told that you are here.'

Israeli explains that a frequent difference between (7a) and (7b) is that the imperfective (7a) is felicitous when the speaker accidentally meets the listener, whereas (7b) is felicitous only in some specific context, such as that the speaker is acting on the information that s/he received and trying to find the listener.

Note that both (7a) and (7b) are equally resultative in terms of a direct communicative result, i.e., that the speaker has internalized the information. In Zel'dovič's terms, we may say that *P'* is the fact that the speaker has acted on the information, an oblique result. Not only is the fact that the speaker has acted on the information not a direct result, but it is also an element of the ground, in that his/her subsequent action has led to the present encounter/speech event and is part of the speaker's knowledge surrounding it.

Examples of subjectification with telic perfective verbs involving a shift in emphasis from the direct result itself to an oblique result can only be considered to involve varying degrees of attenuation, because there is no clear loss of subject control. In (6), the speaker has still placed the call of his/her own volition (discounting exotic scenarios), and in (7b) the indeterminate subject of the clause may or may not have had some interest in influencing the speaker's actions. However, there is clearly an attenuation of subject control in that the oblique results in these examples are not part of the immediate subgoals of the predicates in question, i.e., holding a phone conversation and communicating information respectively.

A piece of circumstantial evidence that subjectification is indeed involved in the cases considered in this section is the fact that in these examples the oblique result has been left unstated. Langacker (1985) argues that subjectively construed entities tend to be left implicit, as explicitly mentioning them puts them on stage. Note in this regard that Russian telic perfective verbs show a strong tendency to be transitive (this is particularly true of resultative verbs), and that the object, which is either the result or an entity metonymically linked to the result, must be explicitly mentioned, apart from the limited possibility of eliding information that has already been mentioned.

4.2. *Subjectification and Russian perfective procedural verbs*

This section considers subjectification first in Russian *po*-delimitatives and then in ironic-resultatives and other intensive-resultative verbs. The account is primarily synchronic, but diachronic issues are considered as well.

4.2.1. Subjectification and po-*delimitatives.* Let us begin with *po*-delimitatives. Isačenko (1960:234–235) characterizes Russian *po*-delimitatives as profiling the relatively limited or indefinite duration of an activity (note that the specification "short/brief" made by Zaliznjak and Šmelev [2000:111] and others is unnecessary; for a cognitive explanation of the interpretation of brevity of *po*-delimitatives, see Dickey [2006:fn. 3]). As mentioned in section 3.3, however, I have argued previously (Dickey [2006]) that Russian *po*-delimitatives do not necessarily foreground indefinite duration. Their functional scope has extended beyond this considerably, so that *po*-delimitatives can function to link an atelic activity with some ensuing consequence (i.e., oblique result), cf. ex. (5a) in section 3.3, or to link an ordinarily goal-oriented (and thus telic) predicate with some oblique result, cf. ex. (5b) in section 3.3; these examples are repeated here as (8a–b).

(8) a. *Abxazskaja milicija slavno* ***porabotala***.
Abzhkazi police gloriously PO-worked
'The Abkhazi police did a terrific job.'

b. ***Popisal*** *i rešil vot vpečatlenija v ètot dnevnik*
PO-wrote.1S anddecided.1S PRESENTATIVE impressions into this diary
pozapisyvat'.
to.write.down.DISTR
'I wrote a little while and decided thus to write down all my impressions in this diary.'

An example demonstrating the reality of some subsequent situation (*P'*) that is not even necessarily an oblique result is given by Zel'dovič (2002: 45, fn. 25):

(9) — *Čto ty budeš' delat'?*
what you.SG will.2SG do.INF
— ***Posplju***.
PO-sleep.PFV.PRS.1SG
— *A ešče čto?*
And still what
— ??*Govorju že tebe* — ***pospolju***.
say.IPFV.PRS.1SG EMPH you.SG.DAT PO-sleep.PFV.PRS.1SG
'— What are you going to do?
— I am going to sleep for a while.
— And what else?
— I'm telling you, I'm going to sleep for a while.'

Zel'dovič observes that in (9) the infelicity of *pospljuPF* 'I will sleep some' in the repeated assertion stems from the fact that *po*-delimitatives, like other Russian perfective verbs expressing a predicate *P*, require that the speaker have in mind another situation *P'* in the contextual set of situations *M*, but in (9) the speaker replies that there is no other situation, only 'sleeping for a while', which violates the aforementioned condition. Note that although the subsequent situation *P'* is unexpressed (which is an indicator that it is subjectively construed), it is part of the conveyed content, which motivates the follow-up question.

It is important to point out that *po*-delimitatives are not resultative predicates (an indication of which is the fact that they are intransitive even when derived from essentially transitive predicates such as *pisat'IMPF* 'write'). Accordingly, I suggest that the use of *po*-delimitatives, e.g., in examples (8a) and (9), evidences a more extreme degree of attenuation. Here the effecting-relationship is that a result has been linked with an atelic activity though the latter is not conventionalized as coding a result (as in the case of *porabotat'*$_{PF}$ 'work a while' in [8a]), or simply that the cessation of the temporally delimited activity enables another situation to be performed by the same agent (as with *pospat'*$_{PF}$ 'sleep a while' in [9]). This is diagramed in Figure 7.

I have kept the term oblique result (OR) for the subsequent situation in this diagram, though this situation can be any subsequent situation. It is worth pointing out that the subjective construal of an attenuated effecting-relationship linking to an oblique result involves the same scanning required for the initial configuration of the perfective (Figure 5), i.e., in accordance with Langacker (1999:153) the subjective effecting-relationship is immanent in the objective effecting-relationship. I see circumstantial support for this in the fact that, as mentioned in section 3.1, in the case of resultative verbs the choice between the imperfective and perfective is by most accounts "subjective" to a limited degree, depending on whether the conceptualizer/speaker conceives and asserts the existence of the result in the event frame or not.

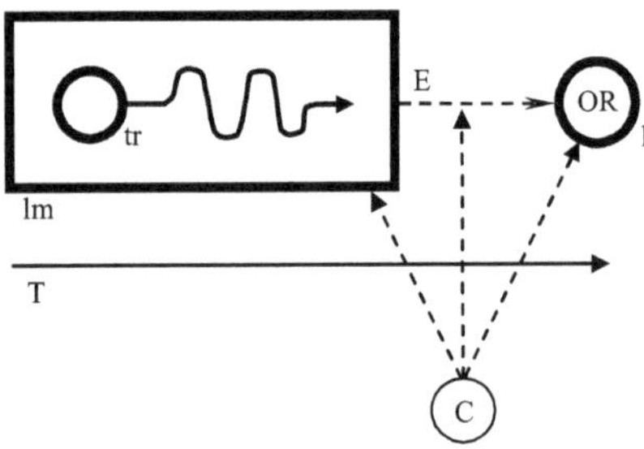

Figure 7. Extreme attenuation of the perfective effecting-relationship: *po*-delimitatives

The question of why *po*-delimitatives, a non-resultative type of predicate, would come to function as if an originally profiled effecting-relationship with a result were attenuated is a complex diachronic question that cannot be discussed adequately here; I will merely speculate briefly. In my view there are two possible answers, which are interconnected. One possibility relies on some particular facts from the history of delimitative *po-* (cf. Dickey [2007] for details). In Old Russian, *po*-delimitatives were telic, i.e., they could profile duration of an atelic activity for the full extent of some specified period of time; around the seventeenth century they lost the ability to profile a full specific duration. This development correlated with the loss by the prefix of its original spatial meanings of surfact-contact and allativity. Other *po*-delimitatives descended from otherwise resultative verbs in *po-*, cf., e.g., Old Russian *povoevati*$_{PF}$ 'conquer [usually distributive]' and Modern Russian *povoevat'*$_{PF}$ 'soldier/wage war for a while'. I consider it possible that the shift from telic verbs in *po-* to atelic *po*-delimitatives involved in one way or another the subjectification of the resultative profile (initial configuration) of the original telic/resultative verbs.

Another possibility involves the diachronic relationship between *po*-delimitatives and determinate motion verbs in *po-*, e.g., *pojti*$_{PF}$ 'go [ingressive]'. Determinate motion verbs in *po-* developed from being able to profile a full spatial trajectory to profiling *ingressive-partial trajectory*, defined as the inception of the motion event in time and some indefinite amount of it (a development which recalls the development of *po*-delimitatives from telic to atelic verbs). Dickey (2007) argues that determinate motion verbs in *po-* played an important role in the rise of *po*-delimitatives in Russian by serving as a semantic model for the new delimitative prefix *po-*. Inasmuch as this is true, it is possible that the unusual profile of the new motion verbs in *po-*, which was ingressive-partial trajectory and in which the destination was optional despite being strongly implied, provided a model for the subjective construal of an oblique result in the new *po*-delimitatives.

These possibilities are not in fact mutually exclusive, and I would suggest that they represent two elements of the same process. In any case, the rise of the new atelic *po*-delimitatives, which began around the seventeenth century, marks a turning point in the development of Russian aspect. I think that the significance of the development of these atelic *po*-delimitatives can be best understood in terms of their addition of the clearly subjective construal of subsequent situations.

4.2.2. Subjectification and intensive-resultatives. Russian intensive-resultatives evidence clear features of subjectification. All of these verbs include an oblique result in the profile, whether they are derived from strictly atelic activity predicates (e.g., *guljat'*$_{IMPF}$ 'stroll') or clearly goal-oriented predicates construed as atelic activities (e.g., *pisat'*$_{2IMPF}$ 'be writing/be a writer'). They never profile the direct, intended result profiled by an ordinarily goal-oriented, telic predicate. Some types of intensive-resultatives do not preclude (some degree of) volition on the part of the subject: referring to the list in (4), they are type (4a), e.g., *otgovorit'sja*$_{PF}$ 'avoid something by talking/talk one's way out of something', type (4b), e.g., *otležat'sja*$_{PF}$ 'lie to the point where one returns to a normal state', type (4e), e.g., *o(b)gljadet'sja*$_{PF}$ 'become accustomed to looking in the dark', type (4f), e.g., *pripit'sja*$_{PF}$ 'get used to a beverage by drinking it', type (4g), e.g., *spet'sja*$_{PF}$ 'reach harmony/accord in singing', and type (4h), e.g., *podol'stit'sja*$_{PF}$ 'gain someone's favor by flattery'. However, only two of these, types (4a) and (4h), assume any significant degree of subject control over the process. The absorptives, resultative-qualificational verbs, and ironic-resultatives likewise preclude subject control. Thus, intensive-resultatives generally involve an attenuation of subject control.

Another way of gauging the degree of subject control in these verbs is to consider whether the profiled result is adverse for the subject or not. Intensive-resultatives profile oblique results ranging from beneficial results to indeterminate results to clearly adverse or harmful results. Of the types in (4), the following involve beneficial results: type (4a), e.g., *otgovorit'sja*$_{\mathrm{PF}}$ 'avoid something by talking/talk one's way out of something', type (4b), e.g., *otležat'sja*$_{\mathrm{PF}}$ 'lie to the point where one returns to a normal state', type (4g), e.g., *srabotat'sja*$_{\mathrm{PF}}$ 'reach harmony/accord in work', and type (4h), e.g., *podol'stit'sja*$_{\mathrm{PF}}$ 'gain someone's favor by flattery'. Others profile an indeterminate result: type (4e), e.g., *obterpet'sja*$_{\mathrm{PF}}$ 'become accustomed to enduring physical or mental pain' and *o(b)gljadet'sja*$_{\mathrm{PF}}$ 'become accustomed to looking in the dark', type (4f), e.g., *prigljadet'sja*$_{\mathrm{PF}}$ 'get used to looking at something' as well as the absorptives, e.g., *zagljadet'sja*$_{\mathrm{PF}}$ 'become absorbed in looking at something'. Note that, as Zaliznjak and Šmelev (2000:117) point out, the absorptives do have a tendency to involve an adverse result. Lastly, many of the types of intensive-resultatives profile an adverse or harmful result: type (4c), e.g., *proferšpilit'sja*$_{\mathrm{PF}}$ 'ruin oneself losing at cards or spending money', type (4d), e.g., *oslyšat'sja* 'hear incorrectly', as well as the resultative-qualificationals, e.g., *izvorovat'sja*$_{\mathrm{PF}}$ 'steal to the point of becoming an incorrigible thief', and the ironic-resultatives, e.g., *dopit'sja* 'drink to the point of an adverse result'. Assuming that people do not ordinarily act to their own detriment, these last types clearly profile an effecting-relationship that is uncontrolled by the subject. Thus, such verbs represent an extreme degree of attenuation comparable to that diagramed in Figure 7.

In my view, the intensive-resultatives, particularly those that profile an adverse or harmful result, represent the most extreme degree of subjectification of the Russian perfective. This view finds circumstantial support in the timing of the rise of such verbs. According to Avilova (1964), who provides commentary on all the verb types in (4) except for (4d, e, g, h), the intensive-resultatives were generally few in number before the time around the end of the eighteenth and the beginning of the nineteenth century, and their productivity increased significantly by the mid-nineteenth century. The absorptives became productive in the eighteenth century, but their productivity nevertheless increased during the nineteenth century, when they came to include "not only verbs of concrete action, but also abstract verbs of bookish language" (Avilova [1964:93]). Avilova (1964:96) also observes that the productivity of prefixal-reflexive verbs expressing an adverse result, i.e., the ironic-resultatives (*do-...-sja*), the resultative-qualificational verbs (*iz-...-sja*) and those in *pro-...-sja*, underwent particularly significant increases in productivity during the course of the nineteenth century. The relatively late appearance of the intensive-resultatives as productive classes of verbs makes sense if these verbs represent the most advanced stage of the subjectification of the Russian perfective aspect, as no result represents a loss of subject control of a situation like an adverse result.

It is important to point out that Langacker's theory of subjectification can make sense of the fact that verbs profiling the attainment of various adverse oblique results have appeared in Russian relatively late and generally account for their synchronic and diachronic nature in a systematic way. To my knowledge, no account of Russian procedural perfectivizing prefixation (synchronic or diachronic) has managed to treat the meanings of the intensive-resultative verbs as anything more than random phenomena.

It is interesting that the Russian perfective appears not to have reached the stage of full subjectification diagramed in Figure 2, but has remained in the attenuation stage. The reasons for this situation lie in the particular nature of Russian aspect as a grammatical category and in the particular path of its grammaticalization. These issues are taken up in the following section.

4.3. *Russian procedurals, subjectification and grammaticalization*

The emergence of Russian aspect as a grammatical category is not a typical case of grammaticalization. Most case studies of grammaticalization (cf., e.g., those in Hopper and Traugott 1993, Lehmann 1995) trace the development of single lexical units into markers of grammatical meaning (a representative case might be the evolution of English *go* from a motion verb to a future auxiliary). Russian aspect (as well as other Slavic aspectual systems) has developed from a lexical derivational category to a grammatical derivational category (cf. in this regard Lehmann [1999], Mende [1999] and Böttger [2004]). Thus, there is no single morpheme that has been grammaticalized as the marker of perfectivity, rather there are several: Mende (1999) lists 12 prefixes that commonly function as purely perfectivizing prefixes in Russian. Moreover, the very same prefixes also function as lexical and procedural prefixes, always with a concomitant perfectivizing effect. For the purpose at hand, this situation means that the distinction between purely perfectivizing prefixation and procedural prefixation is not clear: perfective procedural verbs are perfective verbs, and thus reflect and possibly help shape the semantic category of the Russian perfective aspect.

In my view, this latter point is particularly true of delimitative *po-*. I have argued (Dickey 2005) that Russian delimitative *po-* was instrumental in creating the current category of the Russian perfective, temporal definiteness (or Zel'dovič's [2002] definition outlined in section 3.2 above). Note that *po-* is the only Russian perfectivizing prefix that has lost its original spatial meaning (cf. Dickey [2005: 35] and the references cited there as well as Dickey [2007]), which is an important indication that it has acquired a purely grammatical function. Recall also that, as pointed out in section 3.3, *po-* was responsible for extending the aspect opposition from telic predicates to the class of atelic activities, in the seventeenth century. Inasmuch as a subjective construal of the effecting-relationship was a part of this process, it seems that the appearance of the new Russian *po*-delimitatives, which in my view was a watershed event in the evolution of the semantic nature of Russian aspect, represents a case of an innovative procedural verb reflecting a step in the subjectification of the conceptual content of the Russian perfective. The rise of the new Russian *po*-delimitatives was a very complex lexico-grammatical process, due more or less to the accidental semantic conflation of the old *po*-delimitatives and determinate motion verbs in *po*- (for an attempt at a detailed explanation, cf. Dickey [2007]), and I consider it possible that this process not only reflected but also contributed significantly to the subjectification evident in the Russian perfective today.

I suspect that the situation with the intensive-resultatives is rather different. The history of these verbs is not as complex as that of the *po*-delimitatives.[4] Note that unlike *po*-delimitatives, the intensive-resultatives do profile the attainment of some endpoint by virtue of their telic prefixes (e.g., *za*- 'behind' in the absorptives, *iz*- 'out of' in the resultative-qualificational verbs, *do*- 'up to' in the ironic-resultatives). I suggest that the intensive-resultatives have simply reflected newer morphological possibilities of verbal derivation after the conceptual content of the perfective effecting-relationship was significantly attenuated by the *po*-delimitatives. The derivation of new types of intensive-resultatives has "filled out" the possibilities for various oblique results that are characterized by increasingly lower degrees of subject control. It is for this reason that the appearance of the intensive-resultatives expressing adverse results appears to have been the last major prefixal innovation of the Russian aspectual system. With regard to grammaticalization, the high degree of subjectification evident in Russian perfective verbs indicates a correspondingly high degree of grammaticalization of the perfective aspect and accordingly of the aspect opposition as a whole.

It is important to point out here that with the intensive-resultatives we are not as a rule dealing with the conceptual content of *individual linguistic units, i.e., the various models of derivation, undergoing the subjectification process*. (Compare the case of the development of the *go*-future, where the conceptual content of a single linguistic unit, *go*, undergoes subjectification.) Rather, the individual derivational models themselves represent subjective construals of the original default conceptual content (e.g., *vypit'*$_{PF}$ 'drink up' as the initial configuration versus *popit'*$_{PF}$ 'drink for a while/some' and *dopit'sja*$_{PF}$ 'drink to the point of an adverse result' as progressive stages of attenuation).[5] To

4 This statement refers to the aspectual issues; the precise role of the reflexive suffix *-sja* in the development of such verbs is far from clear.

5 The only exceptions appear to be po-delimitatives, inasmuch as they are partially the product of the subjectification of telic verbs in *po*-, and the ironic-resultatives in *do-...-sja*, which appear to have developed from a neutral oblique-resultative model in *do-...-sja*. This process is itself best viewed as reflecting a progressive loss of subject control in the effecting-relation.

conclude, I would suggest that the history of the various types of Russian perfective procedural derivations tells the story of the subjectification of the Russian perfective much as fossils in sedimentary strata tell the story of evolution.

5. A comparison with other Slavic languages

The last point of consideration is a cross-Slavic comparison: Does Russian evidence a higher degree of subjectification than other Slavic languages? In order to address this question I consider whether the types of perfective procedural verbs discussed in sections 3.3 and 4.2 exist in other Slavic languages. Although a given type of verbal derivation can often be attested if only marginally in most of the Slavic languages, the situation with *po*-delimitatives and intensive-resultatives is fairly clear: they are not equally represented in all of Slavic.

Let us consider *po*-delimitatives first. In terms of the east-west Slavic aspect division outlined in section 1, *po*- has been much more productive as a delimitative prefix in the eastern languages (Russian, Ukrainian, Belarusian, Bulgarian and Macedonian) and Polish than in the western languages (Czech, Slovak, Slovene and Sorbian) and Bosnian/Croatian/Serbian (cf. Dickey and Hutcheson [2003]). Moreover, *po*-delimitatives appear to be slightly less productive in Bulgarian than in East Slavic (Russian, Ukrainian and Belarusian). Thus, the epicenter of delimitative *po*- appears to be East Slavic. If the productivity of *po*-delimitatives in a given Slavic language is correlated with the attenuation of the perfective effecting-relationship, i.e., the subjectification of the perfective aspect, then the languages of the eastern group display a relatively high degree of subjectification of the perfective and the languages of the western group a lower degree, with Polish tending towards the eastern group and Bosnian/Croatian/Serbian patterning clearly with the western group. Further, according to this criterion East Slavic is characterized by a higher degree of subjectification than Bulgarian. This situation corresponds to the facts of aspect usage described by Dickey (2000), according to which aspect is a much more of a lexical category in the languages of the western group, whereas it is much more a discourse-based, referential category in the languages of the eastern group, especially in East Slavic.

The facts regarding intensive-resultatives fit in with this overall picture. The intensive-resultatives are much more productive in Russian (taken as representative of East Slavic; according to Rutkowska 1978 intensive-resultatives are productive in Ukrainian and Belarusian) than in the West or South Slavic languages. The types of intensive-resultatives listed in (4) are not productive in Polish (cf. Rutkowska 1981; isolated verbs of these types in Polish are Russian loans according to Rutkowska 1978). As far as I am aware, none are attested in any other Slavic language. Likewise, according to Rutkowska (1981), resultative-qualificational verbs in *iz-...-sja* do not exist in Polish; according to Isačenko (1960:250) they do not exist in Slovak either. They do not exist in Bosnian/Croatian/Serbian or Bulgarian.

Other intensive-resultatives exist in all of Slavic, but are more productive in Russian. A small number of absorptives exist in all Slavic languages, e.g., Russian *zadumat'sja*$_{PF}$, Polish *zamyślić się*$_{PF}$, Czech *zamýšlet se*$_{PF}$, Bosnian/Croatian/Serbian *zamisliti se*$_{PF}$, Bulgarian *zamislja se*$_{PF}$ 'become absorbed in thought/think hard'. However, it seems that the absorptives that exist across Slavic tend to be predicates of intellectual activity that in reality tend to produce a state of distraction, cf., e.g., Russian *zagljadet'sja*$_{PF}$, Polish *zapatrzyć się*$_{PF}$, Czech *zadívat se*$_{PF}$, Bosnian/Croatian/Serbian *zagledati se*$_{PF}$, Bulgarian *zagledam se*$_{PF}$ 'become absorbed in looking/look hard'. The less likely a process is to produce a state of distraction, the less likely it is to appear in an absorptive verb across Slavic. For instance, *play*, which as a process involving entertainment can produce distraction is unevenly represented in absorptives across Slavic, cf. Russian *zaigrat'sja*$_{PF}$, Bosnian/Croatian/Serbian *zaigrati se*$_{PF}$, Bulgarian *zaigraja se*$_{PF}$ 'become absorbed in play/get carried away playing' (nonexistent in Czech and Polish). Outside of such predicates, absorptives exist only in Russian, cf., e.g., *zafilosovstvovat'sja*$_{PF}$ 'get carried away philosophizing' and *zapisat'sja*$_{PF}$ 'become absorbed in writing'—these verbs have no equivalents in the West or South Slavic languages. I suggest that the productivity of this derivational model with predicates other than those of intellectual activity, as such predicates do not express processes that normally involve distraction, is reflective of a higher degree of loss of subject control in the Russian perfective.

A similar situation exists regarding the intensive-resultative verbs in *do-...-sja*. Such intensive resultatives exist in all Slavic languages, e.g., Russian *dokopat'sja*$_{PF}$, Polish *dokopać się*$_{PF}$, Czech *dokopat se*$_{PF}$, Bosnian/Croatian/Serbian *dokopati se*$_{PF}$, Blg *dokopam se*$_{PF}$, 'discover by digging'. However, the ironic-resultatives, e.g., Russian *dopit'sja* 'drink to the point of an adverse result' have not been developed as a productive class in the West Slavic languages (cf., Isačenko [1960:250] for Slovak, Anan'eva [1998:292] for Czech and Polish; isolated verbs of this type in Polish are Russian loans according to Rutkowska 1978). Ironic-resultatives are also unattested in Bosnian/Croatian/Serbian and Bulgarian.

Thus, the comparative cross-Slavic data indicate that the procedural perfective verbs examined in this study are much more productive in Russian and East Slavic than in the other Slavic languages (though *po*-delimitatives are productive in Bulgarian as well). Again, inasmuch as the development of such procedurals is correlated with the subjectification of the perfective aspect (and in turn with the degree of grammaticalization of aspect), it appears that aspect in

Russian and East Slavic has undergone a greater degree of subjectification than the languages of the western group and even Bulgarian.

6. Conclusion

This article has attempted to apply Langacker's theory of subjectification to the Russian perfective aspect. It has been argued that the use of the Russian perfective in contexts of oblique results as opposed to strictly direct results may be profitably viewed as a case of the subjectification of the effecting-relationship existing between a process and its result. This analysis offers a principled account of the perceived "subjective" or "perspectival" nature of Russian aspect (especially in contrast to aspect in some other Slavic languages such as Czech). As Russian aspect is a lexico-grammatical category, and there are a great number of types of perfective verbs, some of which are associated with individual prefixes and are known as procedural verbs, the approach taken here has not focused on a single linguistic unit and traced its progressive stages of subjectification, but rather attempted to correlate certain kinds of procedural verbs with various degrees of subjectification of the perfective effecting-result. Thus, it has examined individual types of procedural verbs as artifacts of subjectification from a synchronic and diachronic perspective. In particular, it has been argued that Russian *po*-delimitatives represent a considerably attenuated effecting-relationship, and that the intensive-resultative verbs specifically profile the attenuation of subject control of the effecting-relationship, which is most evident in a number of intensive-resultatives that express adverse oblique results.

It has further been suggested that the relatively high productivity of *po*-delimitatives in East Slavic and Bulgarian, as well as the productivity of intensive-resultatives in East Slavic indicate a higher degree of subjectification of the perfective effecting-relationship, and thus the grammaticalization of the aspect category, in the languages of the eastern group than in the western group. Within the languages of the eastern group, aspect in East Slavic displays a higher degree of subjectification and grammaticalization than in Bulgarian. The descriptive picture may also be reformulated in terms of entailment relations: if a language has developed productive intensive-resultatives, it has also developed productive *po*-delimitatives (East Slavic), but not necessarily vice versa (Polish and Bulgarian).

If one allows for the unusual number of forms expressing perfectivity in Russian and the accompanying difficulties in accounting for the grammaticalization of Russian aspect, the semantic development of the Russian perfective appears to be a almost a textbook case of Langacker's process of subjectification. The original, highly lexical aspectual system of Old Russian represents the initial configuration of the perfective, and the contemporary

Russian aspectual system represents a considerable attenuation of the conceptual content of the perfective aspect. Moreover, the subjectively construed effecting-relationship linking with an oblique result was immanent in the objective effecting-relationship linking to a direct result.

An unexpected advantage of a subjectification analysis has in fact turned out to be its flexibility in dealing with the interweaving of the lexical forms and grammatical content of the Russian perfective. Instead of tracing the subjectification of the conceptual content of a single linguistic unit, the analysis simply recognizes the various prefixed procedural verb types as reflecting various stages of attenuation of the effecting-relationship. The established chronology of the appearance of the different types of procedural verbs is what one would expect based on the general theory, as the procedurals expressing the greatest loss of subject control—the intensive-resultatives profiling adverse results—have appeared most recently, becoming productive in the nineteenth century. Applying the principles of subjectification to the types of Russian perfective procedurals exposes them as "fossils" of the semantic development of the Russian perfective. Therein lies one of the greatest advantages of this approach, which is that it brings some order to the types of procedurals that have appeared in Russian, which in the grammatical treatments to date have been listed only as random categories.

References

Anan'eva, N. E. 1998. Problemy sopostavitel'nogo analiza prefiksal'nyx glagolov drevnego perioda (na materiale prefiksal'nyx glagolov s *do-/do-* v drevnepol'skom, drevnečešskom i drevnerusskom jazykax) [Problems of the comparative analysis of prefixed verbs of the ancient era (on the material of verbs prefixed with *do-/do-* in Old Polish, Old Czech and Old Russian]. In A. G. Širokova (ed.), *Sopostavitel'nye issledovanija grammatiki i leksiki russkogo i zapadnoslavjanskix jazykov* [Comparative investigations of the grammar and lexicon of Russian and the West Slavic languages]. Moscow: Moskovskij gosudar'stvennyj universitet.

Avilova, N. S. 1964. Razvitie pristavočnogo glagol'nogo slovoobrazovanija [The development of prefixal verbal derivation]. In V. V. Vinogradov and N. Ju. Švedova (eds.), *Očerki po istoričeskoj grammatike russkogo literaturnogo jazyka XIX veka. Glagol, narečie, predlogi i sojuzy* [Studies in the historical grammar of the Russian language of the nineteenth century. The Verb, adverb, prepositions and conjunctions], 45–104. Moscow: Nauka.

Barentsen, Adrian. 1998. Priznak «sekventnaja svjaz'» i vidovoe protivopostavljenie v russkom jazyke [The feature "sequential connection" and the aspect opposition in Russian]. In Marina Ju. Čertkova (ed.), *Tipologija vida: Problemy, poiski, rešenija* [The typology of aspect: Problems, approaches and solutions], 43–58. Moscow: Jazyki russkoj kul'tury.

Bermel, N.: 1997. *Context and the lexicon in the development of Russian aspect* (Linguistics 129). Berkeley: University of California Press.

Böttger, Katharina. 2004. Grammaticalization the derivational way: The Russian aspectual prefixes PO-, ZA-, OT-. In Walter Bisang, Nikolaus P. Himmelmann and Björn Wiemer (eds.), *What makes grammaticalization? A look from its fringes and its components*, 187–209. Berlin: Mouton de Gruyter.

Carey, Kathleen. 1995. Subjectification and the Development of the English Perfect. In Dieter Stein and Susan Wright (eds.), *Subjectivity and subjectivisation*, 83–102. Cambridge: Cambridge University Press.

Chung, Sandra and Alan Timberlake. 1985. Tense, Aspect, and Mood. In T. Shopen (ed.), *Grammatical categories and the lexicon*, 202–258. Cambridge: Cambridge University Press.

Dahl, Östen. .1985. *Tense and aspect systems*. Oxford: Basil Blackwell.

Dickey, Stephen M. 2000. *Parameters of Slavic aspect: A cognitive approach*. Stanford: CSLI.

Dickey, Stephen M. 2003. "Semelfactive" *-nǫ-* and the western aspect gestalt. *Journal of Slavic linguistics* 10(1): 29–52.

Dickey, Stephen M. 2005. *S-/Z-* and the grammaticalization of Slavic aspect. *Slovene linguistic studies* 5: 3–55.

Dickey, Stephen M. 2006. Aspectual pairs, goal orientation and PO- delimitatives in Russian. *Glossos* 7: http://seelrc.org/glossos/issues/7/ (accessed 18 May 2009).

Dickey, Stephen M. To appear. The varying role of *po-* in the grammaticalization of Slavic aspectual systems: Sequences of events, delimitatives, and German language contact. In Bostjan Dvořak and Elena Gorišneva (eds.), *Collected studies of the Second congress of the Slavic linguistics society*. Berlin: Zentrum für allgemeine Sprachwissenschaft.

Dickey, Stephen M. and Julie Hutcheson. 2003. Delimitative verbs in Russian, Czech and Slavic. In A. Maguire and Alan Timberlake (eds.), *American contributions to theThirteenth international congress of Slavists*, 23–36. Bloomington, IN: Slavica.

Dickey, Stephen M. and Laura Janda. 2009. *Хохотнул, схитрил*: The relationship between semelfactives formed with -NU- and S- in Russian. *Russian Linguistics* 33: 229–248.

Dickey, Stephen M. and Susan Kresin. 2009. Verbal aspect and negation in Russian and Czech. *Russian Linguistics* 33(2): 121–176.

Durst-Andersen, Per. 1993. *Mental grammar. Russian aspect and related issues*. Columbus, OH: Slavica.

Forsyth, James. 1970. *A grammar of aspect. Usage and meaning in the Russian verb*. Cambridge: Cambridge University Press.

Galton, Herbert. 1976. *The main functions of the Slavic verbal aspect*. Skopje: Macedonian Academy of Sciences and Arts.

Holden, Kyril. T. 1990. The functional evolution of aspect in Russian. In Nils Thelin (ed.), *Verbal aspect in discourse*, 131–158. Amsterdam: John Benjamins.

Hopper, Paul J. and Elizabeth Closs Traugott. 1993. *Grammaticalization*. Cambridge: Cambridge University Press.

Isačenko, Alexandr V. 1960. *Grammatičeskij stroj russkogo jazyka v sopostavlenii s slovackim. Čast' vtoraja: Morfologija* [The grammatical structure of Russian in comparison with Slovak. Part two: Morphology]. Bratislava: Izdatel'stvo akademii nauk.

Israeli, Alina. 1996. Discourse analysis of Russian aspect: Accent on creativity. *Journal of Slavic linguistics* 4.1, 8–49.

Israeli, Alina. 1998. Speakers' attitudes, goals and aspectual choices in WH-questions. *Le Langage et l'Homme* 33.1, 55–77.

Israeli, Alina. 2001. The choice of aspect in verbs of communication: Pragmatics contract. *Journal of Slavic linguistics* 9(1): 49–98.

Janda, Laura. 2007. Aspectual clusters of Russian verbs. *Studies in Language* 31 (3): 607–648

Langacker, Ronald W. 1985. Observations and speculations on subjectivity. In John Haiman (ed.), *Iconicity in syntax* (Typological studies in language 6), 109–150. Amsterdam: John Benjamins.

Langacker, Ronald W. 1990. Subjectification. In Ronald W. Langacker, *Concept, image, and symbol*, 315–342. Berlin: Mouton de Gruyter.

Langacker, Ronald W. 1999. Losing control: grammaticization, subjectification, and transparency. In Andreas Blank and Peter Koch (eds.), *Historical semantics and cognition*, 147–175. Berlin: Mouton de Gruyter.

Langacker, Ronald W. 2006. Grammaticization, subjectification, and conceptual archetypes. In Angeliki Athanasiadou, Costas Canakis, and Bert Cornillie (eds.), *Subjectification: various paths to subjectivity*, 17–40. Berlin: Mouton de Gruyter.

Lehmann, Christian. 1995. *Thoughts on grammaticalization*. Munich: Lincom Europa.

Lehmann, Volkmar. 1999. Sprachliche Entwicklung als Expansion und Reduktion [Language development as expansion and reduction]. In Tanja Anstatt (ed.), *Entwicklungen in slavischen Sprachen* [Developments in Slavic languages], 160–254. Munich: Otto Sagner.

Leinonen, Marja. 1982. *Russian aspect,* temporal'naja lokalizacija *and definiteness/ indefiniteness*. Helsinki.

Mende, Julia. 1999. Die Grammatikalisierung des russischen Aspekts [The grammaticalization of Russian aspect]. Tanja Anstatt (ed.), *Entwicklungen in slavischen Sprachen,* 285–325. Munich: Otto Sagner.

Rutkowska, Maria. 1978. Nezauważone rusycyzmy (wschodnioslawizmy) semantyczne w czasownikach polskich XVIII–XX wieku [Unreocognized semantic Russianisms (East-Slavicisms) in Polish verbs of the 18th–20th centuries]. *Z polskich studiów slawistycznych* [From Polish Slavistic studies] 5: 293–299.

Rutkowska, Maria. 1981. *Nie znane polszczyźnie rosyjskie prefigowane derywate odczasownikowe z elementem* -sja *typu* doležat'sja, priest'sja [Russian prefixed deverbal derivatives of the type *doležat'sja, priest'sja* that are unknown in Polish] (Monografie Sławistyczne 43). Wrocław: Komitet Słowianoznawstwa.

Sasse, Hans-Jürgen. 2002. Recent activity in the theory of aspect: Accomplishments, achievements, or just non-progressive state? *Linguistic typology* 6(2): 199–271.

Schuyt, Roel. 1990. *The morphology of Slavic verbal aspect: A descriptive and historical study* (Studies in Slavic and general linguistics 14). Amsterdam: Rhodopi.

Stunová, Anna. 1993. *A contrastive study of Russian and Czech aspect: Invariance vs. discourse*. Ph.D. dissertation, University of Amsterdam. Townsend, Charles. 1975. *Russian word formation.* Columbus, OH: Slavica.

Zel'dovič, Gennadij Moiseevič. 2002. *Russkij vid: Semantika i pragmatika*. Toruń: Uniwersytet Mikołaja Kopernika.

Predicting aspectual choice in modal constructions: a quest for the Holy Grail?

Dagmar Divjak, University of Sheffield

1. Aim

In this paper I will narrate my quest for the variable that best predicts aspect choice in modal constructions in Slavic. Polish plays a key role in this as the language to defy the original hypothesis stating that modality type would determine aspectual choice. Drawing on findings from cognitive linguistics to interpret the results, I will propose a new model for aspect assignment in modal contexts that accommodates Polish and provides a realistic account of the cognitive mechanics involved. Central to this research is the use of copora as an excellent source of authentic language data and of statistical techniques to model the data and make predictions.

1.1. *Background*

Much effort has been put into clarifying the relation between modality and other verbal properties, in particular mood and tense. Until recently the relation between modality and aspect received much less attention, however. For Slavic languages this situation is particularly unfortunate as Slavic languages mark the imperfective versus perfective distinction on all verbal forms. In other words, there is no avoiding aspectual choice, not even in conceptually already complex modal contexts.

Moreover, the hypothesis that directed much of the research states that imperfective aspect prevails in modal constructions or that the imperfective is used to express epistemic modality; perfective aspect would render root modality. This hypothesis does not hold for Slavic data, however: it has long been recognized that, if anything, the perfective would be used more frequently in modal constructions in general (cf. Rassudova 1968, Forsyth 1970) and the imperfective aspect would be preferred in deontic contexts (Durst-Andersen 1992: 144, Padučeva 2006, Šmelev & Zaliznjak 2006, Wiemer ms.).

Yet, conclusive evidence for or against either hypothesis is lacking as linguists involved in the debate, including non-native speakers (Durst-Andersen 1992: 140-151 and 1995: 615-623, for virtually identical accounts), seem to prefer working with (non-)native speaker judgments on a very limited set of data generated and analyzed by means of introspection.

1.1.1. *A data challenge*

Native speakers may well have a good idea of whether or not sentences in their language are grammatical, but their intuitions lack the level of reliability and scope needed to construct a scientific theory of language. For one, native speakers do not necessarily agree on the (un)grammaticality of an item, depending on the context they imagine and the particular dialect they speak. And although such mismatches have a role to play in hypothesis generation, "objective evidence based on experiments that can be replicated and that test falsifiable hypotheses is needed if one wishes to make generalizations about the way in which people ordinarily and automatically engage in cognitive and linguistic processing" (Gibbs 2007: 5)

Croft (1998) cautions that introspection can only be a guide to constraints on possible forms. It cannot, on its own, predict what will in fact occur. This is attested in research on aspect assignment in modal contexts. Padučeva (2006), for example, put forward the claim that the aspectual form of the infinitive under negation would correlate with the concepts of possibility and necessity. Confronting these intuitions with corpus data revealed that only negated deontic modality constrains the aspectual options the infinitive has as predicted, limiting it to the imperfective aspect.

Instead of relying on an introspective analysis of decontextualized sentences I advocate and take a strictly corpus-based, quantitative approach to the analysis of the interaction between aspect and modality. To this end, I assembled a small parallel corpus specifically for this study (see Table 1 in Section 2.1). The concept behind corpus studies should not be alien to Slavic linguists – think of the millions of quotes from the entire Russian literature carefully copied onto cards and stored at the IRJa-RAN in Moscow. Yet, corpus studies are far from mainstream in Slavic linguistics, in particular the type of corpus linguistics that relies on statistical data anlaysis. This is surprising, given that corpus linguistics in its earliest form was practiced in Eastern European linguistic research and careful analyses based on manual counts of the occurrence of a particular form in a limited number of literary works are attested (see Krasnych (1979) to name but one example). Statistical analysis of such data is the next logical, and required, step.

1.1.2. *A methodological challenge*

One of the huge advantages of using corpora is the frequency information that they offer us. Corpus-based studies typically rely to some degree on frequencies in their argumentation, citing, for example, raw numbers or percentages. Yet, despite this reliance on frequencies, confirmatory statistics are conspicuously absent from many corpus-based studies. As a consequence, the authors can strictly speaking not determine whether the frequencies that they have observed are merely a fluke of the data, or capture a regularity that may be expected to exist outside the dataset used. For this, (even) linguists need statistics.

There are many possible reasons why linguists might shun statistics. For one, different from psycholinguists, linguists are unaccustomed to quantitative thinking. Although the tide is turning (see Baayen 2008 and Johnson 2008 for fine examples of textbooks introducing linguists to statistics), most linguists graduate without having sat through a single course in which Greek letters are squared or assigned numerical values, let alone a course in which they are actually taught techniques that are appropriate for the discipline. This is, of course, only made possible by the peculiar data collection and analysis techniques upheld in the field.

Linguists who do rely on corpora should turn to statistics as statistics facilitates summarizing complex numerical data and, if desired, draw inferences. Furthermore, it makes it possible to falsify statements, models and even theories. On what basis can we decide which intuition-based model is better? Who talks prettiest or screams loudest, maybe?

1.1.3. *An interpretative challenge*

That something is "falsifiable" does not mean it is false; rather, that if it is false, then this can be shown by observation or experiment. Karl Popper promoted "sophisticated methodological falsification" with the aim to arrive at an evolutionary process whereby theories become "less bad". When experimental observation makes a theory untenable, a new theory needs to be found that has greater explanatory power, and as a result, can account for the phenomena that falsified the prior theory.

Falsified statements play an important role in this process. Scientific theories are based on a collection of statements, and it is this collection that must be accepted or rejected as a whole: a theory can always be defended by adding ad hoc hypotheses, but, at some point, the collection of ad hoc hypotheses becomes so large that it is unreasonable to support the original theory, and a decision will be made to reject it. This is precisely how corpus research works: "Corpus research in fact, neither denies nor ignores the necessity of interpretations but it takes on a helix-like structure of gradual refinement of interpretations through a repeated confrontation with empirical data. An initial hypothesis, which will more often than not be derived introspectively, is confronted with the corpus data; interpreting the results leads to a more refined hypothesis and more questions, which may then be subjected to further testing – and so on" (Grondelaers et al. 2007: 150).

In this paper, I will show how a statistical analysis of corpus data enabled me to reverse the initial statement on the interaction between aspect and modality, and subsequently falsify that very model using data from Polish, which led to a better-fitting model for Polish and for other languages. The resulting model, at the same time, sheds light on the cognitive mechanisms that have given rise to the pehnomenon.

2. An exploratory comparative study

An exploratory corpus-based, quantitative study (Divjak & Dahl 2007) was carried out to identify the aspectual preferences of dynamic and deontic modality as defined by Nuyts (2006). In what follows I will briefly introduce the key ingredients for this study.

The constructions focused on are positive and negative declarative sentences built around a predicative modal adverbial followed by an infinitive, as illustrated in examples (1) and (2). The former would convey dynamic modality, while the latter would express deontic modality (examples taken from Šmelev & Zaliznjak 2006).

(1) *Zdes' možno perechodit' ulicu*
Here.$_{\text{ADV}}$ can/may$_{\text{.PREDADV}}$ cross.$_{\text{IMPF.INF}}$ street.$_{\text{ACC.F.SG}}$
'You may cross the street here'

(2) *Zdes' možno perejti ulicu*
Here.$_{\text{ADV}}$ can/may$_{\text{.PREDADV}}$ cross.$_{\text{IMPF.INF}}$ street.$_{\text{ACC.F.SG}}$
'You can get across the street here'

The modal word *možno* is a predicative adverbial, i.e. it is an adverb that functions as a verb, hence exists in only one form. It is the infinitive that poses a challenge with respect to aspectual choice. Although both imperfective and perfective infinitives can be used with the modal predicative adverbial *možno*, it is suggested that the aspect of the infinitive co-determines the modal meaning expressed. More precisely, and contrary to what is reported in typological literature (Abraham & Leiss 2008: xii-xiii), it appears to be the case that selecting a perfective infinitive forces a dynamic reading upon the modal situation, whereas choosing an imperfective infinitive pushes the situation towards deontic modality (Durst-Andersen 1992: 144, Padučeva 2006, Šmelev & Zalizniak 2006, Wiemer ms.). Is this really what is going on?

2.1. *Data collection*

As (1) and (2) show, the starting point for the comparative study was the situation in Russian that uses *možno/nel'zja* to express (in)ability, (im)possibility and (non-)permissibility and *nužno* and *nado* to express necessity and obligation. Let us look at one corpus example for each of them.

(3) Да, мы не верим в бога, - чуть улыбнувшись испугу интуриста, ответил Берлиоз, - но об этом **можно говорить**$_{impf.inf}$ совершенно свободно. [MM]
"Correct, we do not believe in God, - Berlioz answered, having smiled (a bit) at the fear of the foreign tourist, - but we **can talk** absolutely freely about that."

(4) Ни в одной лабораторной печи **нельзя** было **получить**$_{pf.inf}$ температуру, при которой бы они плавились, разве что внутри атомного котла. [S]
"There was not a single laboratory oven in which a temperature **could be reached** at which they would melt, except inside an atomic chamber."

(5) Выведите его отсюда на минуту, объясните ему, как **надо разговаривать**$_{impf.inf}$ со мной. [MM]
"Take him outside for a minute, and explain to him how **(one has) to talk** to me."

(6) Когда песок переставал шуршать, **нужно** было **перевернуть**$_{pf.inf}$ книгу и продолжать чтение в обратном порядке, с того места, где остановился, к началу (...) [ChR]
"When the sand stopped rustling, one **had to turn** the book around and continue reading in the reverse direction, from the place where you had stopped to the beginning (...)"

As is clear from the glosses, the infinitives in examples (3) and (5) are imperfective, while the ones in (4) and (6) are perfective. According to the hypothesis put forward by Šmelev & Zaliznjak (2006), among others, (3) and (5) would express deontic modality while (4) and (6) would render dynamic modality.

Given that this situation would reverse the claims about the relation between aspect and modality put forward in the general, typological literature (Abraham & Leiss 2008), I decided to test which of the existing hypotheses about the relation between modality and aspect would hold beyond the very limited set of examples generated and judged with a particular hypothesis in mind. And since it appears that Russian runs counter to what most of the other languages seem to be doing, I found it worth investigating this phenomenon on a larger scale and taking at least an areal perspective. In other words, it is the aim of this study to identify the aspectual preference of deontic and dynamic modality in three Slavic languages, East Slavic Russian, West-Slavic Polish and South-Slavic BCS.

It is well-known to typologists that a decent parallel corpus is needed for this type of investigation as "one cannot simply take any random assortment of texts because the heterogeneity of the corpora would be an obstacle." (Stolz 2007: 102). Since no authoritative parallel Slavic corpus exists, a 1 million word parallel Slavic corpus was compiled specifically for this purpose (see Table 1).

Table 1. Corpus contents

Original	Translation	Translation
(Russian) Bulgakov, M. 1938. *Master i Margarita.*	(Polish) Mistrz i Małgorzata (by Irena Lewandowska & Witold Dąbrowski)	(Serbian) Мајстор и Маргарита (by Milan Čopić)
(Polish) Lem, S. 1961. *Solaris*	(Russian) Солярис (by Dmitrij Bruškin)	(Serbian) Solaris (by Predrag Obućina)
(Serbian) Pavić, M. 1984. *Hazarskij Rečnik.*	(Polish) Słownik chazarski (by Elżbieta Kwaśniewska & Danuta Cirlić-Straszyńska)	(Russian) Хазарский словарь (by Larisa Savel'evaja)

The corpus contains 1 million tokens, with Master i Margarita [MM] accounting for roughly 44,4%, Solaris [S] for 22,2% and Hazarskij Rečnik [ChR] for 33,3% of the corpus. The texts have been scanned, spell-checked, morpho-syntactically annotated, lemmatized and aligned (von Waldenfels 2006) and are now included in ParaSol (www-korpus.uni-r.de/ParaSol), a collection of on-line searchable Slavic (post-war belletristic) texts. Working with a parallel corpus offers many advantages when languages are being compared: "Direct comparability of concrete examples across languages is a strong point of the parallel text method. In the ideal case the same domains, instantiated in the same examples, are represented in the same textual environment with the same degree of emphasis in the same register. That means that, given that the analysis of all examples has been successfully completed, the values for the same features can be determinded by applying the same criteria" (Wälchli 2007: 131-132).

In all, 983 instances were retrieved, starting from Russian. For Russian, there were in all 380 examples, divided over the four afore-mentioned modal predicative adverbials *nado, nužno, možno* and *nel'zja.* For Polish, 12 translational equivalents were found (*dać, dać się, mieć, móc, można, musieć, należy, niepodobno, powinien, sposób, trudno, trzeba*) in 299 sentences, while Serbian yielded 8 (*moci, moguće, morati, smeti, trebati, potrebno, valjati, biti u stanju*) from 304 sentences. Clearly, modality as focused on in this paper is a lexical category.

As recommended, "[...] a full-blown corpus analysis should be carried out separately for each of the various sample languages - and only after their completion, the results of these separate studies can be compared to each other in order to allow for generalizations" (Stolz 2007: 102). Each of the language samples was analyzed separately according to the criteria outlined in the next section.

2.2. *Variables*

The qualifying 821 extractions were tagged for eight properties in all: 1) the language they were written in, 2) the novel they had been extracted from, 3) the author or translator of the work, 4) the modal word exemplified in them, 5) the aspectual range of the infinitive used (imperfective only, perfective only, biaspectual, imperfective & perfective), as well as 6) the actual aspect of the infinitive (imperfective or perfective) used in the extracted sentence. Furthermore, two explanatory variables that had been identified in literature were included: 7) modality type (dynamic vs deontic) and 8) polarity (positive vs negative).

Although most of these properties are self-explanatory, some remarks are in order. First, the fact that the aspectual range was recorded points in the direction of not all extractions being used. Given that I aim to predict the choice of imperfective or perfective aspect, only those examples are taken into consideration with an infinitive that exists in both imperfective and perfective aspect. This was the case in 314 Russian sentences, 240 Polish sentences and 267 BCS sentences.[1]

In Slavic languages, verbal actions come pre-formed, pre-packaged as events, be they perfective or imperfective. In a metaphorical sense, this means that all verbal actions in Slavic have some identity as events beyond and prior to the timeline. They have been given their contours, and are thus objects, objects that can arrive at destinations. In English, events take shape in the timeline, they are not pre-fab objects, which means that they cannot move into their slots. The shapes of events in English are not prior givens, but often interpretable only upon reflection, as a result of factors beyond the verb itself (Janda 2002: 7).

Generally speaking (for details of differences between Russian, Polish and BCS I refer to Dickey 2000), imperfectives signal situations without temporal restriction, i.e. the imperfective focuses attention on the situation's progress, duration or repetition. Imperfectives are thus reserved for situations that are unbounded in time. Perfectives are used to express situations that can be compressed to occupy a single point on the time-line. For example, one-off events, completed events, or events in which beginning and/or ending is stressed

---

1 Restricting the data sample to verbs that occur in both aspects, and for that matter in pairs in which the perfective adds merely the idea of completion to the imperfective, implies that some duratives (i.e. those lacking a perfective partner), all ingressives, all delimitatives and all semelfactives remain outside the scope of this research. For these verbs, there is no imperfective or perfective variant, hence the problem of selecting aspect given a modal context does not pose itself.

would be expressed using a perfective. In other words, perfectives are used to render situations that are bounded in time or restricted to occurring once.

Modality, too, needs a word of explanation as it has been defined differently by different researchers. For the purpose of operationalizing the concept of modality I opted for Nuyts' characterization (2006: 2-5) and work with two subtypes, i.e. dynamic and deontic modality. I prefer this interpretation of modality over many others precisely because it distinguishes between the subject and his/her immediate circumstances on the one hand, and the larger community with its moral, social and legal rules, prescriptions and laws on the other hand. In my view, this division is basic enough to emerge early in infancy and be grounded in our first and most immediate experiences of the world. Rochat and Hespos (1997) report that even neonates discriminate between self and other as they tend to display significantly more rooting responses following external stimulation than following self-stimulation. Moral and conventional regulation takes considerably longer to develop (Turiel 2006), and experts estimate that social understanding only begins to emerge between age 5 and 9. Definitions of modality relying on more fine-grained and thus harder to grasp disctinctions between obligation and necessity or possibility and permission are therefore less likely to relate to early experiences than is a concept of modality based on the self/other versus society distinction.

Dynamic modality (from Greek *dynamis*, meaning 'strength' or 'power') is concerned with the disposition of certain empirical circumstances with regard to the occurrence of some event (Perkins 1983: 34). Following Nuyts (2006, 2-5), I regard the state of affairs (henceforth SoA) as dynamic if it stems from a capacity ascribed to the controlling participant in the SoA. This capacity may be participant inherent or participant imposed. Participant inherent capacities require reference to be made to (participant) internal factors such as ability or capacity, whereas participant imposed capacities imply reference to (participant) external factors, e.g. circumstances. An example of dynamic modality is given in (7). As the example reveals, at this stage no attempt has been made to distinguish between (participant) internal and external factors.

(7) Я постучал — никакого ответа. — "Сарториус!" — крикнул я. — "Доктор Сарториус! Это я, новичок — Кельвин! Мне *нужно* с вами *увидеться*$_{\text{PF.INF}}$, прошу вас, откройте!" [S]
I knocked – no answer. "Sartorius!" – I yelled. – "Doctor Sartorius! It's me, novice Kelvin! I *need to see* you, please, open up!"

Deontic modality (from Greek *deon*, meaning 'duty' or 'that which is binding') is a modality relating to normative or valuational classifications of actions and states of affairs, such as the obligatory, the forbidden, the permitted etc. Following Nuyts, I regard the source of that necessity, obligation, possibility, permission etc. as deontic if an indication of degree of moral or social desirability of the SoA is expressed, as is the case in example (8).

(8) И, конечно, совершенно ужасно было бы даже помыслить о том, что такого человека *можно казнить* $_{\text{IMPF.INF}}$. [MM]
"And, of course, just thinking about the fact that such a man *could be executed* was so awful."

The working hypothesis, borrowed from Slavic linguistic literature, was that both deontic modality and negative contexts would favour imperfective aspect, whereas dynamic modality and positive contexts would favour perfective aspect. This runs counter to typological generalizations (Abraham & Leiss 2008). Hence, if the Slavic version proves to hold, the typological model needs revising.

2.3. *Analysis*

Given that the choice of imperfective versus perfective aspect is a binary choice, a logistic model can be fitted. A logistic model predicts the probability of occurrence of an event; in this case, it assesses the likelihood of encountering an infinitive in a particular aspect (imperfective or perfective) in a particular polar (positive or negated) and modal (dynamic or deontic) context. It does so by examining the relation of a binary dependent variable or response variable to specified independent variables. The dependent variable here is the imperfective or perfective aspect the infinitive occurs in, while the independent variables, a.k.a. explanatory variables are modality type and polarity type.

One crucial assumption logistic regression analyis makes is that all observations are independent. In this sample, however, there is quite some repetition. First, examples are extracted from 3 novels only, written or translated by in all 9 different people on a limited number of modal predicative adverbials. This gives us repeated measurements per subject and item. Furthermore, the number of observations per novel/writer and modal predicative adverbial is uneven.

In addition, not all sources of repetition are of equal interest (at this stage). Since this is the first quantitative study, gaining insight in general tendencies seemed more important than learning more about the details of the system. In other words, I wanted to be able to generalize to the entire "population" of novels and modal verbs, not just study the use of aspect in modal contexts in one particular novel. On the other hand, the parameters included are interesting in themselves and I wanted to study specific levels of modality and polarity.

One way of taking all these facts and factors into account is by fitting a mixed-effects logistic regression model that allows controlling for fixed and random individual differences.

2.4. *Results*

Following Baayen (2008, Section 7.3), I fitted a mixed effects logistic regression, using lmer in R 2.5.0, to the 821 instances that contain an infinitive that exists in both imperfective and perfective, i.e. allow aspectual choice. The aim was to predict the propensity of modality type and polarity type to co-occur with imperfective or perfective aspect. Different models were fitted and compared using ANOVA. The final model incorporates novel and modal word as random effects (interest is in the general population, not in the behaviour of aspect in modal contexts in one novel, or aspectual choice with one modal verb) and modality type and polarity as fixed effects (here, interest in the particular levels distinguished, i.e. deontic versus dynamic modality and negative vs positive). The results of the best performing model are summarized in Table 2.

Table 2. Comparing models across languages

Russian	Polish	Serbian
a modal adverb followed by a perfective infinitive is not typically used to express deontic modality [estimate = -5.4567, p= 6.95e-11]	a modal adverb followed by a perfective infinitive is not typically used to express deontic modality [estimate = -2.1838, p= 1.5e-06]	a modal adverb followed by a perfective infinitive is not typically used to express deontic modality [estimate = -2.8217, p= 3.53e-09]
a modal adverb followed by a perfective infinitive is found when the modal statement is positive [estimate = 3.8689, p= 0.000807]	a modal adverb followed by a perfective infinitive is found when the modal statement is positive [estimate = 0.7439, p= 0.05308]	a modal adverb followed by a perfective infinitive is found when the modal statement is positive [estimate = 1.3420, p=0.000362]
Estimated scale [0.9864484] C index of concordance [0.8670398] Somer's D [0.7340796]	Estimated scale [0.991989] C index of concordance [0.7405442] Somer's D [0.4810883]	Estimated scale [0.980616] C index of concordance [0.8037842] Somer's D [0.6075684]

In words: a model with language and modal word as random effects and modality type plus polarity as fixed effects revealed that, in all three Slavic languages studied, 1) it is significantly less likely to find a modal adverb followed by a perfective infinitive when deontic modality is expressed than it is to find an imperfective infinitive; 2) it is significantly more likely to find a perfective infinitive when the modal statement is positive than it is to find an imperfective infinitive. The outcome is exactly what Slavic linguists had predicted.

Although both modality and polarity show up as significant predictors of the choice of a particular aspect for the infinitive in all three languages, the diagnostics reveal that the model fits the Russian data best. The estimated scale of variation for Russian, [0.9864484], is very close to the ideal of [1]; in other words, there is only slightly less variation in the data than the model would expect. The C index of concordance between predicted probabilities and observed responses equals [0.8670398]. When C takes on the value 0.5 the predictions are random, and when it takes on the value 1 the predictions are perfect. A value above 0.8 indicates that the model has real predictive capacity, as it can be interpreted to mean that 80 percent of the time a randomly selected individual from one group has a test value larger than that for a randomly chosen individual from the other group. Despite the fact that both true and false positives are taken into account when calculating C, the overall concordance rate of 86.7% is still considerably higher than the 79.9% of utterances that would be classified correctly if no explanatory variables would be taken into account, and instead the perfective would always be selected (Johnson 2008: 254-255). Thirdly, Somer's D, the rank correlation between predicted probabilities and observed responses, ranges between 0 (randomness) and 1 (perfect prediction); in general a correlation is considered large if it takes on a value over 0.5. The obtained [0.7340796] thus indicates a large correlation between predicted probabilities and observed responses.

For Polish, in particular the value of the C index of concordance is disconcerting. C equals [0.7405442], which is below the 80% threshold. This performance is particularly poor given that the percentage of correctly predicted cases would be about 75%, merely by selecting perfective infinitives in all cases, and not including any predictor variables. Likewise, Somer's D stands at [0.4810883], falling just short of 0.5, which signals a large correlation.

BCS takes up a position between Russian and Polish with a respectable Somer's D of [0.6075684] and C at [0.8037842], which is 3% up from the 77% that would be classified correctly if the perfective would always be selected.

2.5. *Interpretation*

The initial outcome suggested that the "lexical" meaning of modality (dynamic vs deontic) as well as polarity (positive vs negative) predict aspectual choice in modal constructions quite well, at least for Russian and to a lesser extent Serbian. This finding reverses the claims made in the general linguistic literature while confirming the corrections proposed by Slavic linguists. Modality type seems a good predictor, but how good is it?

Judging from the C index of concordance, between 13% and 26% of all cases are incorrectly predicted in a model that assigns aspect on the basis of modality type and polarity. For Russian, the number of incorrectly predicted cases stands at 13.3%, for BCS it is 20%, yet for Polish it amounts to 26%.

Interestingly, the fact that the model would fit Russian best is expected given Dickey's (2000) division of the Slavic aspectual world: with the Russian aspectual system focused on definiteness in time, the imperfective expresses "qualitative temporal indefiniteness", i.e. lack of assignability to a single, unique point in time, which fits well with the "general timeless applicability" of deontic modality. Polish and Serbian being transitional zones between the Eastern and Western systems, they likewise display the pattern observed for Russian, albeit to a lesser extent. Polish, although predicted to be more similar to Russian than Serbian, seems to deviate in particular from the expected aspectual pattern: Somer's D reveals only a medium rank correlation between predicted probabilities and observed responses while the obtained C index of concordance remains below the 0.8 threshold. Although it would be possible to argue that modal contexts are an addition to the 7 situations considered in Dickey (2000) and that it is this modal context that makes the Polish scales tip towards the Western end, the possibility should be considered that Polish needs a better model. Maybe modality is not the best predictor of aspectual choice?

3. A better model for Polish

If modality is not what predicts aspectual choice in modal contexts best, then what is? In the literature on aspect in non-finite contexts, several other variables have been put forward, such as the semantics of the modal word (Padučeva 2006) or the level of control the agent has over the action (Šmelev & Zaliznjak 2006, Wiemer ms.), to name but a few relevant parameters. In the next sections, I will explore how increasing both the size of the dataset and the number of variables influences the fit of the model for Polish.

3.1. *More data, more variables*

In order to arrive at an adequate model of aspect assignment in modal constructions including Polish, the corpus sample for all languages was extended. For Russian, all in all 516 examples were analyzed, for BCS 367 and for Polish an additional 160 examples were included, yielding 400 examples.[2]

2 At QITL 2008 (Helsinki, Finland) the question was asked whether the fact that the data are "filtered" through Russian skews the results. As a test case I looked at Polish *można* for which there are 269 examples in the entire corpus, 121 of which occur in the sample used for this study. Out of those 269 examples, 55 come from *Solaris*, and 22 of these *Solaris* examples are included in the sample. So the question is: does *można* as used in the originally Polish novel *Solaris* differ as far as choice of aspect according to modal

At the same time the number of modal predicative adverbs was decreased from 12 to 7, i.e. *można* ('can, may'), *móc* ('can, be able'), *musieć* ('must, have to'), *mieć* ('have'), *należy* ('should, ought'), *powinien* ('should, ought'), *trzeba* ('should, ought, must, it is necessary').

Moreover, additional properties were taken into account, relating to the meaning of aspect, the meaning of the modal word and the relation between the subject and the infinitive action it carries out. Let us briefly look at these new variables in turn.[3] First, a variable relating to the semantics of the modal word was added. This variable captures the meaning of the modality type expressed, i.e., possibility, permissibility, necessity, obligation, ability, volition and prediction. In (9) through (12), examples are given of the first four semantic types in turn, the latter three occurring only rarely in the corpus.

(9) Za drugim razem zakończył u nich swoją drogę, jako że do Chazarów **można uciec**$_{PF.INF}$, lecz od nich uciec się nie da. [ChR]
'For the second time his journey ended with them, as though one could run to the Chazars, but not away from them.'

(10) **Litować**$_{IMPF.INF}$ **możesz się** nad sobą, nie nad nią. Ona zawsze będzie miała dwadzieścia lat. Nie udawaj, że tego nie wiesz! [S]
'You can feel pity for yourself, not for her. She will always be 20 years old. Don't pretend you would not know that!'

(11) Tylko **nie trzeba sobie**, drogi sąsiedzie, **zaprzątać**$_{PF.INF}$ **głowy** wielkimi planami, niech mi pan wierzy!
'Just don't, dear neighbour, be infatuated with big plans, take my word for it.'

(12) Jeszcze, jeszcze, królowo Margot - szeptał Korowiow, który znalazł się obok niej – **trzeba oblecieć**$_{PF.INF}$ sale, aby czcigodni goście nie czuli się zaniedbywani. [MM]
'More, more, queen Margot, whispered Korowiow, who was next to her – you have to fly round the halls, so that the venerable guests would not feel neglected.'

Next, the degree of control the subject has over the infinitive action was inlcuded. Three different degrees were distinguished: high, medium, low, and an example for each level is given in (13) through (15). The tests used in tagging for "control" are inspired by the Vendlerian stativity tests (Vendler

situation is concerned from the other occurrences of *można*, i.e. in Polish translations of a Russian and Serbian novel? Given that authorship was found not to influence aspectual choice in a mixed effects model, a simple chi-squared test was run on the data. It revealed that there is no significant difference between the way in which *można* is used in the originally Polish novel Solaris to render modality and the way in which *można* is used for this same purpose in Polish translations of Russian and Serbian novels.

3 To ensure a native-like interpretation of the examples, the dataset was double-tagged by Adrianna Jakóbczyk, an MA student from Warsaw University, who is a native-speaker of Polish.

1967). In (13) agent control is considered high since joining a group passes the imperative test that can be seen to include the "persuade to V" and "force to V" test and combines with the adverb-tests "to V deliberately" or "to V carefully". Weakly controllable verbs such as 'meet' in (14) pass the imperative tests but do not pass the adverb tests, e.g. you can order someone to find something but you cannot find something carefully or deliberately. Verbs like 'perish' in (15) that do not pass either test are tagged "non-controllable".

(13) Jedna z nich [...] będzie eskortowała skazanych [..] na Nagą Górę, a kiedy już tam przybędzie, **ma się przyłączyć**$_{PF.INF}$ do kordonu ochraniającego jej szczyt. [MM]
'One of them will be escorting the condemned [...] to Golgotha, and upon arrival will have to join the cordon securing its top.'

(14) W państwie chazarskim **można spotkać**$_{PF.INF}$ uczonych Żydów, Greków czy Arabów, którzy dobrze znają jego przeszłość [ChR]
'In the Chazar Empire one could meet educated Jews, Greeks, and Arabs, who knew its past quite well.'

(15) Ocean nie zachowuje się agresywnie i **zginąć**$_{PF.INF}$ w jego plazmatycznych odmętach **może** tylko ten, kto się o to szczególnie stara ... [S]
'The ocean does not behave aggressively and only the one who really tries can perish in its plasmatic depths.'

Thirdly, I added a parameter that relates directly to the meaning of aspect and encodes the aspectual type rendered. Examples (16) and (17) illustrate the difference between what I will call a generic and a specific situation. In (16), a generic situation is expressed as the speaker lists what one should *never* do over the phone. In (17), on the other hand, the expressed situation clearly refers to specific patients, i.e. us versus them.

(16) Przez telefon nie **należy** nikomu **ubliżać**$_{IMPF.INF}$. Przez telefon nie należy kłamać. [MM]
'Over the phone one should never offend anyone. Over the phone one should never lie.'

(17) Mnie z tobą **można zabić**$_{PF.INF}$. A ich nie? Nie radzę ci próbować. Straszne widowisko! [S]
'To kill us? And not them? I don't suggest you try. A terrible spectacle!'

This variable has been hinted at in the literature. Murphy (1965: 136) points out that, in Russian, a distinction between imperfective and perfective usage of infinitives in contexts that express ability can be drawn between a rule with general applicability and permission given for one occasion only. Similar ideas are expressed by Dahl & Hedin (2000) for Russian and Greek, Padučeva (2006) and Wiemer (ms.). Let's see what including these variables does for the Polish model.

3.2. *Excellent results*

A new mixed effects logistic regression model was fit to the corpus data in order to reveal the variable or set of variables that has the highest predictive power for aspect assignment in modal constructions. The results of the best performing model (again with Novel and Modal Word as random effects) are summarized in Table 3.

Table 3. A new model for Polish

Polish (old)	**Polish (new)**
it is significantly less likely to find a modal adverb followed by an perfective infinitive when deontic modality is expressed [estimate = -2.1838, p= 1.5e-06]	it is significantly more likely to find a modal adverb followed by a perfective infinitive when dynamic modality is expressed [estimate = 1.0474, p= 0.00955]
it is marginally significantly more likely to find a perfective infinitive when the modal statement is positive [estimate = 0.7439, p= 0.05308]	it is significantly more likely to find a modal adverb followed by an imperfective infinitive when a generalization is expressed [estimate = 3.6962, p= < 2e-16]
Estimated scale [0.991989] C index of concordance [0.7405442] Somer's D [0.4810883]	Estimated scale [0.9748724] C index of concordance [0.9016152] Somer's D [0.8032304]

Although type of modality remains a significant contributor to aspectual choice, the fact whether the option, permission, order etc. has been given in a generic or specific way outperforms the type of modality in predicting the choice of aspect for the infinitive. From the two factors that prove important in a predictive statistical model of aspect realization in Polish modal contexts, Modality type and SoA Applicability, the latter does better, as is revealed by z that stands at [2.592] for Modality Type and [10.260] for SoA Applicability. Of interest is also the fact that the other added variables do not seem to play any role, i.e. the semantics of the modal word and level of agent control over the infinitive action. Polarity, too, seems to have become superfluous.

Different from the old model, this model does fit the Polish data extremely well. C, the overall concordance rate is up from 74% to 90.1%. This 90.1% is also 18% better than the 72% of utterances that would be classified correctly if the perfective would always be selected. Somer's D, the rank correlation between predicted probabilities and observed responses, equals 80.3%, a 32.2% increase from the previous model, indicating a high degree of correlation between predicted probabilities and observed responses. Clearly, aspect assignment in Polish modal contexts needed a better model than the one originally proposed, a fact revealed by the diagnostics alerting to the poor fit of the original statistical model.

4. Theoretical implications

This study illustrated that quantitative corpus-linguistic methodologies are capable of honoring the multifaceted nature of a linguistic phenomenon. The initial, parallel, outcome of the study suggested that the "lexical" meaning of modality (dynamic vs deontic) as well as polarity (positive vs negative) predict aspectual choice in modal constructions quite well, at least for Russian and to a lesser extent Serbian. This finding reverses the claims made in the general linguistic literature while confirming the corrections proposed by Slavic linguists. Yet, an in-depth study of Polish revealed that other variables might be better at predicting aspectual choice: the "grammatical" meaning of aspect as captured by the parameter specific vs generic outperforms the "lexical" meaning of modality when it comes to predicting aspectual choice, making polarity superfluous, not only in Polish, but also in Russian and BCS (Divjak 2009).

This outcome comes as no surprise, although the similarities between aspect on the one hand and modality on the other hand are far from conspicuous. Recall that, in Polish and other Slavic languages, imperfectives signal situations without temporal restriction, while perfectives are used to express situations that can be compressed to occupy a single point on the time-line. As far as modality is concerned, deontic modality is defined as regulating existence for everyone, always and everywhere while dynamic modality is concerned with a particular situation or a participant in that situation. Deontic modality regulates existence for everyone, always and everywhere, hence expresses a meaning that is similar to the prototypical interpretation of the imperfective as encoding events without focus on temporal restrictions, e.g. as repeated events or with focus on their process or duration. Dynamic modality, on the other hand, is concerned with a particular situation or a participant in that situation, hence shares characteristics with the prototypical interpretation of perfectively coded events as presenting situations as relating to a specific point in time, e.g. as one-off events or as focusing on their time-limits. Going from grammatical, aspectual meaning directly to lexical modal meaning requires the cognitive mechanism of analogical mapping between domains. Mapping between domains requires the intermediate step of extracting properties and abstracting over these, revealing their commonalities. Figure 1 visualizes this mapping.

Expressed in words, Figure 1 visualizes that the core of unmarked, imperfective meaning, i.e. highlighting the actual process, duration or repetition of an event, is used in a deontic situation that applies always, everywhere and to everyone. The core of the marked perfective meaning is employed to signal a situation that is being limited by or to certain participants or circumstances, as is typical of dynamic modal situations. In other words, what imperfective aspect and deontic modality have incommon is their preoccupation with the temporally unmarked, generally applicable situation-as-type. Perfective aspect and dynamic modality, on the other hand, deal with situations-as-tokens that are temporally marked, specific to a subject or circumstances.[4]

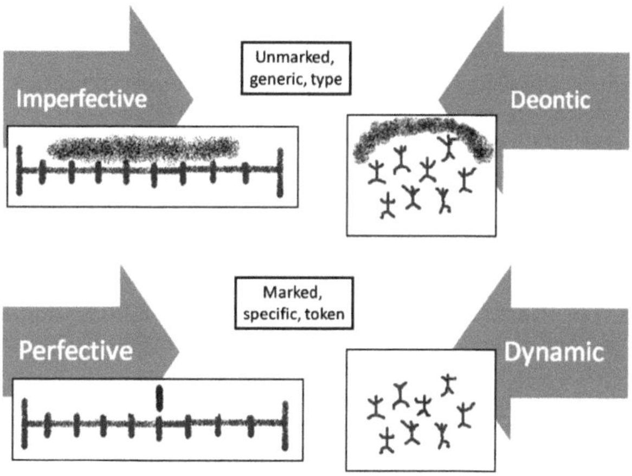

Figure 1. Analogical mapping between domains

Strikingly, the generic vs specific parameter captures the commonalities between aspect and modality that facilitate the reasoning by analogy underlying the mapping between these two domains. By abstracting over the prototypical meaning of each aspect and modality type, the generic vs specific situation parameter acts as a cognitive stepping stone. Since it is closer to both aspectual and modal situations than aspect and modality are to each other, it predicts aspectual choice in modal contexts best.

The generic vs specific parameter can be considered the result of highly complex analogical comparison that gives rise to abstract "rules" through schema abstraction (Gentner & Medina 1998). As I argued in Divjak (2010), "faced with the need of having to make an aspectual choice in a modal context, speakers of Russian were guided by the deep similarities between the two domains. This in turn has given rise to the very strong tendency to use the imperfective aspect to signal deonticity, and the perfective to render dynamic modality".

4 The type vs token interpretation of aspect is borrowed from Hedin (2000: 228) who presents a "functional description of aspect based on the view that imperfective and perfective represent two ways to refer to situations. The imperfective is used when reference is made to situations as types, when they are considered in a non-temporal perspective as abstractions not existing in time but corresponding to the denotative content of some verbal expression. The perfective, on the other hand, is used when reference is made to situations as tokens, as instantiations of situations in time."

After all, in cognitive grammar (Langacker 1987: 445-447), rules are analogies that have become entrenched in the linguistic system.

References

Abraham, Walter & Elisabeth Leiss. 2006. Cross-linguistic aspect/Aktionsart-modality links', Workshop organized at the SLE 2006 meeting in Bremen.

Abraham, Walter & Elisabeth Leiss. 2008. *Modality-aspect interfaces: implications and typological solutions*. Amsterdam: John Benjamins.

Baayen, R.Harald. 2008. *Analyzing Linguistic Data: A Practical Introduction to Statistics using R*. Cambridge: Cambridge University Press.

Croft, William. 1998. Linguistic evidence and mental representations. *Cognitive Linguistics* 9 (2): 151-173.

Dahl Östen & Eva Hedin. 2000. Current Relevance and Event Reference. In: Östen Dahl. *Tense and Aspect in the Languages of Europe.* Berlin: Mouton de Gruyter, 385-401.

Dickey, Stephen. 2000. *Parameters of Slavic Aspect: A Cognitive Approach.* Stanford: Center for the Study of Language and Information.

Divjak, Dagmar & Östen Dahl. 2006. Quantifying the interaction between modality and aspect in Slavic languages, paper presented at the ICLC-10, July 16, Kraków, Poland.

Divjak, Dagmar. 2009. Aspect and modality in Slavic. An exceptional interaction, paper presented at SALC-2, June 10, Stockholm, Sweden.

Divjak, Dagmar. 2010. Corpus-based evidence for an idiosyncratic aspect-modality interaction in Russian (working title). In D. Glynn and K. Fisher (eds.). *Usage-based Cognitive Semantics. Corpus-Driven methods for the study of meaning.* Applications of Cognitive Linguistics Research, Berlin: Mouton de Gruyter.

Durst-Andersen, Per. 1992. *Mental Grammar: Russian Aspect and related Issues.* Columbus, OH: Slavica Publishers, 140-151.

Durst-Andersen, Per. 1995. Imperative frames and modality. Direct vs. indirect speech acts in Russian, Danish, and English. *Linguistics and Philosophy*, 18: 611-653.

Forsyth, J. 1970. *A Grammar of Aspect. Usage and Meaning in the Russian Verb.* Cambridge: Cambridge University Press.

Gentner & Medina. 1998. Similarity and the development of rules. In: *Cognition* 65: 263-297.

Gibbs, R.W.Jr 2007. Why cognitive linguists should care more about empirical methods. In: Monica Gonzalez-Marquez, Irene Mittelberg, Seana Coulson and Michael J. Spivey (eds.). *Methods in Cognitive Linguistics*. Amsterdam: Benjamins, pg. 2-18 [Human Cognitive Processing 18].

Grondelaers, Stefan & Geeraerts, Dirk & Speelman, Dirk. 2007. A case for a cognitive corpus linguistics. In: Monica Gonzalez-Marquez, Irene Mittelberg, Seana Coulson and Michael J. Spivey (eds.). *Methods in Cognitive Linguistics*. Amsterdam: Benjamins, pg. 149-170 [Human Cognitive Processing 18].

Hedin, Eva. 2000. The type-referring function of the imperfective. In: Östen Dahl (ed.). *The Languages of Europe. Tense and Aspect.* Berlin: Mouton de Gruyter, 27-264.

Janda, Laura A. 2002. The conceptualization of events and their relationship to time in Russian. *Glossos* 2 [http://www.seelrc.org/glossos/issues/2/janda.pdf], 10 pg.

Johnson, Keith. 2008. *Quantitative methods in Linguistics*. Kirjastus: Wiley-Blackwell.

Krasnych, V.I. 1979. О наиболее употребительных глаголах, сочетающихся с инфинитивом. In: *Русский язык для студентов-иностранцев. Сборник методологических статей.* Moskva: Russkij Jazyk, 74-86.

Langacker, Ronald W. 1987. *Foundations of cognitive grammar. Volume I. Theoretical prerequisites.* Stanford; Stanford University Press.

Murphy, A.B. 1965. *Aspectival Usage in Russian.* Oxford: Pergamon Press.

Nuyts, Jan, 2006, Modality: Overview and linguistic issues. In: W. Frawley (ed.). *The expression of modality,* 1-26. Berlin: Mouton.

Padučeva, E. 2006. Modality, Negation and Aspect: the case of the Russian *možet* and *dolžen*, paper presented at the 39th Annual Meeting of the SLE, Bremen August 30 – September 2, 2006. [downloaded from www.lexicograph.ru/files/modality_ aspect_negation_ hnd.doc, page last accessed on June 13, 2008]

ParaSol. Parallel Corpus of Slavic and Other Languages, online available at http://www.uni-regensburg.de/Fakultaeten/phil_Fak_IV/Slavistik/ RPC/index.html.

Perkins, M. R. 1983. *Modal Expressions in English.* London: Frances Pinter & Norwood, NJ: Ablex.

R Development Core Team. 2007. R: A language and environment for statistical computing. R Foundation for Statistical Computing, Vienna, Austria, URL http://www.R-project.org.

Rassudova, O. P. 1968. *Upotreblenie vidov glagola v russkom jazyke.* Moskva: Izdatel'stvo Moskovskogo Universiteta.

Rochat, P. & Hespos, S. J. 1997. Differential rooting response by neonates: Evidence for an early sense of self. *Early Development and Parenting,* 6 (3 & 4), 105–112.

Šmelev, Alexej & Anna Zaliznjak. 2006. Aspect, Modality and closely related categories in Russian, paper presented at the Inaugural Meeting of the Slavic Linguistic Society in Bloomington, IN, September 8-10, 2006.

Turiel, Eliott. 2006. The Development of Morality. In Nancy Eisenberg (ed). *Handbook of Child Psychology: Social, emotional, and personality development.* 6th edition: 6. Oxford: John Wiley and Sons, pg. 789- 857.

Vendler, Zeno. 1967. Verbs and Times. In: Z. Vendler. *Linguistics in Philosophy.* Ithaca: Cornell University Press, 97-121.

Von Waldenfels, Ruprecht. 2006. Compiling a parallel corpus of Slavic languages. Textstrategies, Tools and the Question of Lemmatization in Alignment. In: Brehmer, B., Ždanova, V., Zimny, R. (Eds). *Beiträge der Europäischen Slavistischen Linguistik (POLYSLAV) 9.* München, 123-138. (Die Welt der Slaven. Sammelbände 27).

Wiemer, Björn. 2006. *Aspect choice in modal and pragmatic contexts: pieces of a puzzle in Russian and other Slavic languages.* Manuscript.

CORPUS

Bulgakov, Michail. 1938. *Master i Margarita* [MM] (Polish) *Mistrz i Małgorzata* (by Irena Lewandowska and Witold Dąbrowski) (Serbian) *Мајстор и Маргарита* (by Milan Čopić)

Lem, Stanisław. 1961. *Solaris* [S] (Russian) *Солярис* (by Dmitrij Bruškin) (Serbian) *Solaris* (by Predrag Obućina)

Pavić, Milorad. 1984. *Hazarskij Rečnik* [ChR] (Polish) *Słownik chazarski* (by Elżbieta Kwaśniewska and Danuta Cirlić-Straszyńska) (Russian) *Хазарский словарь* (by Larisa Savel'evaja)

Cognitive morphology and the architecture of case in Slovak

Christoph Rosenbaum, Matej Bel University in Banská Bystrica
Wolfgang Schulze, Ludwig-Maximilians-Universität, München

1. Introduction

Case syncretism is a phenomenon common to many often fusional languages that belong to the dependent marking type (here: case marking). Traditionally, syncretism is seen as a formal process starting from the assumption that each of the case forms involved in the syncretism should be encoded by a distinct morphology in at least one part of the paradigm (see, e.g., Comrie 1991). This formal aspect can be related to processes within the conceptual domain on the claim that a formal process usually invokes corresponding processes in those conceptual domains that are related to the formal expression of a morpheme. In the present paper, we will examine some aspects of this hypothesis that is rooted in the framework of Cognitive Morphology. Still, we wish to stress that it is not our goal to review the vast literature on case syncretism and its motivations, nor do we aim at spelling out in full detail the framework of Cognitive Morphology as applied in the present paper. In fact, our goal is far less ambitious, and perhaps, in some parts, at the same time too general. Quite in accordance with the scope of the present volume, we intend to concentrate on processes of case syncretism, and, more globally, on the architecture of case paradigms in one Slavic language, namely Modern Slovak. This language has not yet been extensively referred to, neither in Cognitive Linguistics nor in Language Typology, and hence the expert in Slavic linguistics reading this contribution will perhaps deplore the fact that we occasionally address issues of Slovak linguistics that belong to the common agenda of many such experts. However, going into every detail, in particular those of a comparative nature, from a 'Slavic perspective' would have expanded the paper unnecessarily. Likewise, cognitive linguists will possibly ask for a more comprehensive review of assumptions related to the linkage of morphological expression and morphological concepts, especially in the light of the ongoing debate in Construction Grammar and Cognitive Grammar. Here, we take a rather simple stance that nonetheless has its foundations in a broader framework called Radical Experientialism.[1] One of the claims of Radical Experientialism (and other related frameworks in Cognitive Linguistics) is that formal linguistic variation always entails variation on the conceptual level, which is

1 See Schulze (1998) for a first outline, Schulze (2009) for some details.

quite in accordance with the well-known motto "everything has meaning".[2] In this sense, we argue that most of the many types of paradigmatic variation and case syncretism observable in Modern Slovak reflect shifts in the corresponding conceptual domains, mainly related to the categorial layer of 'virility'. The fact that Slovak continues to elaborate this feature in itself is nothing new and observable in other Slavic languages, too. Still, reviewing its interaction with case selection, formal aspects of the morphological paradigm, and shifts in the semantics of gender categorization may help to sustain older observations and to interpret them in the light of Cognitive Morphology.

The paper is organized as follows: In section 2, we give a very brief draft of Cognitive Morphology with emphasis on case/gender/number marking. In section 3, we discuss the notion of Case, Number, and Gender, before turning to the overall architecture of the corresponding Slovak case paradigms in section 4. Section 5 concentrates on one of the hotspots of Slovak morphology, namely the dimension of (male) animateness. In section 6, we give a brief summary illustrating the make-up of Slovak case paradigms from the perspective developed in this paper.

2. Cognitive Morphology

As for most other Slavic languages (with the exception of e.g. Bulgarian), Slovak morphosyntax is dominated by a paradigm of functional markers that show the fusion of Case, Gender, and Number properties.[3] By itself, this is not really surprising, because the basic architecture of this paradigm goes back to at least Late Indo-European strategies to mark nominal expressions. In other words: In order to 'explain' the Slovak set of markers, one might simply refer to older stages of the language and to those of its preceding variants (Slavic, Indo-European), claiming that the functional scope of the Slovak paradigm is nothing but a diachronically motivated, merely learned feature of the language. This assumption, however, goes against standard hypotheses concerning the synchronic coupling of linguistic form and cognitive procedures (or: 'events'). Accordingly, any linguistic form is said to be part of a symbolic construction (*linguistic* sign) that pairs a *signifiant* domain (articulation) with a *signifié* domain (conceptualization):

2 Adopting a phrase by Bryan S. Turner (Turner 1990:3).

3 See Janda/Clancy (2002) for an overview on case in Russian from a cognitive perspective. Note that in this paper we do not want to monitor extensively the vast literature on case/gender/number in Slavic.

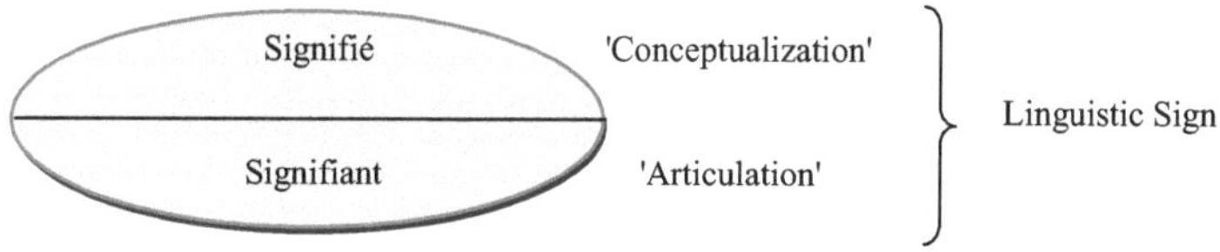

Figure 1. The basic structure of a linguistic sign

Naturally, the concrete types of pairing expression (or: articulation) and conceptualization has to be learned by every individual of a given language community, but it is reasonable to assume that the 'coining' of such pairs helps to entrench conceptual 'spaces' and to keep them synchronically 'alive' (or: active). In this sense, we assume that articulation sequences can be (metaphorically speaking) described as 'nails' that fix parts of a flexible and by itself instable conceptual entity, as illustrated in Figure 2:

Figure 2. The entrenchment of a conceptual entity

Most importantly, we argue that except for the representation of names, the 'nails' rarely fix the totality of a conceptual entity: Rather, conventionalization and thus language use means that only parts of it are fixed. This domain can be called 'active zone' - taking up a term coined by Langacker (2000:62ff.). In other words: A conceptual entity is usually 'more' than what is expressed linguistically. Still, the 'active zone' is not always the same for a given linguistic unit. Rather, it is reasonable to hypothesize that its contextualization may result in shifting processes that narrow down or expand the active zone of such an entity (a process related to conceptual blending, see Fauconnier / Turner 2002). Likewise, variation on the articulation side (allotypes) may indicate a shift with respect to the 'location' of the active zone (AZ), as illustrated in Figure 3 (α = articulatory pattern):

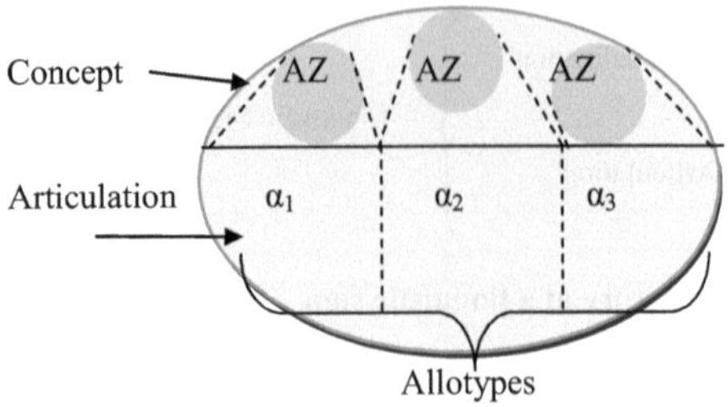

Figure 3. Articulatory allotypes and shifts within the conceptual domain

A key assumption in Cognitive Linguistics claims that there is no basic structural difference between lexical units used to pair articulation and concept and grammatical units pairing articulation and a more general domain of conceptualization, namely category and function. Here, we have considered two crucial points: First, any lexical unit (except perhaps for names) is embedded in one or more categorial spaces, no matter whether these spaces are organized in a disjunctive, prototypical, or radial way (C = concept, α = articulation):

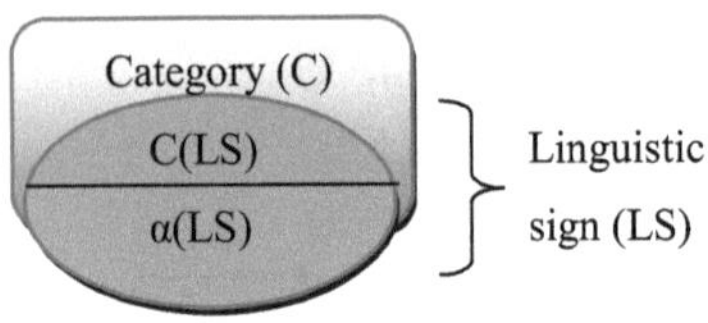

Figure 4. The categorial background of a linguistic sign

A grammatical element associated with a given lexical entity usually refers to the category at issue, not to the conceptual domain of the linguistic sign itself.

Second, the notion of 'morpheme' as expressed in, e.g., the phrasing "the smallest unit of meaning" needs to refined. We argue that a 'morpheme' is a non-autonomous linguistic sign that elaborates a given linguistic sign (or its phrasal frame) as for its cotext and context, or that varies that *signifié* domain of the linguistic sign. Note we define cotext as the textual surroundings of a given linguistic unit, whereas context refers to extra-linguistic knowledge and use (also compare Dressler 1989 for a more detailed discussion). A co(n)textualizing morpheme thus operates in terms of inflection (a), whereas morphemes that vary the *signifié* domain corresponding to what is conventionally called derivation (b) (M= morpheme):

a.

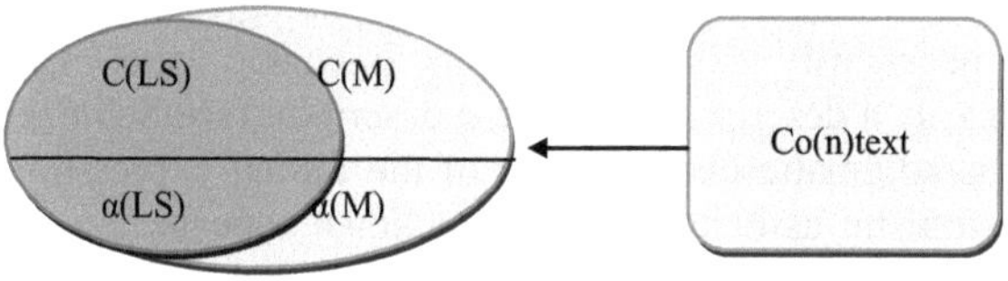

Slovak: *býva-* *-jú* {Given Third Person Plural referent
dwell 3PL:PRES Time frame: Present

b.

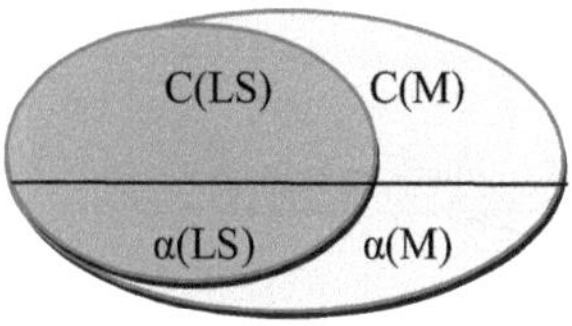

Slovak: *úči-* *-teľ*
teach- Nomen Agentis
'Teacher'

Figure 5. Inflection and derivation

As has been said above, both procedures that 'gestalt', i.e. form/make up a lexical unit refer to the categorial domain of the unit. Accordingly, the conceptual value of a morpheme has to be compatible with at least one of the categorial aspects of the lexical unit, or - in other words - a given morpheme echoes categorial properties of the unit. In addition, the morpheme contorts the conceptual domain of the lexical unit towards the domain represented by the morpheme, especially if it has inflection, that is, co(n)textualizing properties:

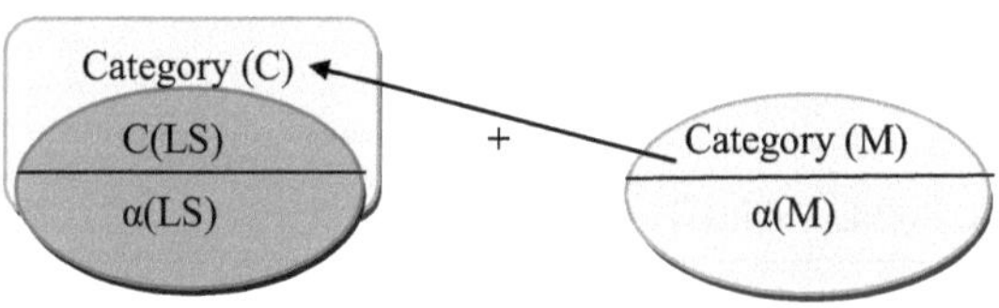

Figure 6. Categorial mapping

An example is (Slovak):

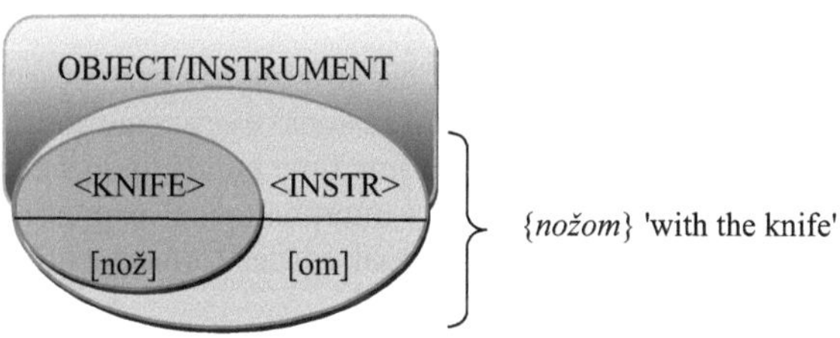

Figure 7. Categorial mapping in nož-om 'with the knife'

It should be noted that the cognitive value of a morphological linguistic sign (morpheme) is not necessarily expressed by its linguistic label. Linguistic categorization is primarily a descriptive artifact, a descriptor (see Schulze 1998), stemming from the paradigmatic classification of the formal expression plus a more or less impressionistic association with functional aspects (e.g. *genitive*, *dative*, *accusative*). In fact, linguistic categories are (first and foremost) nothing but heuristic tools for accessing the categorial layer as such, compare:

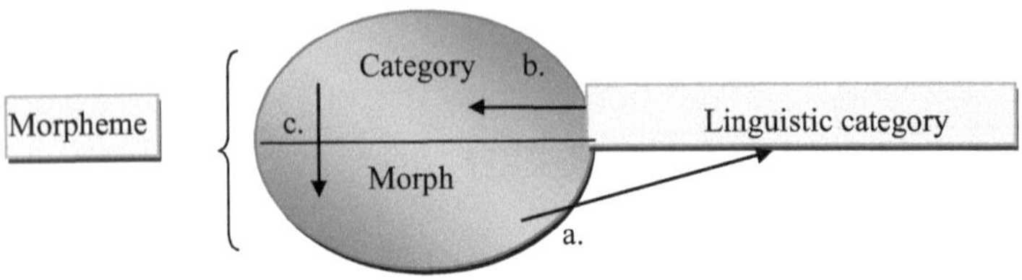

Figure 8. Linguistic and conceptual categories

We hence cannot expect that linguistic values like those mentioned above are cognitively 'real'. Rather, we often have to translate the corresponding labels into 'cognitive' terms. In the present paper, cognitive labels are marked by capital letters, whereas linguistic labels are indicated by small letters.

In Slovak, the fusional characteristics of noun inflection requires that the articulatory domain of a morphological linguistic sign (morpheme) can be coupled with rather heterogeneous categorial layers that nevertheless echo categorial properties of the 'same' referential unit. Consider for instance the Slovak term *v mestách* 'in the cities':[4]

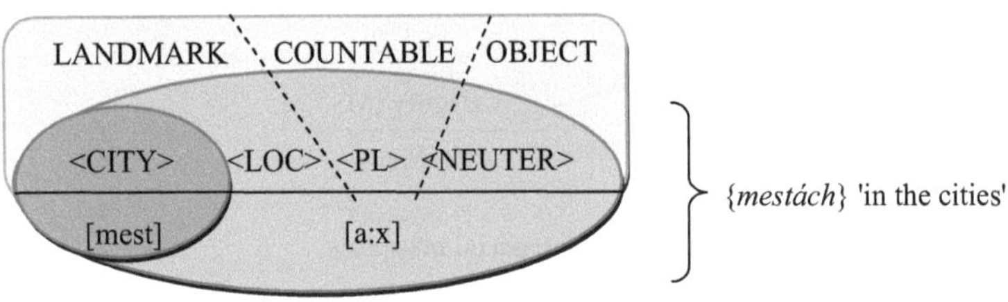

Figure 9. Categorial mapping in (v) mest-ách 'in the cities'

The morphological structure in Figure 9 comes close to what has been described for active zones in Figure 3. The assumption that a fusional morpheme addresses different active zones of a categorial layer at the same time builds upon a central claim of Cognitive Linguistics, namely that *each* morphological expression is coupled with *one* more or less complex conceptual space. Consequently, the

4 The preposition *v* 'in' is ignored in the given context.

categorial 'subdomains' (or: individual active zones) symbolized by a given morpheme must be conceptually related in order to be collectively paired with a formal expression (LOC = Locative, PL = Plural, N = Neuter):

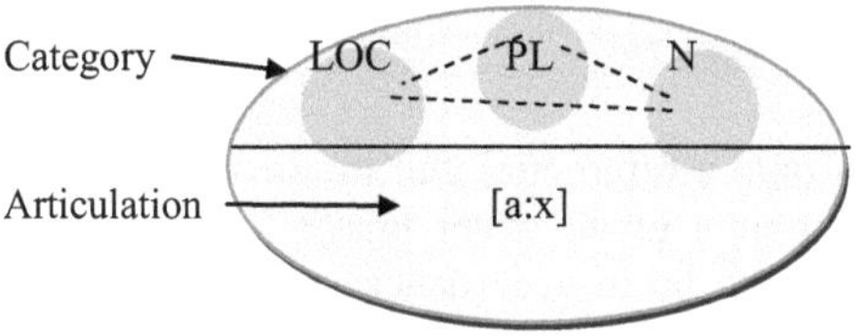

Figure 10. *The* basic categorial network of the Slovak morpheme *-ách*

In the next section of the paper, we will have a brief look at the types of categorial network established by the inflectional morphology of Slovak nouns before turning to the question of animateness as one of the central categorial domains in the subsequent sections.

3. Case, Number, and Gender in Slovak

The linguistic categories of case, number, and gender are usually considered as a categorial cluster that is typically associated with (linguistically speaking) nominal expressions.[5] However, a closer look reveals that these categories are rather heterogeneous in nature: Case represents a functional domain that is strongly relational in nature. That is to say, the case normally originates from the verbal valence given in a phrasal expression (clause, sentence). It copies verbal (that is relational) properties onto the nominal expression (termed relational echo here) or (by itself) serves as a verb-like relator:[6]

(1) *Slov-o robí chlap-a*
word-N:SG:NOM make:PRES:3Sg man-M:SG:GEN/ACC
'The word makes the man.'

(2) *Žen-a vidí mačk-u v záhrad-e*
woman-F:SG:NOM see:PRES:3Sg cat-F:SG:ACC in garden-F:SG:LOC
'The woman sees the cat in the garden.'

5 The following discussion of Slovak case morphology has profited especially from Ružička (1966), Pauliny (1963, 1981, 1983a, 1983b, 1989, 1997), Pauliny/Ružička/Štolc (1968), and Mistrík (2003).

6 Note that with interlinear glosses, we normally refer to linguistic categories, that is (in the given case) to labels stemming from both typology and the descriptive grammar of Slovak.

The corresponding formulae are (𝔑 = referential unit, → = Relator, AGR = Agreement):[7]

(3) 𝔑:CASE:A →/AGR:A 𝔑:CASE:O
𝔑:CASE:A →/AGR:A 𝔑:CASE:O → 𝔑:CASE:LOC

Case thus copies relational properties onto a referential unit in accordance with the role that is assigned to a referent in a given event image (*scene*). Most importantly, such a role assignment has to be in accordance with the inherent semantics of the referent at issue, more precisely with some aspects of its categorization. If the categorial domain of a referential concept does not match the prototypical domain of a grammatical role, shifts with respect to case assignment may indicate this atypical relation. Figure 11 gives an example:

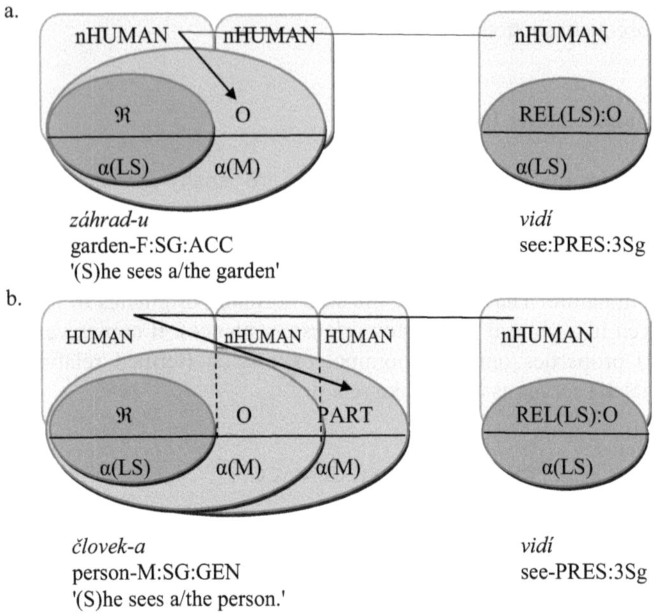

Figure 11. An instance of O-Split [8]

7 We use the following labels to indicate grammatical relations: S (Subjective), A (Agentive), O (Objective), IA (Indirect Agentive ~ Instrumental), IO (Indirect Objective). See Schulze (2000), Schulze/Sallaberger (2007) for details.

8 The word order in both sentences is arranged in order to the graphics: The unmarked positional preference would be *vidí záhradu / človeka*.

This figure reads as follows: An event image (meronymically expressed by the 'verbal' relator, see Schulze (forthcoming)) prototypically calls for a referential unit in O function marked for the feature [non-human]. The corresponding case echo would be represented by a morpheme {O;α}, as shown in (a). However, if the categorial domain of the referent at issue does not match the categorial prerogatives of the verbal relator, the echo-morpheme (Case) is changed resulting in a morpheme that is more compatible with the category of the referent, here {O;α} > {PART;α}, as shown in (b). Here, the categorial domain of the Partitive (PART) is activated because of the fact that it is construed as being more compatible with the category <HUMAN> than the prototypical O-domain. The schema in Figure 11 also illustrates that we are not dealing with a process of replacement. Rather, the original O-domain is still present, but contorted towards a Partitive that is now blended with features of the O-domain (O>PART).[9]

Gender, on the other hand, is a categorial domain that is immediately embedded in the referential frame, which means that it is conditioned not by the (verbal) relator, but by the referential unit itself. Hence, we have to deal with a classificatory device that adds a categorial notion to a given referent and/or to entities associated with this referent.[10] In Slovak, the classificatory system is – to a large part - overt, which means that the classificatory device occurs on the trigger itself as well as on qualifying entities such as attributes. In addition, Gender is also used as some kind of referential echo - that is, as an Agreement device in past tense verb forms (covert). The Slovak (and Slavic) gender system is marked for the fact that the covert (agreement) system is categorially reduced, distinguishing gender only in the singular. Nevertheless, one may proceed from the verbal agreement in order to isolate the most unmarked gender morphemes, compare Table 1:

Table 1. The basic Gender suffixes of Slovak (Nominative)

GENDER/CASE										GENDER/AGR			
OVERT						COVERT							
SG			PL			SG			PL	SG			PL
M	F	N	M	F	N	M	F	N		M	N	F	
-Ø	*-a*	*-o,-e*	*-i/-y,-e* *-ovia*	*-i,-e*	*-a*	*-y/-Ø*	*-a*	*-e*	*-e/-i/-í*	*-Ø*	*-a*	*-o*	*-i*
-a	[palat]			*-i,-e*									

The agreement suffixes copy Gender/Number features of a given referential unit in S or A function, that is, a noun in its nominative case. From this we may assume that prototypically, Gender has the following markers:

9 See Schulze (1998, 2000), and Dewell (2000) for the underlying interpretation of the {O;IO} domain as a conceptual landmark.

10 See Doleschal (1995) for gender association in Russian.

Table 2. The prototypical gender markers of Slovak

	M	F	N	PL
Covert:	-*Ø* (~ -*a*)	-*a*	-*o* (~ -*e*)	-*i*
Overt:				-*i* [M,F]; -*a* [N]

Still, it is not possible to claim that Gender is just added to the feature of Case (or *vice versa*). The fusion of Gender and Case features suggests that we are dealing with a categorial cluster that interlinks properties of Gender and Case. In the given context, it is the relation to categorial properties of the referential entity that are of greatest interest. Above it was claimed that relational echoes (Case) may reflect categorial properties of the referential unit to which they are added. Gender, in fact, does very much the same, compare:

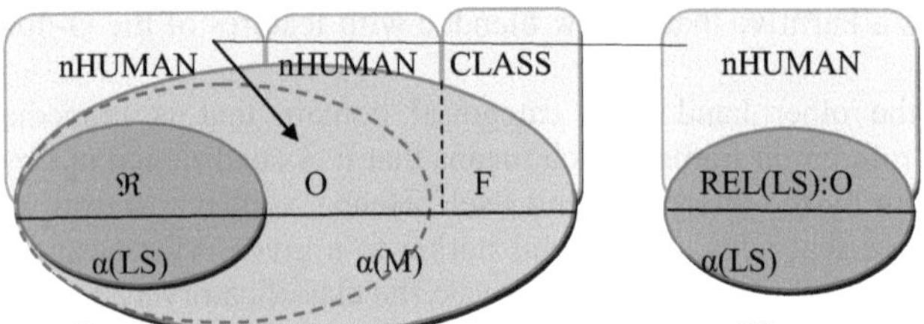

záhrad-u *vidí*
garden-F:SG:ACC see:PRES:3Sg
'(S)he sees the garden'

Figure 12. The interaction of Case and Gender

In Slovak, Gender is a mixture between motivated classification and what is awkwardly called grammatical gender. Here, we cannot dwell upon the question whether the three classes of Slovak are based on pure semantics, i.e., a conceptual type of classification. Still, the fact that the three-gender system is in concurrence with a conceptually motivated type of categorization distinguishing [<MALE>]<HUMAN> from others suggests that the underlying paradigm, namely the Case/Gender/Number cluster, too, is liable to be processed semantically.

Before turning to this point in the subsequent sections, we should consider the third type of categorization strategies, namely Number. Number, again, is a reference-based strategy that starts from the categorization of referential units in terms of what has been called 'Singular Object Nouns' (SON), 'Set Nouns', and 'Sort Nouns' (see Rijkhoff 2002). In Slovak, the presence of an independent plural paradigm suggests that, in most cases, we are dealing with the SON-type of nouns. In this case, plurality prototypically refers to the feature of countability, adding the notion of 'more than one unit'. The categorial background is that of individuation, or - in other terms - to the construction of referential entities as individual units that can be grouped together to form 'plural' clusters. Its relation to the category of Gender is not obvious: The covert paradigm of Gender marking in Slovak (agreement) illustrates that the Gender feature is deleted in the plural. Here, we are left with the impression that Gender is restricted to the categorization of individuated referential entities. Still, overt Gender marking in Slovak suggests (as is true for the model of late Indo-European) that countability was (or still is) construed as a feature that depends primarily from Gender categorization, compare:

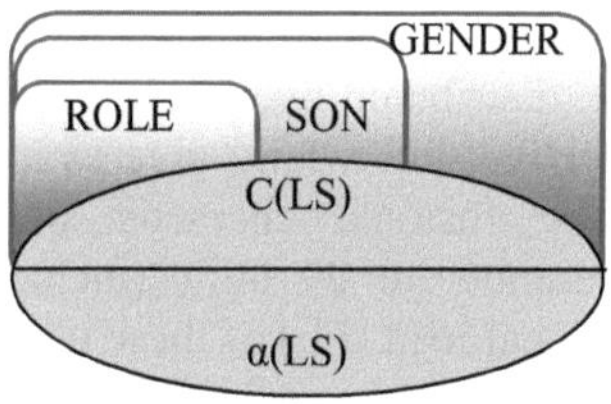

Figure 13. The prototypical hierarchy of Gender/SON/ROLE categorization with Slovak nouns

As has been said above, reference towards these categorial domains is basically fusional in Slovak. Hence, it is difficult to tell whether the hierarchy indicated in Figure 13 has iconic matches in the serialization of the corresponding morphology. From a systematic point of view, morphemes that are marked for derivational rather than for inflectional features are expected to be closely associated with the lexical stem and are hence placed nearer to this stem. However, this hypothesis is not always corroborated by typological data. There are many instances which illustrate that Number features can heavily interact with Case features. For instance, a group plural may qualify for the A role to a lesser extent than individuated referents. The same holds for Gender categorization: In some cases, Gender is rather derivational in nature; in other cases, it is clearly related to the profile of a referent with respect to its role in an event image (as is true for Slovak). Nevertheless, the following conceptual ordering of categorizing and echo morphemes can be suggested for Slovak:

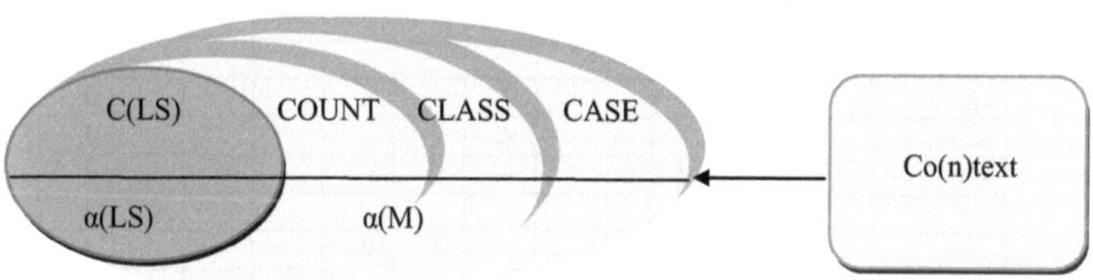

Figure 14. The conceptual linearization of the inflectional cluster with Slovak nouns

In Slovak, Gender classification thus dominates both the COUNT domain and the CASE domain. Although much of the Gender system is merely an inherited and thus functionally less pronounced, learned paradigm, the language illustrates that some semantic aspects of Gender classification still are or are even becoming increasingly more relevant. Here, the most salient feature is that of (male) animateness.[11]

4. The architecture of the Slovak Case paradigm

If we neglect the controversies concerning the existence of a Vocative case, we can describe six case categories for Slovak, distributed over three genders and two numbers. From an agglutinative perspective, this would give us a total of 6*3*2 (= 36) morpheme sequences. The fusional character of Slovak morphology has reduced this theoretical number to 29 units (including -Ø, specific paradigms are neglected). The functional load (FL) of these units ranges from ten to three (the mean would be 4.68), whereas the individual features of a category can be represented by six to eighteen units (indicated by 'degree of variation, VAR' in the following table):

Table 3. Functional load and degree of variation for Slovak inflectional devices (nouns)

	Gender			Number		Case						FL
	M	**F**	**N**	**Sg**	**Pl**	**N**	**A**	**G**	**D**	**L**	**I**	
-e	√	√	√	√	√	√	√	√	√	√		**10**
-i	√		√	√	√	√	√	√	√	√		**9**
-Ø	√	√	√	√	√	√	√	√				**8**
-u	√	√	√	√			√	√	√	√		**8**
-y[12]	√	√		√	√	√	√	√				**7**
-a	√	√	√	√		√		√				**6**
-om	√		√	√	√				√		√	**6**
-í		√		√	√			√		√		**5**
-ia			√		√	√	√	√				**5**
-ami	√	√	√		√						√	**5**
-á ~ -a			√		√	√	√					**4**
-ách		√	√		√					√		**4**
-ám		√	√		√				√			**4**
-iach		√	√		√					√		**4**
-iam		√	√		√				√			**4**
-ie			√	√		√	√					**4**
-o			√	√		√	√					**4**
-ho	√			√			√	√				**4**
-m	√			√						√	√	**4**
-ov	√				√		√	√				**4**
-ovi	√			√					√	√		**4**
-iami			√		√						√	**3**
-ím			√	√							√	**3**
-iu			√	√					√			**3**

11 See, for instance, Janda (1999) for a discussion of 'virility' in Slavic.
12 From a phonetic point of view, *-i* and *-y* merge in Standard Slovak.

-ou		√		√							√	**3**
-mi	√				√						√	**3**
-mu	√			√					√			**3**
-och	√				√					√		**3**
-ovia	√				√	√						**3**
VAR	**16**	**12**	**17**	**17**	**18**	**11**	**11**	**10**	**9**	**9**	**6**	
	M	**F**	**N**	**Sg**	**Pl**	**N**	**A**	**G**	**D**	**L**	**I**	
	Gender			Number		Case						FL

The functionally most loaded morpheme is *-e,* which serves to encode any of the given categories except for the instrumental case. On the other hand, *-ovia* has become highly specialized for the cluster {aM;Pl;Nom}, just as is true for *-ou* {F;Sg:Instr}. As for the degree of variation, the plural category and the neuter category show the greatest number of morphological alternations.

Naturally, one has to take into account the fact that the distribution of the morphemes listed above is also controlled by features of the given lexical stem, resulting in the well-known inflectional classes of Slovak. As a result, certain types of allomorphism as well as of homonymic pairings have to be described:

Table 4. The elaboration of Case morphology with Slovak nouns (Number cumulated)

Nominative			
	M	F	N
-a	√	√	√
-á			√
-e	√	√	√
-i	√	√	
-ia			√
-ie			√
-o			√
-Ø	√	√	
-ovia	√		
-y	√	√	

Accusative			
	M	F	N
-a	√		√
-á			√
-e	√	√	√
-ho	√		
-i		√	
-ia			√
-ie			√
-o			√
-Ø	√	√	
-ov	√		
-u	√	√	
-y	√	√	

Genitive			
	M	F	N
-a	√		√
-e		√	
-ho	√		
-i		√	
-í			√
-ia			√
-Ø		√	√
-ov	√		
-u	√		√
-y		√	

Dative			
	M	F	N
-ám		√	√
-e		√	
-i		√	
-iam		√	√
-iu			√
-mu	√		
-om	√		
-ovi	√		
-u	√		√

Locative			
	M	F	N
-ách		√	√
-e	√	√	√
-i	√	√	√
-í			√
-iach		√	√
-m	√		
-och	√		
-ovi	√		
-u	√		√

Instrumental			
	M	F	N
-ami	√	√	√
-iami			√
-ím			√
-m	√		
-mi	√		
-om	√		√
-ou		√	

The table shows that the accusative case is marked for a relatively high degree of variation (12 morphemes), followed by the nominative, genitive (10), dative (9), the locative (8) and the instrumental (7). If we assume that morphological variation within a functional category is at least partially motivated by semantic factors, we can relate this pattern to a certain preference for manipulating the O-domain (expressed by the accusative case), whereas the instrumental is rather fixed. The picture can be further refined by reducing the number of allomorphs, thus neglecting stem-specific variants:

Table 5. The morpheme bases of Slovak noun inflection

	-a- base	-á- base	-e- base	-ho- base	-i- base	-o- base	-zero- base	-ov- base	-u- base	-m- base	-ch- base	
Nominative	$-a_1$	*-á*			$-i_1$	*-o*	$-Ø_1$	*-ovia*				(6)
Accusative	$-a_2$	*-á*		*-ho*	$-i_1$	*-o*	$-Ø_1$	*-ov*	$-u_1$			(8)
Genitive	$-a_2$		$-e_1$	*-ho*	$-i_2$		$-Ø_2$	*-ov*	$-u_1$ $-u_2$			(8)
Dative			$-e_2$		$-i_3$			*-ovi*	$-u_2$	*-Vm(V)*		(5)
Locative			$-e_2$		$-i_3$			*-ovi*	$-u_2$	*-m*	*-Vch*	(5)
Instrumental										*-(V)m(i)*		(1)

Table 5 also illustrates the major types of case syncretism in Slovak. It becomes clear that the central domains of syncretism involve the following domains: {NOM/ACC}, {ACC/GEN}, and {DAT/LOC}. A minor type is {INSTR/DAT} that is present only with *-om* and that also includes Number features. The major syncretism patterns can be clearly related to Gender categorization. Accordingly, the {NOM/ACC} syncretism is typical of non-animate and plural female referents, whereas the {ACC/GEN} syncretism is confined to the category of (male) animate referents. Both patterns are at least in part derived from a common Slavic pattern, see below section 4. The third syncretism {DAT/LOC} is present with male animate and female referents (singular only). In addition, note that this syncretism also occurs with some inanimate nouns (naM and N) ending in a velar, e.g. *v hrachu* 'in (the) pea', *v kožuchu* 'in the fur coat', *v poriadku* 'in order', *v jazyku* 'in language', *na brehu* 'at the coast', *v batohu* 'in the bundle', compare Stanislav 1959:29). The syncretism possibly has to do with the tendency to avoid palatalization, a feature that - as we will see - strongly correlates with animateness.

Finally, quite a number of naM-terms have a genitive in *-u* (instead of *-a*), especially if the referent has abstract or collective semantics, such as *hmyz* 'bugs, vermin', *vrh* 'hit', *cukor* 'sugar', *piesok* 'sand' etc. Occasionally, the syncretistic form *-u* (GEN/DAT) is used to distinguish countable 'possessors' from collective ones (e.g. *papier-a* 'of a piece of paper' vs. *papier-u* 'of/from paper'). With some other naM-terms, the exact conceptual motivation for using the *-u*-genitive (= dative) is not fully transparent, e.g. *breh* 'coast, hill', *cesnak* 'garlic', *druh* 'sort, type', *hrad* 'castle', *sad* 'garden', *stan* 'tent', *vinohrad* 'vineyard', *vlak* 'train', *vrch* 'peak, mountain peak'. In section 4, it will be argued that the syncretism probably results from the transfer of dative properties onto the genitive domain, as is true for, e.g., German feminine singular nouns. In this sense, nouns of this class would simply lack a true genitive (which would be *-a*), using the dative instead. We may assume that this pattern is another clue for the stabilization of the feature [animate] in Slovak, singling out *-a* as a marker of basic aM nouns (excluding the *hrdina*-type, see below), as opposed to naM nouns becoming marked by the *-u*-genitive.

The following table summarizes the main types of syncretism (minor types are neglected):

Table 6. The basic types of case syncretism in Slovak (nouns)

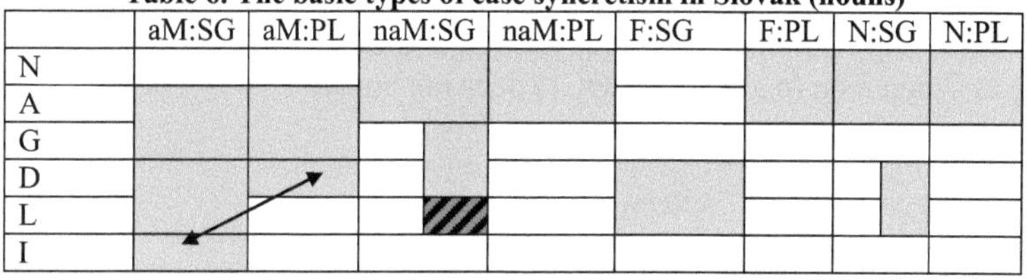

We may again condense this pattern by ignoring parallel Number behavior:

Table 7. The basic types of case syncretism and Gender in Slovak (nouns)

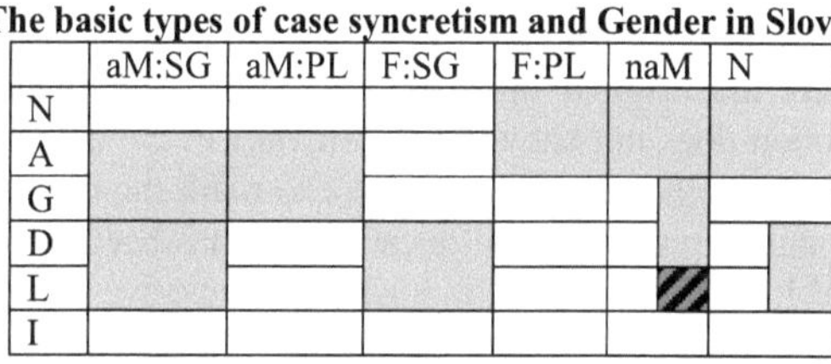

This table entails a crucial point: The animate male category (aM) shows up as the most central category with which each of the two other types share one (different) feature. Here, we must distinguish the following two types: F:SG is embedded in the syncretistic strategies of aM, whereas the complex {naM/N/F:PL} is related to it via the behavior of the accusative:

(4)

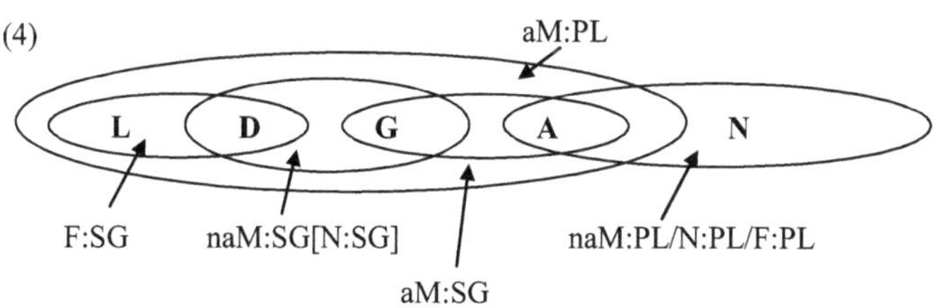

From this we may assume that F:SG shares certain conceptual features with aM, whereas the cluster {naM/N/F:PL} has a common motivation different from aM and F:SG. These aspects will be elaborated in more detail in the following section.

5. Animateness in Slovak

As have been said above, the functional value of the individual case forms is conditioned by echoing properties related to the verbal node, be it a full verb or a verboid, for instance a preposition. The basic pattern of relational behavior (see Dixon 1994, Schulze 2000) is Accusative in nature: It is marked for the parallel encoding of S and A functions as opposed to the O function (S=A;O). The remaining relational primitives IA, IO, and LOC are encoded separately, as long as syncretism (e.g. {DAT;LOC}) does not apply:

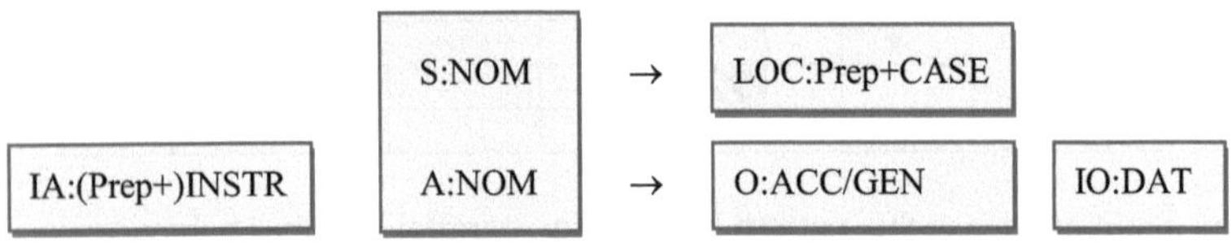

Figure 15. The prototypical coding of grammatical relations in Slovak

The table neglects the original dimension of Possession (genitive *sui generis*) because this function does not show up (in Slovak) in the context of grammatical relations (see below for the use of the genitive to mark the O function). The table, however, also entails a problematical question: As has been illustrated above, the cluster {F:PL;naM;N} does not distinguish an accusative from the nominative. Within theories of Case, at least two different positions can be taken: The 'splitting' perspective argues that we need to start from the maximal differentiation within the paradigm of case forms, which would be F:SG in Slovak. F:SG and aM are the only categories that formally mark the S=A function in opposition to the O function. The conclusion would be that if a language opposes a (say) nominative to an accusative in one subparadigm, it also opposes these case forms in subparadigms that show formal merger. The 'lumping' tradition argues the other way round: Accordingly, it starts from the assumption of either a macro-category or a prototypical category, that are instantiated by specific paradigms:

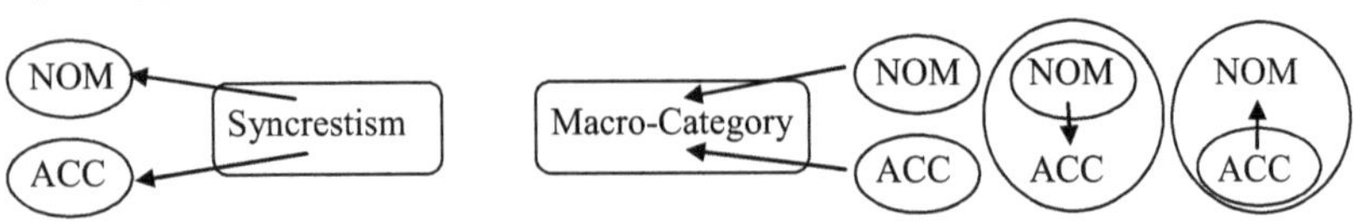

a. Splitting b. Lumping (1) c. Lumping (2)

Figure 16. Splitting and Lumping the nominative/accusative domains

In order not to complicate the matter, we will take a 'splitter' perspective in the remainder of this paper. However, this perspective is modified in so far that a referent in function X is said to modify this function if the referent becomes marked by an element that qualifies for a different function. In the given context this means that a referent functionally marked by the accusative domain acquires properties of the nominative domain if its case marking merges with that of the nominative:

(5)

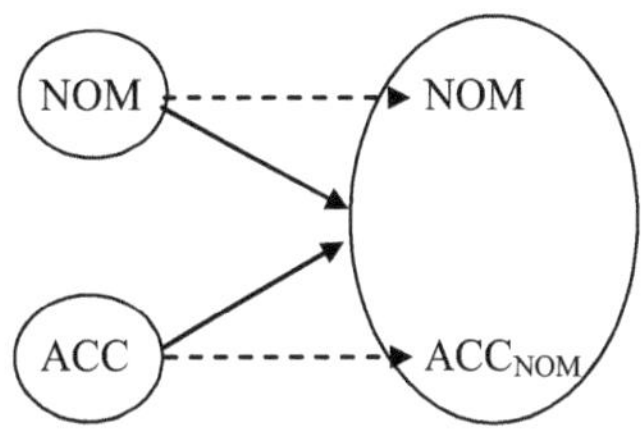

In this sense, the following two sentences differ semantically:

(6) a. *muž vidí mesto*
man:aM:SG:NOM see:PRES:3Sg town:N:SG:ACC(>NOM)
'The man sees the town.'
b. *muž vidí mačk-u*
man:aM:SG:NOM see:PRES:3Sg cat-F:SG:ACC
'The man sees the cat.'

In (6.a), the semantic information entailed in the original accusative morpheme has been bleached and (in part) transferred to positional factors. (6.a.) can thus be paraphrased, for instance, by 'the man (Nominative) sees [what is] a town (Nominative)', whereas (6.b) can read: 'the man (Nominative) sees [towards] a cat (Accusative)'.

The fact that the nominative/accusative syncretism is confined to the cluster {F:PL;naM;N} leads to the assumption that the elements of this cluster have something in common that deprives them of a morphologically marked accusative. As for the sub-cluster {naM;N}, this common denominator can be described in terms of the feature [-anim], although exceptions may occur: *dieťa* 'child', *knieža* 'prince, duke' (also aM), *zviera* 'animal', *kura* 'chicken', *prasa* [13] 'young pig', *mláďa* 'whelp, puppy etc.', *šteňa* 'lap dog', *bábä* [14] 'small boy, page, tot', *húsa* 'gosling', *dvojča* 'twin', *dievča* 'girl', *decko* 'young person up to the age of fifteen (often slightly pejorative)'. Here, we cannot elaborate in detail the categorial frame of referent [+anim] marked for the neuter gender. However, the data just presented hint at the fact that most neuter nouns [+anim] refer to animate

---

13 Also used to simply denote a domestic pig. A 'young pig' would then be *prasiatko*.

14 Rare. The standard form is *bábo*.

concepts characterized by features of smallness, youth, or pejorativeness. This set of features is typical of an intermediate sub-class that mixes categorial aspects of animateness and inanimateness. As a result, the special marking of affectedness (or, of the O function) is not as relevant as for fully 'animate' referents. The fact that the sub-gender F:PL is also included in this pattern is probably due to the feature of plurality, which typically reduces the discreteness of a referent and hence subsumes it under a class marked for features of collectivity.

The domains of aM and F:SG share the feature of differentiating the O domain from the S=A domain. However, the two sub-classes do not take the same option to mark the O domain. Referents included in the F:SG domain overtly mark the O function with the help of the corresponding morpheme *-u*, as long as the noun does not end in a consonant (type: *dlaň* 'palm etc.', *kosť* 'bone'[15]). Hence, the *-u*-morpheme is strongly associated with the concept of affectedness (and with directionality). Still, the overall presence of non-animate referents in this category precludes the proposal of a semantic motivation for all the class members. Obviously, the feminine gender is marked for both semantic features (with respect to animate referents) and paradigmatic, mainly phonological features, resulting from the paradigmatization of the *a*-vowel of the nominative. The following chart summarizes the distribution of S=A vs. O coding in the classes mentioned so far:

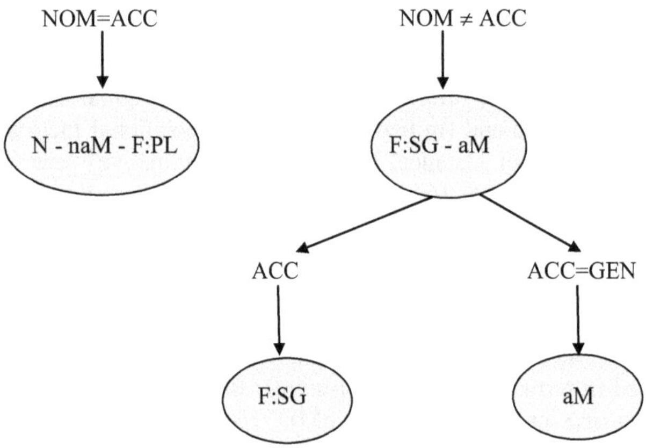

Figure 17. The classification of S;A;O coding in Slovak

15 Recent changes in the two paradigms *dlaň* (F:SG, N:PL -e) and *kosť* (G:SG; N:PL -i) nicely illustrate the relevance of semantic features in the architecture of the Slovak case paradigm: The *kosť*-paradigm tends to specialize for 'abstract' referents marked by *-osť*, whereas all other consonant-final feminine nouns are subsumed under the *dlaň* -paradigm.

By itself, the feminine gender is thus neutral with respect to the domain of animateness. However, it entails features that set it apart from the category of animateness with 'male' referents as expressed in the aM gender. One crucial point is the lack of palatalization that is present with aM nouns ending in *-k*, *-g* and *-ch* in the nominative plural, compare *žiak* → *žiaci* 'pupil, disciple', *mních* → *mnísi* 'monk', *hroch* → *hrosi* (~ *hrochy* [16] ~ *hrochovia*) 'hippo', 'sea cow', *valach* → *valasi* 'Walachian' (see below). This fact relates the feminine gender to the non-animate sub-class of masculine nouns. However, the feminine gender is not totally unrelated to the animate class of masculine gender (aM). Here, the most important point is given by the presence of the morpheme *-u* with the so-called *hrdina*-paradigm ('hero') that shows *-u* both in the genitive and the accusative. Although we are dealing with two different morphemes from a historical perspective, it is reasonable to assume that the two morphemes are processed at least in 'semantic assonance', see below.

Slovak has highly elaborated the category of male animateness. As has been remarked before, we cannot oppose it to a class of 'inanimateness', but only to a class of referents that are not animate males (compare the stemma given in Rubach 1993:84). In this sense, aM is a marked class, as opposed to the unmarked rest, which is not sensitive for this distinction. The class aM includes two sub-classes, namely a) male human beings and b) animals. However, whereas sub-class (a) is motivated for the feature [male], the sub-class (b) is cross-cut by the feminine class. Thus only a portion of animal referents belongs to the 'animate' class, while others are included in the 'intermediate' unmarked class [feminine]. Both genders are used to mark the set of *epicoena* (see Brugman 1897:12). The following tables illustrate the basic distribution (based on 79 non-exotic generic animal terms; animal terms classified as 'neuter' are not included):

The overall distribution is aM (animals) 56 %, F (animals): 36 %. The diagrams support the impression that non-exotic animals are preferably marked by the masculine gender,[17] although we cannot describe a full-fledged motivation for the individual class allocations. Nevertheless, the diagrams, as well as terms classified as 'feminine' such as *hus* 'goose', *koza* 'goat', *mačka* 'cat', *sviňa* 'pig', *kuna* 'marten', *líška* 'fox', and *myš* 'mouse' (and even *ťava* 'camel') make it impossible to claim that the aM class is the general class for male animate beings. This may also be deduced from the fact that many animal concepts are animate in the singular, but both animate and inanimate in the plural, depending on the degree of anthropomorphization. This also holds for concepts like *robot,* which can be inflected both as an animate concept ('human-

16 Usually categorized as an inanimate plural.

17 A similar tendency can be described for Russian, where all uninflected animal terms are masculine by default.

like robot') or as an inanimate concept ('robot' as a tool). Also compare the well-known distinction *kohútik* 'rooster' (animate) vs. *kohútik* 'water tap' (inanimate).

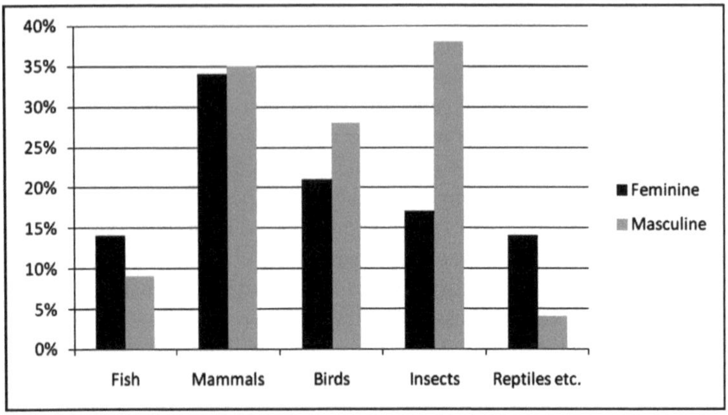

Diagram 1. The distribution of animal terms (percentage of class)

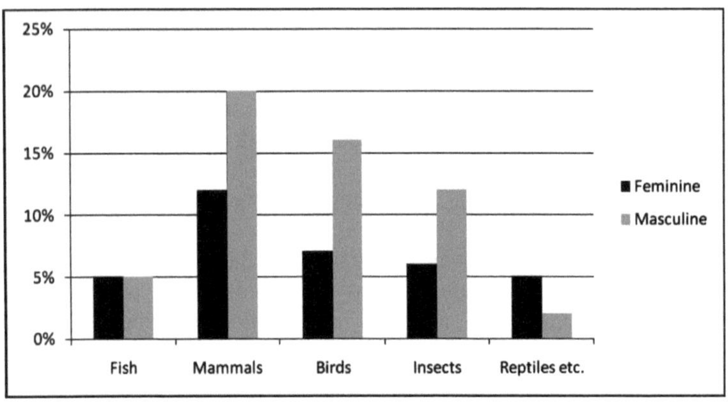

Diagram 2. The distribution of animal terms (percentage of total)

On the other hand, Slovak continues an (ultimately perhaps Indo-European) tradition to mark certain male referents for the feminine gender.[18] This technique, conventionally called 'natural gender' (see Strunk 1994), is present e.g. in Latin *poeta*, *nauta*, *incola*, *agricola*, *collega*, *conviva*, Greek *polítēs*

18 More precisely, such nouns are marked for feminine case, but have masculine agreement. It may be that we have to deal with a blend of time-stable 'male' properties (expressed by agreement/concordance) and the temporal qualification of the given referent (expressed by the feminine gender).

'citizen', *ornithothēras* 'bird hunter', Lithuanian *válkata* 'vagabond' (compare Vykypěl 2006; often pejorative), Latvian *pļāpa* 'gossiper' (often pejorative) etc. Slavic equivalents start with Old Church Slavonic, e.g. *bližnika* 'relative', *vinopinica* 'wine drinker', *vladyka* 'ruler', *drevodelja* 'carpenter', *starěišina* 'elder', *sluga* 'servant', *sěč'ca* 'headsman', *ubiica* 'murderer', *čarodějca* 'magician', and *junuša* 'youngster' (see Leskien 1969 [1910]:78-79, Brugmann 1889). With respect to the languages mentioned above, many such terms figure as *nomina agentis*[19]. In Slovak, this pattern has become fixed in the so-called *hrdina*-paradigm. Many of these nouns continue the old *nomen agentis* marker (*-ca*) as in *výherca* 'winner', *záujemca* 'client, interested party', *sudca* 'judge, arbiter', *radca* 'mentor', *darca* 'donator', *dozorca* 'controller', *dôchodca* 'retiree', *obranca* 'champion, defender', *ochranca* 'guardian', *predchodca* 'predecessor', *sprievodca* 'conductor, guard', *vládca* 'ruler, monarch'. Another pattern is given by nouns ending in *-ta* (loans and native terms), including e.g. *anarchista* 'anarchist', *gitarista* 'guitarist', *socialista* 'socialist', *husita* 'Hussite', *prednosta* 'chief, manager', *mešťanosta* 'alderman', *starosta* 'mayor'. Other terms are, for instance: *sluha* 'servant', *družba* 'bride's male attendant', *šarha* 'horse doctor', *tata* 'daddy', *tlčhuba* 'babbler' (Ružička 1966:86). At least in part, we have to consider the morphology of the *hrdina*-class as merely a phonologically determined variant of the standard *chlap*-class, as can be seen from the following example:

(7) *Aj po tragick-ej smrt-i*
even after tragic-F:SG:LOC death-LOC

Tomáš-a Bať-u
Tomáš-aM:SG:GEN Baťa-aM:SG:GEN
'Even after the tragic death of Tomáš Baťa...'

Here, the compound expression *Tomáša Baťu* refers to one single (name) concept, which means that in the given case, we cannot assume a semantic difference between the *-a* and the *u*-genitive. Likewise, the well-known syncretism of the genitive and the accusative with (animate) *hrdina*-nouns sufficiently illustrates that the basic conceptual architecture of the *hrdina*-paradigm is in analogy with the standard animate *chlap*-paradigm. This also holds for the third sub-paradigm (animate males), the *pony*-paradigm ('nag, pony') that is marked for a different type of singular genitive/accusative morpheme (*ponyho*)[20].

Nevertheless, other parts of the *hrdina*-paradigm suggest that there may be some kind of conceptual peculiarity present in this paradigm. One point is the

19 Also compare the Old Armenian *nomen agentis* marker *-ič'* that stems from a feminine marked suffix **-ik-jā*, see Meillet (1936:79).

20 In the given context, we neglect a discussion of the unusual L/I-syncretism (*ponym*) that stands in opposition to D *ponymu*.

morpheme of the nominative plural (*-ovia*) that replaces the nominative singular morpheme *-a* (*hrdinovia*). This morpheme may also occur with *chlap*-nouns both for phonological and semantic reasons. Most importantly, the *-ovia*-plural is used with concepts that are part of the kinship terminology, such as *otec* 'father', *syn* 'son', *kmotor* 'godfather' and the in-law terms *švagor* 'brother-in-law', *svokor* 'father-in-law', and *zať* 'son-in-law'. However, the associative plural ('family of X') is not included (compare *Štefan-ovc-i* 'the family of Stephen'). Here, the *-ovia*-plural marked a non-associative, individuated plural, treating names like Singular Object Nouns, e.g. *Štefan-ovia* 'persons called Stephen'). In addition, the *-ovia*-plural frequently occurs with diminutive names and kinship terms ending in *-o*, as in *dedo* 'grandfather' → *dedovia*, *Kubo* 'Little Jacob' → *Kubovia*. Also note *chlapčisko* 'cad, churlish boy' → *chlapčiskovia* (Stanislav 1956:50).[21]

More or less free alternation between the *-i*-plural (*chlap*-type) and the *-ovia*-plural (*hrdina*-type) is documented, among others, for terms like *posol* 'ambassador', *muž* 'man', *cár* 'Tsar', *bán* 'margrave' *buržuj* 'bourgeois', *apoštol* 'apostle'.

The third variant, *-ia* (*chlap*-paradigm) occurs, among others, with referents that are part of a social or regional group. In addition, it is typical of terms like *rodič* 'parent', *brat* 'brother', *dedič* 'descendant', *manžel* 'husband', *sused* 'neighbor', *hosť* 'guest', *človek* 'person', and *žid* 'jew' (Stanislav 1956:43ff).

Contrary to the singular domain, the plural of the *chlap*- and *hrdina*-paradigms differs only with respect to the nominative case. The fact that some *chlap*-terms allow both markers or even all three (*-i*, *-ovia*, and *-ia*) suggests that the *hrdina*-paradigm in fact is confined to the singular, whereas we would have to describe a set of (formally speaking) allomorphs for the nominative plural of one single paradigm:

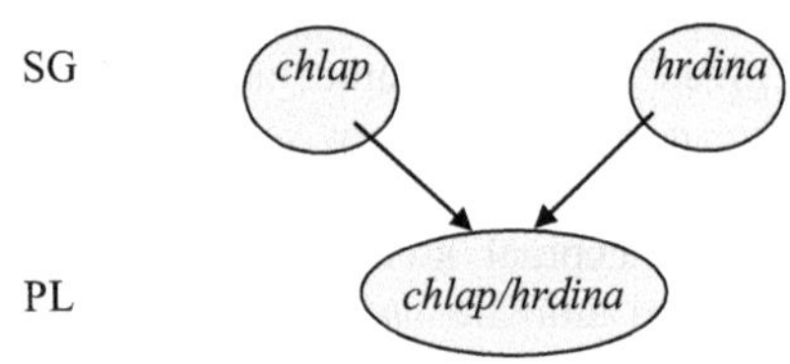

Figure 18. The merger of the animate chlap/hrdina-paradigms in the plural

Nevertheless, the presence of the set of plural allomorphs (nominative) suggests that, conceptually speaking, different domains (or: active zones) of the given referent may be addressed. Taking up the semantics of those referents that are marked for *-ovia,* one might argue that in most instances, we are dealing with referents who play a particular 'role' in their social environment, either in terms

21 Also N:PL *chlapčiská.*

of profession or in terms of social relations. In other words: The plural refers to referents marked for an extrinsic or non-native property, whereas the standard *-i*-plural alludes to the intrinsic or 'given' properties of a referent. Naturally, this hypothesis has to be seen in the light of the underlying diachronic processes. It is generally assumed that *-ovia* has resulted from the merger of two plural forms, namely Old Church Slavonic *-ove* (plural of *ŭ*-stems, see Comrie/Corbett 2002:86, Stanislav 1956:43,50) and *-ję* (plural of *-jā*-stems). The merger is said to have occurred in Middle Slovak times[22] spreading to terms of the *chlap*-paradigm (*o*-stems) in order to keep the lexical base more distinct especially for those nouns that end in *-k*, *-g,* and *-ch*: As has been said above, such nouns undergo palatalization, which conditions lexical stem variants (*žiak* → *žiaci* 'pupil, disciple'). Hence, *-ovia* would have been a means to avoid the fusion of lexical and morphological information. This hypothesis also entails the assumption that - beginning in Middle Slovak - morphological transparency became a relevant issue. In addition, we may argue that the avoidance of forms like *žiaci* can be iconically related to the avoidance of plural forms that look like lexical plurals, often including the notion of set nouns (collectives). The use of *-ovia* would then have strengthened the character of Slovak nouns as Singular Object Nouns (SON).

The hypothesis mentioned here is based on the assumption that the segment *-ov-* has been reanalyzed as a 'meaningful' element, expressing a conceptual layer that entails the notion of 'individuation within a group'. In order to fix the semantics of the element *-ov-*, it seems appropriate to oppose it to the *-ia*-plural described above. The morpheme *-ia* is conventionally related to the old feminine singular *–ьja,* as in Old Church Slavonic *kъnęž-ьja* 'dukes' and *bratr-ьja* 'brothers' (usually collective, compare Leskien 1969:269). In Slovak, *-ьja* became *–ä,* being diphthongized later on (> *-ia*).[23] Accordingly, *-ia* once had a stronger notion of group or collective marking, perhaps induced by the collective *a*-neuter (plural).[24] The *-ov*-segment would then have again reduced this feature in favor of individuation:

---

22 In the Middle/Central Slovak dialect. The first documentation of *-ovia* (1589) comes from a text from Slovenská Ľupča (*svedkovia* 'witnesses', see Stanislav 1956:51). Elsewhere, *-ove* was in use.

23 The original Old Slovak form *-ä* is preserved in the variants of Lower Orava and Gemer: *bratä*, *židä*, *roďičä*, *pohoňičä* (Stanislav 1956:43).

24 Also see Gorškova/Xaburgaev (1997:220): “Jarkim primerom izmenenija grammatičeskogo značenia formoobrazujuščego affiksa javljaetsja ispol´zovanie drevnix sobirateľnyx obrazovanij dlja oboznačenia množstvennosti. Častotnosť form tipa *brat'ja*, *djad'ja*, *knjaz'ja* (< *knjazija*) pri ukazanii na sovokupnost´ lic (po sravneniju s formami množstvennogo čisla *brati*, *djadě*, *knjazi* s tem že real´no-čislovym značeniem) (...).” Obviously, this development already started in Common Slavic.

(8)

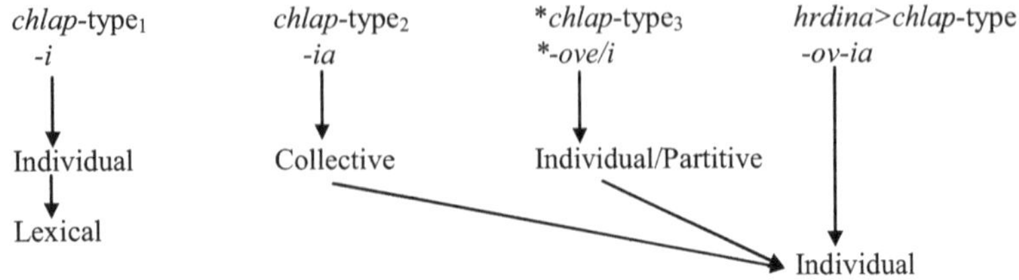

The fact that the seemingly allomorphic distribution of the three animate plural markers is confined to the nominative case is by itself not surprising: Obviously, the question of whether a referential entity in S=A function is conceived as a group (Set Noun) or an ensemble of individuals (Singular Object Noun) is especially relevant with this (agentive) function, whereas in the oblique cases (encoding the O, IO, IA and LOC functions) this differentiation is not seen as being decisive, at least in Slovak.

The opposition -*i* ~ -*ia* ~ -*ov-ia* thus rests upon two major processes: First, the reinterpretation of the original feminine singular ending -*ia* as some kind of collective marker, and second, the reanalysis of the segment -*ov*- as a distributive plural marker. The second process is intimately related to the overall-scope of the morpheme -*ov* in Slovak that is present in the following case forms:

Table 8. The basic distribution of -ov-morphemes in Slovak

	aM		naM	
	SG	PL	SG	PL
NOM		*-ov-ia*		
ACC		*-ov*		
GEN		*-ov*		*-ov*
DAT	*-ov-i*			
LOC	*-ov-i*			

In part, the underlying pattern was already elaborated in Old Church Slavonic. The main point is the paradigmatic shift of many *ŏ*-stem nouns that assimilated properties of the *ŭ*-stem paradigm (Leskien 1969:82-85). The stem-final segment -*ŭ*- that showed up only in those cases that did not encode the singular {S;A;O} cluster (or, in other words: that served to mark indirect and locative functions) became reanalyzed as part of the case forms:

(9)

Late Indo-European			Old Church Slavonic	Slovak
SG				
N	**sŭn-ŭ-s*	→	*syn-ъ*	*syn*
A	**sūn-ŭ-m*	→	*syn-ъ*	*syn-a*
G	**sūn-ŏu̯-s*	→	*syn-u*	*syn-a*
D	**sūn-**ŏu̯**-ĕi̯*	→	*syn-**ovi***	*syn-**ovi** ~ syn-u*

L	*sūn-ŏu̯-Ø	→	*syn-u*	*syn-***ovi** *~ syn-u*
I	*sŭn-ŭ-mĭ	→	*syn-ъmь*	*syn-om*
PL				
N	*sŭn-**ŏu̯**-ēs	→	*syn-***ove**	*syn-***ovia**
A	*sūn-ŭ-ns	→	*syn-y*	*syn-***ov**
G	*sūn-**ŏu̯**-ŏm	→	*syn-***ovъ**	*syn-***ov**
D	*sūn-ŭ-mŭs	→	*syn-ъmъ*	*syn-om*
L	*sūn-ŭ-sŭ	→	*syn-ъxъ*	*syn-ox*
I	*sŭn-ŭ-mīs	→	*syn-ъmi*	*syn-mi*

The paradigm illustrates that *-ov* became a distinct morpheme especially in the genitive plural. Here, two conceptual features were blended: a) the partitive notion of the genitive plural, b) the prevalent feature of (male) animateness related to such high frequent terms as OChSl *synъ* 'son' and *volъ* 'oxen'. The element *-ov* thus became associated with a conceptual cluster that entailed the semantics [+male(animate);+partitive]. The partitive semantics was functionally reduced especially with the dative singular, because here, the element was followed by the segment *-i* leading to a more complex form that accounted for this specific dative function.[25]

Adding the feature of SON-marking isolated above for *–ovia,* we may hypothesize that the segment *-ov(-)* is marked for the following categorial domains (aM = male animate):

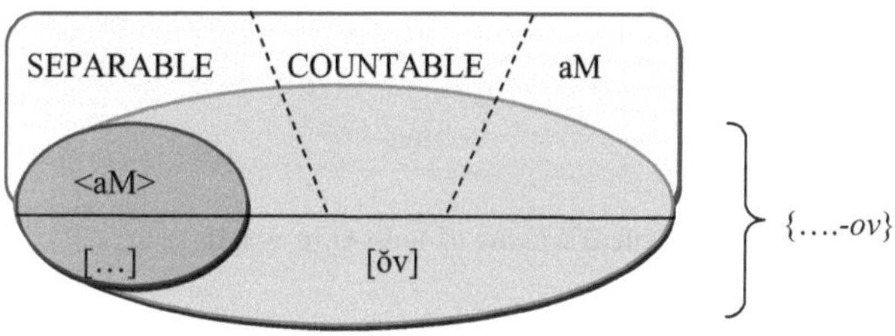

Figure 19: Categorial mapping of the case segment *-ov*

The fact that already in Old Church Slavonic, the dative singular *-ovi* is present with some *ŏ*-stems denoting animates (males) such as *vinarь* 'vintner', *gospodъ* 'lord', *mǫžь* 'man'. *duxъ* 'soul, spirit', *bog* 'God', *cěsarь* 'Caesar, emperor' suggests that the conceptual features mentioned in Figure 19 were active already by that time. However, the adaptation of the genitive plural morpheme *-ov* (*ŭ*-stems) by *ŏ*-stem nouns as documented in some Old Church Slavonic sources

25 The morpheme *-ovi* competes with *-u* that is typical of naM nouns and some generic animate terms such as *člověk* 'man, person', *pán* 'Sir' (in apposition), *boh* 'god', *čert* 'devil', *otec* 'father', *syn* 'son' (often in a religious context), *pes* 'dog', *vlk* 'wolf'.

(see Leskien 1969:83-84 for details) illustrates that the feature 'aM' is secondary as opposed to the feature 'separable' (> partitive). In fact, the sources show that as early as Old Church Slavonic, *-ov* (genitive plural) could be used with inanimate referents, too. Hence, we can describe the following path for the conceptual shift of the *-ov*-morpheme:

(10) SEPARABLE (→PARTITIVE) → aM → SON/COUNTABLE

The starting point of the reanalysis of *-ov* would then have been its partitive notion which perfectly meets the conditions to develop a conceptual split within the O-domain ('patient'). Here, we cannot take up this well-known issue in all its details. Conceptually speaking, we are dealing with a metaphorization of the ablative/partitive domain (see Schulze 2009; Nikiforidou 1991 for a general overview).[26] 'Taking away' a part *from* something in order to address it (in replacement of the whole) is a typical instance of social deixis that may single out as an honorific marker. Hence, saying 'I see Christ' would be less respectful than saying 'I see a part of Christ', compare the scheme in Figure 20:

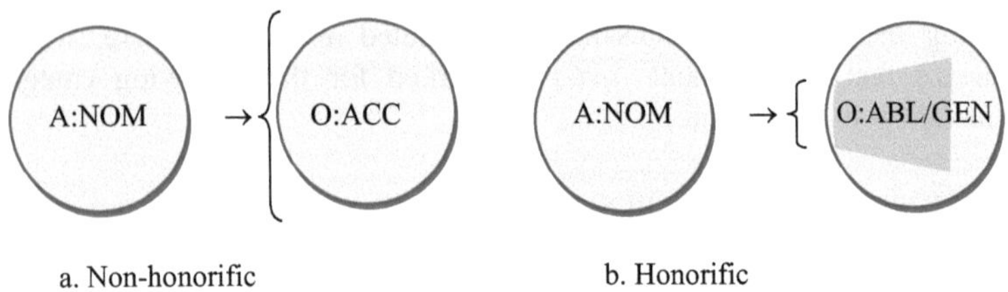

Figure 20. A prototypical scheme of Split-O in Slavic

The split gradually evolved from reference towards person concepts of high esteem to male humans and then male animates, turning the originally pragmatic split (Fluid-O in terms of Dixon 1994 and Schulze 2000) into a more or less semantically motivated split (Split-O). The split is also present in the singular, where the *a*-morpheme of the old *ŏ*-inflection clearly shows its derivation from on older ablative (**-od* > *-ăd*), compare the paradigm in (11):

---

26 Noonan/Mihas (2008:3) strongly argue in favor of a sprachbund-related phenomenon: "There seems little doubt that this is a feature of the European Sprachbund. Its rarity outside seems to mean that this particular grammaticalization path is not often utilized for the creation of new genitives." However, the assumed rarity of this syncretism outside Europe is not fully corroborated by the facts, see e.g. Schulze (2009).

(11)

	Late Indo-European		Old Church Slavonic	Slovak
SG				
N	**u̯l̥k^{u}-ŏ-s*	→	*vlъk-ъ*	*vlk*
A	**u̯l̥k^{u}-ŏ-m*	→	*vlъk-ъ*	*vlk-**a***
G/ABL	**u̯l̥k^{u}-ŏ-**ăd***	→	**vlъk-**a***	*vlk-**a***
D	**u̯l̥k^{u}-ŏ-ĕī*	→	*vlъk-u*	*vlk-u* (or else *-ovi*)
L	**u̯l̥k^{u}-ŏ-i̯*	→	*vlъc-ě*	*volk-ovi*
I	**u̯l̥k^{u}-ŏ-mĭ*	→	*vlъk-omъ*	*vlk-om*
PL				
N	**u̯l̥k^{u}-ŏ-s*		*vlъc-i*	*vlc-i ~ vlk-y*
A	**u̯l̥k^{u}-ŏ-ns*	→	*vlъk-y*	*vlk-ov*
G	**u̯l̥k^{u}-ŏ-ŏm*	→	*vlъk-ъ*	*vlk-ov*
D	**u̯l̥k^{u}-ŏ-mŭs*	→	*vlъk-omъ*	*vlk-om*
L	**u̯l̥k^{u}-ŏ-sŭ*	→	*vlъc-ěxu*	*vlk-och*
I	**u̯l̥k^{u}-ŏ-ŏi̯s*	→	*vlъk-y*	*vlk-mi*

The *hrdina*-paradigm (stemming from the feminine inflectional pattern of *a*-stems) takes up the same technique but from a different point of departure: The Slovak suffix *-u* (GEN/ACC) goes back to the Old Church Slavonic feminine accusative singular (*-ǫ* < **-ām*), replacing the genitive morpheme Old Church Slavonic *-y ~ ję*, Slovak *-y ~ -e* etc. The question of why the *hrdina*-paradigm did not follow the way of the *chlap*-paradigm by starting from the genitive (GEN/ACC ***hrdiny*) is difficult to answer. Most likely, the process has to do with the inherent semantics of the *hrdina*-terms (extrinsic agentivity). Accordingly, concepts related to professions etc. (see above) would not have been accessible to primary strategies of honorific-partitive marking. This hypothesis may be further corroborated if we are allowed to interpret the dative singular variant of *chlap*-nouns (see footnote 22) not merely as a homophone. It is a well-known fact that in some languages, a less 'agentive' (or: male animate) category tends to encode possession differently from the more 'agentive' class, at least on a synchronic level. Compare for instance German:[27]

(12) a. *d-as Buch d-es Mann-es*
DEF:N:SG:NOM book DEF:N:SG:GEN man-SG:GEN
'The book of the man'

b. *d-as Buch d-er Frau*
DEF:N:SG:NOM book DEF:F:SG:GEN/DAT woman
'The book of the woman'

For German, it is reasonable to assume that the case form *d-er* in (12.b) is nothing but the dative that also occurs in the analytic possessive construction (13):

27 Also compare Krifka (2009).

(13) *d-er Frau ihr Buch*
DEF:F:SG:DAT woman her book
'The book of the woman.'

Taking this analysis as an analogy, we can hypothesize that the *u*-genitive of *hrdina*-terms is not the accusative (alone), but at least induced by the old *u*-dative of the *ŏ*-stems. In this sense, the *hrdina*-paradigm would be partially motivated in analogy with the already mentioned GEN/DAT syncretism on some naM-terms (-*u*). If this assumption is true, the *hrdina*-syncretism of genitive and accusative has to be read differently from that of the *chlap*-paradigm, compare the prototypical patterns:

(14) Chlap: *vidím chlap-**a***
I see [part of]$_{GEN}$ *the man.*

Hrdina: *vidím hrdin-**u***
I see [to(wards)]$_{*DAT}$ *the hero.*

Chlap: *dom chlap-**a***
The house [of]$_{GEN}$ *the man*

Hrdina: *dom hrdin-**u***
The house [to/for]$_{*DAT}$ *the hero*

At a later stage, the functional domains would have merged in favor of a re-conceptualization of the *hrdina*-strategies to mark possession and hence Split-O.

6. Conclusions

In the present paper, we have attempted to give an outline of the architecture of case paradigms in Slovak from the perspective of Cognitive Morphology.[28] Cognitive Morphology treats morphological units as variants of linguistic signs (or: symbolic constructions) that are marked for the symbolic linkage of an expression domain (*signifiant*) to a conceptual domain (*signifié*) that is categorial in nature and that relates the conceptual domain of an adjacent lexical unit to its co(n)text (here, only inflectional patterns are at issue). A morphological linguistic sign may be explicit with respect to one categorial domain of its host, resulting in what is conventionally called agglutination; or, it may integrate several such categorial domains into one 'covering'

---

28 See, among many others, Wierzbicka (1988), Niemi/Laine/Tuominen (1994), Bybee (1985, 2001), Gaeta (2004).

category (fusion). We have argued that such 'fusional' morphemes do not represent separate categorial domains in *one* form, but complex categorial entities that are profiled according to one or more of the sub-domains in terms of active zones. As for the theme of this paper, this concerns the three major conceptual domains Gender, Number, and Case. Moreover, it is argued that some types of syncretism especially within the layer of Case do not reflect homonymic expressions but result from the blending of two or more elsewhere functionally distinct domains, based on the extension of one of the domains towards the others.

Basically, the general architecture of the Slovak case paradigm continues patterns of its ancestor language (Common Slavic). However, Slovak has elaborated the virility dimension that dominates a larger part of the case paradigm. This process has affected nearly the whole case paradigm and has resulted in the reanalysis of some of the morphemes as markers relevant for this dimension. The most important morphemes and morphophonemes are:

Table 9. The basic morphological strategies relevant for the animateness dimension in Slovak

-a	Intrinsic aM	Singular	Partitive/Possessive	→ Split-O
-i [+ pal., if possible]	Intrinsic aM	Plural	S=A	
-ia	Social aM	Plural	S=A	
-ov	aM	Plural	Partitive/Possessive	→ Split-O
-ovi	Ex-/Intrinsic aM	Singular	LOC/IO	
-ovia	Extrinsic aM	Plural	S=A	
-u₁	not intrinsic aM/N	Singular	LOC/IO>POSS	→ Split-O
-u₂	F	Singular	O	

The underlying types of syncretism argue in favor of a reduced case paradigm in Slovak. Accordingly, aM:SG would encompass four cases, naM:SG three to five, F four to five, and N three to five cases:

Table 10. A rearrangement of Slovak case patterns (singular, disregarding the instrumental)

<table>
<tr><th></th><th colspan="2">aM (chlap)</th><th colspan="2">aM (hrdina)</th><th colspan="2">naM</th><th colspan="2">N</th><th colspan="2">F</th></tr>
<tr><td>NOM</td><td colspan="2">-Ø</td><td colspan="2">[-a] / -Ø</td><td colspan="2" rowspan="2">-Ø</td><td colspan="2" rowspan="2">-o ~ -e</td><td colspan="2">[-a] /</td></tr>
<tr><td>ACC</td><td colspan="2" rowspan="2">-a</td><td colspan="2" rowspan="2">-u</td><td>-u</td><td>-Ø</td></tr>
<tr><td>GEN</td><td>-a</td><td rowspan="3">-u</td><td>-a</td><td rowspan="3">-u</td><td colspan="2">-i [~ -e]</td></tr>
<tr><td>DAT</td><td rowspan="2">-ovi</td><td rowspan="2">-u</td><td rowspan="2">-ovi</td><td rowspan="2"></td><td></td><td></td><td colspan="2" rowspan="2">-e [~ -i]</td></tr>
<tr><td>LOC</td><td>-e</td><td>-e</td></tr>
</table>

In the plural, the syncretism goes the same direction: Here we have aM:PL with four cases, naM:PL with four cases, N and F with five cases:

Table 11. A rearrangement of Slovak case patterns (plural, disregarding the instrumental)

	aM (*chlap*)		aM (*hrdina*)	naM	N	F
NOM	*-i*	*-ia*	*-ovia*	*-y*	*-a ~ -e*	-y ~ -e
ACC	*-ov*					
GEN					*-Ø ~ -í*	
DAT	*-om*				*-am*	
LOC	*-och*				*-ach*	

All types of syncretism can be interpreted in terms of polysemic extension of one of the underlying case functions, although we cannot always be sure which of the individual subdomains was the starting point of these extensions. The following processes can be described as:

(15)

Function1		Function2		Function3
NOM	→	ACC		
GEN	→	ACC		
DAT/LOC	→	LOC/DAT		
DAT/LOC	→	LOC/DAT	→	GEN

The relevance of these blending processes can also be seen from the accompanying paradigms as expressed (among others) by the inflection of attributes and pronouns, compare Tables 12 and 13:

Table 12. The basic pattern of attributive case marking in Slovak

	Masculine		Neuter	Feminine	Plural	
	aM	naM			aM	not [aM]
N	*-ý*	*-ý*	*-é*	*-á*	*-í*	*-é*
A	*-ého*			*-ú*	*-ých*	
G				*-ej*		
L	*-om*					
D	*-ému*				*-ým*	
I	*-ým*			*-ou*	*-ými*	

Table 13. The basic paradigm of personal/anaphoric pronouns in Slovak

	I	you	he	it	she	we	you:pl	they[+aM]	they[-aM]
NOM	*ja*	*ty*	*on*	*ono*	*ona*	*my*	*vy*	*oni*	*Ony*
ACC	*ma*	*ťa*	*ho*		*ju*	*nás*	*vás*	*Ich*	
GEN									
LOC	*mne*	*tebe*	*ňom*		*nej*			*Nich*	
DAT			*mu*		*jej (nej)*	*nám*	*vám*	*Im*	
INSTR	*mnou*	*tebou*	*ním*		*ňou*	*nami*	*vami*	*Nimi*	

Nevertheless, it has to be born in mind that some of the syncretisms mentioned above are partly due to phonetic processes, not to conceptual blends. Others may reflect older stages of the language, which means that they are merely conventionalized mergers stabilized during learning (and teaching). The rather heterogeneous picture given in section 5 suggests that the rearrangement of case morphology has not come to an end yet. Obviously, the Slovak case paradigm is in a stage of transition that still allows the existence of idiosyncratic allo-variants. Still, we can assume that such a stage will be followed by another one, where the strategies lurking through the current paradigms will become stabilized and fully entrenched in the cognition of Slovak speakers.

References

Brugmann, Karl 1889. Das Nominalgeschlecht in den indogermanischen Sprachen. *Internationale Zeitschrift für allgemeine Sprachwissenschaft* 4, 100-109.

Brugmann, Karl 1897. *The Nature and Origin of the Noun Genders in the Indo-European Languages.* New York: Charles Scribner's Sons.

Bybee, Joan L. 1985. *Morphology: A study of the relation between meaning and form.* Amsterdam/Philadelphia: Benjamins.

Bybee, Joan L. 2001. *Phonology and Language Use.* Cambridge: Cambridge University Press.

Comrie, Bernard 1991. Form and Function in Identifying Cases. In: Frans Plank (ed.). *Paradigms: The Economy of Inflection,* 41-56. Berlin: Mouton de Gruyter.

Comrie, Bernard, Greville G. Corbett 2002 [1993]. *The Slavonic Languages.* London: Routledge.

Dewell, Robert B. 2000. Case Meaning and Sequence of Attention Source: Landmarks as Accusative and Dative Objects of the Verb. In: Ad Foolen and Frederike Van der Leek (eds.). *Constructions in Cognitive Linguistics: Selected Papers from the Fifth International Cognitive Linguistics Conference*, 47-65. Amsterdam: Benjamins.

Dixon, Robert M.W. 1994. *Ergativity.* Cambridge: Cambridge University Press.

Doleschal, Ursula 1995. Genuszuweisung im Russischen. Ein kognitives Assoziationsmodell. In Uwe Junghanns (ed.). *Linguistische Beiträge zur Slawistik aus Deutschland und Österreich. Wiener Slawistischer Almanach Sonderband* 37:75-96.

Dressler, Wolfgang U. 1989. Prototypical differences between inflection and derivation. *Zeitschrift für Phonetik, Sprachwissenschaft und Kommunikationsforschung* 42:3-10.

Fauconnier, Gilles, Mark Turner 2002. *The Way We Think. Conceptual Blending and the Mind's Hidden Complexities.* New York:Basic Books – Perseus Books Group.

Gaeta, Livio 2004. Thoughts on Cognitive Morphology. In Gertraud Fenk-Oczlon, Christian Winkler (eds.). *Sprache und Natürlichkeit*, 107-127. Tübingen: Narr.

Gorškova, Klavdia V. and Xaburgaev, Georgij A. 1997. *Istoričeskaja grammatika russkogo jazyka.* Moscow: Vyssaja škola.

Janda, Laura A. 1999. Whence virility? The rise of a new gender distinction in the history of Slavic. In Margaret H. Mills (ed.). *Slavic gender linguistics*, 201-228. Amsterdam/Philadelphia: John Benjamins

Janda, Laura A. 2002. Cognitive hot spots in the Russian case system. In: Michael Shapiro (ed.). *Peircean Semiotics: The State of the Art*, 165-188. New York: Berghahn Books.

Janda, Laura A., Steven J. Clancy. 2002. *The Case Book for Russian.* Bloomington: Slavica.

Krifka, Manfred 2009. Case syncretism in German feminines: Typological, functional and structural aspects Patrick O. Steinkrüger, Manfred Krifka, Manfred (eds.). *On Inflection*, 141-172. Berlin: Moutin de Gryter.

Langacker, Ronald 2000. Grammar and Conceptualization. Berlin/New York: Mouton de Gruyter.

Leskien, August 1969 [1910]. *Handbuch der altbulgarischen Sprache*. Heidelberg: Winter.

Meillet, Antoine 1936². *Esquisse d'une grammaire comparée de l'arménien classique*. Vienne: Impr. des Pp. mékhitharistes.

Mistrík, Jozef 2003. Gramatika slovenčiny. Bratislava: Slovenské pedagogické nakladateľstvo.

Niemi, Jussi, Matti Laine, Juhani Tuominen 1994. Cognitive morphology in Finnish: Foundations of a new model. *Language and Cognitive Processes*, Vol.9,3: 423-446.

Nikiforidou, Kiki 1991. The Meanings of the Genitive: A Case Study in Semantic Structure and Semantic Change. *Cognitive Linguistics* 2:149-205.

Noonan Michael / Elena Mihas 2008. Areal Dimensions in Case Syncretism: Ablatives and Genitives. Paper presented at the ALT conference in Paris, Sept. 2007, and at the University of Hamburg, May, 2008 [http://www.uwm.edu/~noonan/Ablatives&Genitives.pdf]

Olsen, Birgit A. 1999. *The Noun in Biblical Armenian. Origin and Word-Formation - with special emphasis on the Indo-European heritage*. Berlin/New York: Mouton de Gruyter.

Pauliny, Eugen 1963. *Fonologický vývin slovenčiny*. Bratislava : Vydavateľstvo Slovenskej akadémie vied.

Pauliny, Eugen 1981. *Slovenská gramatika. Opis jazykového systému*. Bratislava : Slovenské pedagogické nakladateľstvo.

Pauliny, Eugen 1983a. *Dejiny spisovnej slovenčiny od začiatkov po súčasnosť*. Bratislava : Slovenské pedagogické nakladateľstvo.

Pauliny, Eugen 1983b. *Historický a nárečový základ spisovnej slovenčiny*. Banská Bystrica : Pedagogická fakulta, 1983.

Pauliny, Eugen 1989. *Vývin slovenskej deklinácie*. Bratislava : Slovenský spisovateľ, 1989.

Pauliny, Eugen 1997. *Krátka gramatika slovenská*. Bratislava : Národné literárne centrum.

Pauliny, Eugen, Jozef Ružička, Jozef Štolc 1968. *Slovenská gramatika*. Bratislava: Slovenské pedagogické nakladateľstvo.

Rijkhoff, Jan 2002. *The Noun Phrase* Oxford: Oxford University Press.

Rubach, Jerzy 1993. *The lexical phonology of Slovak*. Oxford: Oxford University Press.

Ružička, Jozef (red.) 1966. *Morfológia slovenského jazyka*. Bralistava: Vydavateľstvo Slovenskej akadémie vied.

Schulze, Wolfgang 1998. *Person, Klasse, Kongruenz. Fragmente einer Kategorialtypologie des einfachen Satz in den ostkaukasischen Sprachen*. Vol. 1-2. Munich: Lincom.

Schulze, Wolfgang 2000. Towards a Typology of the Accusative Ergative Continuum - The Case of East Caucasian. *General Linguistics* 37 (2000): 71-155.

Schulze, Wolfgang 2009. A new model of metaphorization: Case semantics in East Caucasian. In: Klaus-Uwe Panther, Linda L. Thornburg, Antonio Barcelona (eds.) *Metonymy and Metaphor in Grammar*. Amsterdam/Philadelphia: Benjamins (in press).

Schulze, Wolfgang und Walther Sallaberger 2007. Grammatische Relationen im Sumerischen. *Zeitschrift für Assyriologie* 97 (2007):163-214.

Stanislav, Ján 1956. *Slovenská historická gramatika*. III. diel. Skloňovanie. Bratislava: Slovenské pedagogické nakladateľstvo.

Strunk, Klaus 1994. Grammatisches und natürliches Geschlecht in sprachwissenschaftlicher Sicht. In: Venanz Schubert (Hrsg.). *Frau und Mann - Geschlechterdifferenzierung in Natur- und Menschenwelt.* Wissenschaft und Philosophie 10, 141-164. St. Ottilien: EOS Verlag.

Turner, Bryan S. 1990. *Theories of Modernity and Postmodernity*. London: Sage.

Vykypěl, Bohumil 2006. Eine Bemerkung zum Genus. *Acta Linguistica Lithuanica* 54/2006: 95-100

Wierzbicka, Anna 1988. *The Semantics of Grammar.* Amsterdam: John Benjamins.

The notion of affectedness in expressing interpersonal functions

Mirjam Fried, Czech Academy of Sciences, Prague

1. Introduction

Dative-marked nominals cover a lot of semantic territory in Czech. Many are governed by specific predicate classes and linked to various argument roles (experience, recipient, goal, etc.) depending on the meaning of the predicate. The examples in (1) show just one role type, one that is associated with a number of 2-place predicates and semantically most directly related to the issues discussed in this paper; the datives and their English equivalents in the translations will be in boldface throughout the paper:[1]

(1) a. *já **vám** budu chodit po večerech pomáhat*
1SG:NOM 2PL:DAT be:FUT:1SG come:INF onevenings help:INF[2]
'in the evenings I'll be coming to help **you**' [PMK]
b. *trend, **kterému** pomohl především*
trend which:DAT:SG:M help:PST:SG:M especially
{*silný příliv zahraničního kapitálu*}[3] [SYNEK]
'a trend **which** was helped particularly {by a strong flow of foreign investment}

The dative-marked arguments in (1) are affected by the event denoted by the predicate, but in a less direct, perhaps mental, way as compared to the accusative-marked patient (cf. also Croft 1992, Kemmer 1993 on this distinction). We can label this role as 'indirect affectedness'.

Many dative NPs, however, are added as extra elements that must be integrated into the clause structure by a mechanism other than a simple head projection. Some illustrative examples of this type are in (2). The verbs *otevřít* 'open:TRANS' (2a), *ochladit se* 'turn.cooler:INTRANS' (2b), or *poranit* 'injure' (2c) have no dative-marked argument in their valence (i.e., their lexical meaning does not require it) but such a participant can be added.

(2) a. *Tento balón otevřel **lidstvu** oblohu.*

1 Unless otherwise noted, the examples come from the Czech National Corpus, both spoken (oral2006, oral2008, PMK, BMK) and written (SYN, SYNEK). By their very nature, the forms discussed in this paper are primarily a feature of conversational language but they are attested in written texts as well.

2 Abbreviations in the glosses: SG/PL 'singular/plural', NOM 'nominative', DAT 'dative', ACC 'accusative', LOC 'locative', M/F/N 'masc./fem/neuter', FUT 'future', PST 'past', PRES 'present', IMP 'imperative, IRREAL 'irrealis', NEG 'negative', RF 'reflexive', (IN)TRANS '(in)transitive', DM 'discourse marker'.

3 When larger context is helpful for better understanding of an example or for easier exposition, the text will be added in curly brackets but left without glossing.

This balloon :NOM:SG:M open:PST:SG:M mankind:DAT:SG:N sky:ACC:SG:F
'This balloon opened up the sky **for mankind**.' [SYN]

b. *{jo na noc pojedete. to byste měli fajn,}*
kdyby se ***vám*** *ochladilo viď*. [oral2006]
if:IRREAL RF:ACC 3PL:DAT turn.cooler:PST:SG:N DM
'{Oh, you'll be traveling overnight. It'd be good} if **you** had cooler weather, wouldn't it?' [lit. 'if the weather turned cooler for you']

c. *Další výstřel* ***mu*** *však těžce*
next:NOM:SG.M shot:NOM:SG:M 3SG:M:DAT however seriously
poranil nohu. [SYN2006pub]
injure:PST:SG:M leg:ACC:SG:F
'The second shot, however, seriously injured **his** leg.'

Semantically, the participant's role in the reported event is related to the role type in (1), but in a narrower interpretation of affectedness: the referent of these extra elements is cast as having a kind of 'interest', broadly understood, in the outcome of the reported event while not being in control of the event. Consequently, and in contrast to (1b), the datives in (2) require a human referent. I will refer to the role as Dative-of-Interest (DI; cf. also the notion *dativus (in)commodi* in more traditional terminology). The relationship between (1) and (2) has been discussed and justified elsewhere (e.g. Fried 1999a) and I will take it for granted in the present analysis.

The central concern of this paper is yet another set of patterns, partially illustrated by the example in (3):

(3) *{přijdu si večer v pátek domů, povečeřím, a}*
Začal ***ti*** *mě bolet zub.*
start:PST:SG:M 2SG:DAT 1SG:ACC hurt:INF tooth:NOM:SG:M
{tak sem si hbitě vzala prášek a ono to přestalo} [PMK]
'{Friday evening I get myself home, have a little supper, and} [**get this**], my tooth starts hurting. {so I quickly popped in a pill and it stopped}

The dative in (3) shares one crucial property with (2) in that it is also an extra element, not one required by the head verb (here *bolet* 'hurt'). It further shares with (2) the requirement that the referent be an animate entity (and more specifically, a human). But it also differs from both (1) and (2) in an important way. In (1) and (2), the dative marks an entity that participates in the narrated event – either on its own, as in (1) and (2a-b), or as a possessor of another affected entity, as in (2c). In contrast, the referent of the pronominal dative *ti* in (3) does *not* participate in the state of having a toothache. Instead, the dative indexes the hearer as a participant in the present *discourse*, not in the reported event itself.

This much is commonly acknowledged in the literature on 'ethical datives' in various languages (e.g. Poldauf 1962, Leclere 1976, Berman 1982, *Mluvnice češtiny* III 1986, Borer & Grodzinski 1986, Authier & Reed 1992, Janda 1993,

Dabrowska 1997, King 1997). I will argue, however, that the pattern shown in (3) is not the only variant of an 'interactional' dative and that the distribution of all such datives in authentic usage is licensed by a relatively complex set of constraints that have not been addressed in any detail yet. For reasons that will become clear in the exposition, I will refer to these patterns collectively as 'datives of empathy' (DE), in order to avoid any misunderstanding that might be brought about by the term 'ethical dative'. The latter seems to mean somewhat different things to different analysts and at best would cover only a subset of the data discussed in this paper.

The present analysis is framed by the following general question: what is the nature of the linguistic knowledge that allows the Czech speakers to produce and interpret the discourse-based datives with native-like fluency? Put differently, does the distribution of DEs reflect a recurrent, conventional grammatical pattern, or is their use just a matter of context-dependent inference or interpretation based on extralinguistic knowledge? The material also raises broader questions of general theoretical interest. In particular, we must consider (i) the relationship between the conceptual aspects of a linguistic form and its pragmatic function and (ii) the role of interactional factors in shaping linguistic structure. I will argue that the use of DEs indeed involves grammatical knowledge, manifesting a conventional association between a specific morphosyntactic behavior and certain well-defined pragmatic factors, as well as a link to a particular semantic relation conceptualized as a type of affectedness. The gist of the analysis amounts to a proposal for positing an extension of the semantic notion of affectedness from a 'semantic dative' to an 'interactional dative', as a cognitively plausible transfer from the experiential domain of event types to the domain of communicative strategies and specifically speaker-hearer relations. I will show that both types of datives can be organized in a single network of related but distinct grammatical forms that occupy overlapping conceptual spaces.

The nature of the data and the generalizations to be drawn from the observed patterns call for an analytic model that allows us to treat grammatical forms as relatively complex signs in which form and meaning/function constitute further indivisible units of linguistic structure, i.e. grammatical constructions in the sense of Construction Grammar (CxG; e.g. Fillmore 1988, 1989, Fried & Östman 2004). This approach is framed by the following hypotheses, which form the conceptual and architectural basis of CxG: speakers rely on relatively complex grammatical patterns – CONSTRUCTIONS – for producing and understanding linguistic expressions; linguistic expressions reflect the effects of interaction between CONSTRUCTIONS and the words that fill them in actual expressions; and CONSTRUCTIONS are organized into networks of overlapping patterns related through shared properties.

Two fundamental features of this approach will be relevant to the present analysis. (i) Constructions (the basic units of analysis) are multidimensional objects in which various *types* of features are integrated in a single, further indivisible, form-meaning/function pairing. The form pole may consist of any combination of syntactic, morphological, or prosodic features and the meaning/function pole may refer to lexical semantics, pragmatics, and/or discourse structure; particular combinations will depend on the construction. And (ii) the relationships among constructions are identified through inheritance links, which capture the sharing of properties across related constructions and simultaneously ensure representational parsimony.

All of these features of the constructional model will be relevant in the analysis of the material at hand. I will first isolate the properties of each DE variant in section 2, proposing a constructional treatment of their defining characteristics. In section 3 I will suggest a way of linking the variants in a network of related expressions that can be mapped onto a conceptual space of indexing speaker-hearer relations; each DE variant will be treated as a distinct expression of a specific discourse-deictic function, occupying a particular portion of the conceptual space. Section 4 will summarize my conclusions.

2. Variants of DE

2.1 *General DE properties*

Let us start by surveying the relevant data. DE seems to come in three different flavors, one of which is illustrated in (3) above. Slightly different uses of DE are shown in (4) and (5). For reasons that will become clear in the course of the paper, I will refer to the three patterns as Contact DE (3), Subjective DE (4), and Distancing DE (5); for now let us just note that the labels are motivated by the pragmatic function associated with each variant.

(4) *No to je* ***mi*** *ale náhoda*! [SYN]
DM it is 1SG:DAT but coincidence:NOM:SG:F
'**My**, what a coincidence!'

(5) *Dej* ***si*** *mu přes hubu, jak chceš,*
give:IMP:2SG RF:DAT 3SG:DAT on mouth:ACC:SG:F as want:PRES:2SG
mně je to jedno. (Janda 1993: 103)
3SG:DAT is that:NOM:SG:N one:NOM:SG:N
'[**Help yourself**] and beat him up [**if it makes you feel good**] – as you want, I don't care.'

The examples (3–5) share certain properties that set them apart from other Czech datives, whether governed (1) or ungoverned (2). The morpho-syntactic idiosyncracies (cf. Fried 1999b for arguments and full discussion) and functional status of the DEs can be summarized as follows:

- They are only pronominal in form and if a pronoun has both a full and a clitic form, the DE can only be in the clitic form. The clitics tend to fall into a particular slot within the second-position clitic cluster, distinct from other dative clitics, although corpus data also show some degree of variable placement (to be addressed in this paper).
- Semantically, they resemble ungoverned datives (2), which always mark human referents with some interest in the reported event (DIs). However, DEs do not behave like syntactic arguments in that they cannot be extracted (in questions or relative clauses).
- They can co-occur with other dative-marked nominals in the same sentence without violating the bi-uniqueness condition; this is exemplified by the verb *dát* 'give' in (5), which brings its own dative NP as a recipient (*mu* 'to him').
- They serve a discourse-deictic function in speaker-hearer relations. For now, let us just observe that the Contact DE is directed from the speaker to the hearer, the Subjective DE appears to be fully speaker-oriented, and the Distancing DE is primarily also speaker-centered, albeit in a more complicated way.

Another striking feature of the DEs is their textual distribution. They all are associated with emotional speech, whether the speaker makes an appeal to the listener (3), or whether he asserts his own emotional state about a piece of discourse – with respect to himself (4) or to another entity (5). Consequently, these datives are restricted to informal spoken discourse (including its written renditions) and typically occur in emotional, intimate genres in particular. They also necessarily involve features of sub-standard speech or colloquialisms at all levels – phonetic, morphological, and lexical. As, for example, King (1997: 73) observes about the Contact DE, its use would be impossible in a formal situation, such as in the welcoming of an important official guest. A possible form of such a text is shown in (6); the formal language and deferential tone clearly mark the social distance between the speaker and the addressee. The form in (6) contrasts with (7), in which the insertion of DE *vám* '2PL:DAT:POLITE' results in a pragmatically odd utterance; the relevant sections in both examples are underlined, the contextual inappropriateness in (7) is indicated by the symbol '#'.

(6) *Dovolte mi, pane ministře, abych Vás jménem městské rady přivítal v našem městě. <u>Máme velkou radost,</u>*
have:PRES:1PL big:ACC:SG:F joy:ACC:SG:F
že jste přijal naše pozvání, a doufáme, že se Vám bude u nás líbit.
'Allow me, Mr. Secretary, to welcome you in our town, on behalf of the City Council. <u>We are very pleased</u> that you accepted our invitation and hope that you will enjoy your stay.'

(7) *Dovolte mi, pane ministře, abych Vás jménem městské rady přivítal v našem městě. #<u>Máme **vám** velkou radost,</u>*

have:PRES:1PL 2:PL:DAT:POLITE big:ACC:SG:F joy:ACC:SG:F
že jste přijal naše pozvání, a doufáme, že se Vám bude u nás líbit.
'Allow me, Mr. Secretary, to welcome you in our town, on behalf of the City Council. #[**Hey, listen,**] we are so pleased that you accepted our invitation and hope that you will enjoy your stay!'

Based on the observations presented so far (and mostly familiar from existing analyses), we can formulate a preliminary generalization as follows. There is a cluster of properties that characterize all the DE variants and identify DEs as a formally and functionally distinct family of expressions that mark certain kinds of interpersonal relations. The features shared across the variants are listed in (8):

(8) a. form: clitic/unstressed pronoun
dative case
b. semantics: 'interest'
c. textual: informal, emotional register; (semi)private conversation
d. pragmatics: interpersonal function

However, in order to account fully for all the data in (3–5), the general contours of the DE family must be fleshed out in more detail, since each subtype involves some idiosyncratic behaviors and constraints and as such constitutes a distinct member of the family.

2.2 *Contact DE*

Judging from the corpus evidence, this DE variant is most frequent of the three; in addition to the introductory example in (3), we can also consider the attestations in (9), to get a relatively full range of the interpretive nuances:

(9) a. {*ňákou chvíli sme eště vo tom mluvili, vona vodešla a*}
já ***vám*** *sem vo tom přemýšlela i v noci*
1SG:NOM 2PL:DAT AUX:1SG about it:LOC think:PST:SG:F also at night
'{we kept talking about it for a while longer, she left and} **you [know what]**, I kept thinking about it even at night' [PMK]
b. {*ráno sme spolu dneska stály u pokladny v diskontu, tak aby řeč nestála, říkám: „Štěpánka už má za sebou druhý spalničky a, a to?" a vona říká:*}
„ježiš já ***vám*** *ani nevim."* [oral2006]
Jesus 1SG:NOM 2PL:DAT not.even NEG:know:PRES:1SG
'{this morning we were standing at the cashiers in the market and so to make conversation I say: "Štěpánka's been through two cases of measles by now, huh, and, and y'know? " and she said:} "Jesus, I don't even know, **[believe it or not]**."'
c. {A: *říkám: „ale to je holka, kam se Erika za ní hrabe" hele, ...*
B: *to neni pravda.*}

A: *Erika **ti** měla kouli z chemie.*
E:NOM 2SG:DAT have:PST:SG:F [grade]F:ACC:SG:F from chemistry
{A: 'I say: "now that's [one gem of] a girl, Erika can't hold a candle to her", look...'
B: 'that's not true'}
A: '[**hey listen**], Erika got an F in chemistry.' [oral2008]

As we can see from the English translations, to render the meaning of the Contact DE (C-DE) in English is not easy, certainly not by using a single equivalent in all contexts. It is generally accepted that the function of the dative is to secure the listener's attention. But a careful analysis of the distribution in natural discourse suggests that C-DE serves not only to draw the hearer in by issuing a direct appeal; it appears to be used on the assumption that the hearer either is not (fully) aware of what is going on in relation to the narrated event or that he will find it in some ways surprising (and, hence, interesting). The speaker's intention in using the DE is to emphasize the presumed newsworthiness of what is being said, either because it somehow goes against what can be expected, or because the speaker appeals to the hearer's interest in the speaker's role in a given piece of discourse and hopes to elicit a sympathetic reaction. The surprise reading is quite apparent in (9b-c); the empathy-seeking modulation may sometimes simply follow from the lexical meaning conveyed, as in reporting the pain in (3), but it emerges as very salient especially in contexts where the speaker admits some inadequacy. In such contexts, the use of DE may even imply a mildly apologetic flavor. It may be no coincidence that C-DE often collocates with *nevím* 'I don't know', as we see in (9b) or (10) below; in fact, in the sample of 96 DEs with *ti* (which is the total of C-DE *ti* tokens in the oral2008 corpus), full 20% are of the form *já ti nevím* '[believe it or not], I don't know'.

A precise characterization of this DE variant raises several issues that require a closer study: its relationship to vocatives; its combinability with other nominals that index 2nd person referents; and its linearization habits, especially in relation to the template that normally determines the ordering of multiple second-position clitics in Czech. I will start with the comparison to vocatives, which, however, is closely related to the referential issues.

C-DE is limited to 2nd person referents only; this follows from its pragmatic function of making a direct appeal to the hearer. The latter is a property shared with vocatives, but the two forms cannot be simply lumped together as two formal variants of (roughly) the same functional category. For one thing, the presence of one does not preclude the presence of the other; they can co-occur, as illustrated in (10).

(10) *já **ti** Klárko, nevim, {já si to možná vzpomenu někdy}*
1SG:NOM 2SG:DAT K.:VOC NEG:know:PRES:1SG

'**you [may not believe it]**, Klárka, I just don't know {maybe it'll come back to me later}' [oral2008]

For another, there is referential evidence that (10) is not simply an issue of some kind of double marking of the discourse-deictic function of direct address: unlike the vocative, C-DE is sensitive to the identity of the participants in the narrated event. Specifically, C-DE cannot co-occur with other 2nd person referents in the same clause, as illustrated in the following example. The presence of DE *ti* in (11) below precludes the presence of a thematic dative with a 2nd person referent, while any other referent is possible (cf. also King's 1997: 75 speculation that multiple datives in one clause must have distinct referents). In the corpus example (11a), we have two dative nominals expressed pronominally: the dative that marks the possessor of the patient in the kissing event (*líbat někomu*$_{DAT}$ *ruce* 'kiss someone$_{DAT}$'s hands'), followed by a C-DE. The possessor in (11a) happens to refer to the speaker (*mi* 'to me'), but the referent could be anybody, except the hearer, as shown in (11b):

(11) a. *Von mi **ti**, představ si, začal*
3SG:M:NOM 1SG:DAT 2SG:DAT imagine:IMP:2SG RF:DAT start:PST:SG:M
líbat ruce [SYN]
kiss:INF hand:ACC:PL:F
'**Y'[know]**, just imagine, he started to kiss my hands.'

b. *Von*$_i$ *mu*$_j$*/jim/*ti/*vám **ti**, představ si, začal líbat ruce*
3SG/3PL/*2SG/*2PL:DAT 2SG:DAT
'**Y'[know]**, just imagine, he$_i$ started to kiss his$_j$/their/*your(sg/pl) hands.'

This cannot be a matter of form, i.e. a phonetic prohibition on the repetition of *ti ti* (or *vám vám*, in the plural). Such combinations are unproblematic and the *ti ti* sequence can easily occur since the word *ti* is also a form of the demonstrative pronoun 'that:NOM:PL:M:ANIM'. Moreover, we see in (11b) that the sequence *vám ti* is not a coherent possibility either, even though there is no haplology involved. Finally, the prohibition on accumulating multiple hearer references in a single clause is not limited to the dative case. It seems equally awkward to use C-DE in combination with other nominals whose referent is the 2nd person, regardless of the event role (and, hence, case marking) of that nominal. Thus, for example, C-DE does not occur in imperative clauses (12a) and would be at best very awkward in other grammatical patterns; the corpus example in (12b) shows a 3rd pers. accusative patient (*ho* 'him'), which could not be replaced with the accusative *tě* 'you:ACC:SG'. Notice that this is in obvious contrast to the use of the vocative, which would be equally natural in both (12a) and (12b), in the latter regardless of *ho* or *tě* as the object:

(12) a. **Podej **ti** tátovi kladivo*!
hand:IMP:2SG 2SG:DAT dad:DAT:SG:M hammer:ACC:SG:N

*'[**Let me tell**] **you**, hand dad the hammer!'

b. *já už sem **ti** ho/??tě neviděla*
1SG:NOM already AUX:1SG 2SG:DAT 3SG:M/2SG:ACC NEG:see:PST:SG:F
teďkon tak dlouho [oral2008]
now so long
'**you [believe it or not]**, I haven't seen it [=a TV program] / ??you for such a long time now'

In accounting for these restrictions, we must appeal to the quintessential pragmatic function of C-DE, which presupposes a certain degree of surprise and empathetic attitude on the part of the hearer; this is particularly evident in (12) but it underlies all the clashes. Propositions describing activities that necessarily involve the hearer as one of the event participants must result in a semantic conflict if the hearer is simultaneously expected to be surprised by the reported events.

It is important to emphasize that none of these constraints apply to Czech vocatives, which are truly extra-clausal elements; a vocative could appear freely in all of the above examples. It is clear that C-DE, while a pragmatic device, is at least partially integrated in the semantic structure of the clause. In this respect, the DE resembles real arguments, for which the same restriction holds: a single referent cannot be cast simultaneously in multiple roles (without a reflexive configuration of participants). C-DEs are special only in that the role-sensitivity carries over to discourse roles as well (speaker, hearer) and is not limited to the domain of event roles (agent, patient, etc.). This partial semantic integration must be a function of the dative form, which, unlike the vocative, is inherently and robustly associated with a particular semantic content. The content – or the speakers' knowledge of it – evidently does not just disappear when the form gets extended to another function in a new domain.

The semantic persistence points toward at least certain remnants of the DE's relationship to other datives (in the spirit of Hopper's 1991 idea of 'layering' as a concomitant feature of language change), those expressing event roles. At the same time, syntactically, C-DE seems to be asserting itself as an independent category, sufficiently defined by its distinct pragmatic function. The independence manifests itself not just in the fact that the C-DE does not have the status of a full-fledged argument, but also through its unexpected linearization habits. Prosodically, all the DEs are enclitics, whether inherent (*ti, mi, si*) or contextually conditioned (*vám, nám* in their unstressed variant). As such, they are expected to appear in a particular slot in the clitic cluster (Fried 1994: 171-172). The core part of the cluster consists of clitic auxiliaries followed by pronominal arguments in a particular order: [aux – reflexive – dative – accusative]. The corpus material suggests that the distribution of the C-DE follows three different patterns relative to this cluster.

- Pattern 1. The dative slot splits into two positions, to accommodate multiple dative clitics in a sentence. The more common configuration is one with the

DE preceding an argument dative, as exemplified in (13a) below; the dative *mi* in this example is added as an extra DI to the verb *sežrat* 'to gobble up', patterning after the examples in (2). The cluster is enclosed in brackets [ ] in (13).

- Pattern 2. Less frequently, DE follows the argument dative (11a).
- Pattern 3. The DE appears immediately to the left of the cluster, i.e. preceding the auxiliary, as in (9a) or even more clearly in (13b).

(13) a. *von [**ti** mi jí] celou*
3SG:M:NOM 2SG:DAT 1SG:DAT 3SG:F:ACC whole:ACC:SG:F
sežral [oral2008]
gobble.up:PST:SG:M
'he, **[can] you [believe it]**, gobbled up the whole [thing] on me'
b. *{von začne řvát a}*
*já **ti** [bych mu] takovou*
1SG:NOM 2SG:DAT 1SG:IRREAL 2SG:M:DAT such:ACC:SG:F
dala pěstí [oral2008]
hit.by.fist:PST:SG:F
'he starts howling and [**I'll tell**] **you**, I'd so [like to] let him taste my fist'

Argument datives are *not* attested in the position described in Pattern 3. The minimal pair in (14) demonstrates the difference. The (canonical) post-auxiliary placement in (14a) allows both a C-DE (i) and a DI reading (ii) of the dative, while (14b), with the (unexpected) pre-auxiliary placement, results in the C-DE reading only. This difference can be confirmed by the choice of a follow-up. A felicitous continuation in (14a) but not in (14b) could be 'So I'm going to read it to you now – are you ready?', which unambiguously picks out the argument reading in (ii).

(14) a. *Tohle [jsem **ti**] ještě nečetla.*
this:ACC AUX:1SG 2SG:DAT yet NEG:read:PST:SG:F
(i) '[**Believe it or not**] I haven't yet read this [one].'
(ii) 'I haven't yet read this [one] **to you**.'
b. *Tohle **ti** [jsem] ještě nečetla.*
this:ACC 2SG:DAT AUX:1SG yet NEG:read:PST:SG:F
(i) '[**Believe it or not**] I haven't yet read this [one].'
(ii) *'I haven't yet read this [one] **to you**.'

The distributions summarized in Patterns 1-3 can be, again, motivated by the pragmatic function: if the main purpose of using C-DE is to secure the hearer's attention and, possibly, a reaction, it makes sense that the speaker would signal such an intention early in the utterance. A general leftward tendency, whether inside or outside the cluster, is noticeable in the corpus data.

However, there are also signs that C-DE may be dissociating itself from the formal status of a second-position clitic altogether and becoming a relatively independent unit; prosodically still an enclitic, but not tied to the second-position cluster. Evidence of this dissociation comes from what appears as a 'free' placement anywhere in the sentence (except sentence-initially, due to its persisting enclitic status): in the middle (15a) or even at the end of a clause (15b), sometimes also inside a phrase (nominal or adjectival); the empty brackets [_] indicate the position where *ti* could be canonically expected as a true second-position clitic.[4]

(15) a. *já* [_] *měla* ***ti*** *pocit,*
1SG:NOM have:PST:SG:F 2SG:DAT feeling:ACC:SG:M
{*že se ta plachetnice láme*} [oral2008]
'[**hey listen**], it felt as if the sailboat was breaking up'

b. *tak* [_] *mu kouká pod nohy* ***ti****, jesi*
so 3SG:M:DAT look:PRES:3SG under feet 2SG:DAT if
ňákej neukopneš náhodou, [oral2008]
some:ACC:SG:M NEG:kick.off:PRES:2SG by.chance
'so she looks where he puts his feet **[imagine],** [to see] if one doesn't happen to kick one [=stone] out of its place'

```
cat       pro_pers.                                        Contact DE
prosody [ stress  -          ]                  inherit DI construction
        [ (2P-) enclitic)    ]
prag    [ discourse-frame  [ PRIVATE CONVERSATION ] ]
        [                  [    register  informal ] ]
        [ involvement      'empathy-seeking contact with Hearer' ]
morphol. [ pers.  2  ]
         [ case  DAT ]
sem      [role  Interest]
```

Diagram 1. Contact DE construction

It is beyond the scope of this study to investigate in detail the conditions for such placements; it will be necessary to address this behavior in the general context of pragmatic particles, their functions and distribution, in order to gain a full, comprehensive picture of the C-DE's status in relation to other discourse-pragmatic devices. For the purposes of the present study, it suffices to simply

4 The apparent absence of C-DE in the sentence-initial position is interesting in the light of the fact that other second-position clitics can be, occasionally, found at the very beginning of an utterance in conversational Czech. This asymmetry deserves further investigation, possibly on the hypothesis that the DE's persistent enclitic status has to do with its pragmatic function. Exploring this hypothesis must be left to future research.

note that through these idiosyncracies, C-DE further confirms its status of a distinct linguistic category.

In order to represent this category, we must posit a construction that contains the features outlined in (8) but narrows down their effect. The construction must specify the phatic function of establishing a particular type of contact with the hearer and it must mark the discourse role of the dative as 'hearer', since this association does not follow from any features of the dative itself, nor from anything in the structure of the surrounding clause. The construction must also indicate the DE's attenuated status as a second-position clitic. A possible formalization of this cluster of co-occurring properties is in Diagram 1. In order to keep the representation as uncluttered as possible I am using a simplified short-hand for the more elaborate CxG notation; including all the details of the formalism would be more distracting than informative (the interested reader is referred to Fried & Östman 2004 for the details of the full formalism). Diagram 1 specifies the *prag*(matics) of C-DE, which consists of the type of discourse, register and the contact function (the latter glossed simply as 'empathy seeking' in the *involvement* statement). These features, together with the restrictions in *cat*(egory), reference to *2nd pers.*, and the *prosody*, are shown in boldface, to indicate that these are features specific to C-DE, unpredictable from anything else in the form. The italicized statements represent features that are shared with (*inherited* from) the Dative-of-Interest construction, which licenses argument expression in sentences of the type in (2). The inherited features are not idiosyncratic to C-DE and, strictly speaking, we do not need to list them explicitly in the representation; I include them in the interest of clearer exposition.

2.3 *Subjective DE*

The Subjective DE (S-DE), introduced in (4) and further illustrated in (16) below is a more controversial, harder-to-identify example of an interactional dative, although it is as such briefly mentioned, without further commentary, in the Czech academic grammar (*Mluvnice češtiny* II 1986: 102).

(16) a. {*Tak počkej – to máme burzovní makléř, známý umělec, taxíkář a teplouš.*} *No to je* ***mi*** *teda bratrstvo*. [SYN]
DM it is 1SG:DAT DM brotherhood:NOM:SG:N
'{Let's see – we've got here a stock-trader, a famous artist, a cabdriver, and a gay.} I mean, quite a brotherhood **[if you ask] me**.'

b. *Hehe, to je* ***mi*** *fórek*... {*- A jak mi vysvětlíš, dědku, že tahle*
haha it is 1SG:DAT joke:NOM:SG:M
ženička nabízí všecko a nic za to nechce?} [SYN]

'Haha, that's a fine little joke, [**what**] **I** [**am hearing**]... {- And how are you going to explain, you old fart, that this little woman is offering everything and doesn't want anything in return?}

Some researchers (e.g. King 1997) have questioned its existence altogether on the grounds that it can easily be confused with argument-marking datives and, thus, misidentified. The confusion involves particularly DIs, illustrated in the initial examples in (2a-b) and here repeated in (17a-b), since they, too, are not required by the verb. A brief digression on the DI's meaning and use is thus now in order.

(17) a. *Tento balón otevřel **lidstvu** oblohu.*
this balloon:NOM:SG:M open:PST:SG:M mankind:DAT:SG:N sky:ACC:SG:F
'This balloon opened up the sky **for mankind**.' [SYN]
b. *{jo na noc pojedete. to byste měli fajn,}*
*kdyby se **vám** ochladilo viď*. [oral2006]
if:IRREAL RF:ACC 3PL:DAT turn.cooler:PST:SG:N DM
'{Oh, you'll be traveling overnight. It'd be good} if **you** had a cooler weather, wouldn't it?' [lit. 'if the weather turned cooler for you']

As already mentioned, the DI is very commonly added to all kinds of predicates – transitive (17a), intransitive, atransitive (17b) – as a way of introducing an additional participant in the narrated event; the participant is interpreted as having some 'interest' in the event denoted by the verb. In this semantic role, we would not expect it to be limited to 1st person referents and the examples in (17) demonstrate that no such restriction applies. This observation is in sharp contrast to the copular sentences in (4) and (16), which are necessarily restricted to the 1st pers. sg. pronoun. Moreover, the dative pronoun in (4) and (16) does not mark an event participant: the coincidence (4), the motley group of people (16a), or the presumed absurdity (16b) do not exist for the speaker's benefit (or to his detriment, as the case may be); the speaker just expresses his subjective, emotional reaction to the report of their existence. Consequently, the 1st pers. sg. dative in these examples cannot be replaced by any nominal that would have to be interpreted as an event participant, rather than a discourse participant and specifically the speaker. An attempt at such a substitution by a 3rd person referent is shown in (18), in contrast to (4); it is difficult to even provide a coherent interpretation:

(18) **No to je* ***jim/Evě*** *ale náhoda*!
DM it is 3PL/Eva:DAT but coincidence:NOM:SG:F
*'Hey, what a coincidence **for their/Eva's** sake!'

Yet, some analysts – both for Czech (Grepl & Karlík 1986, Janda 1993) and for comparable phenomena in other languages (Dabrowska 1997 for Polish, Borer & Grodzinski 1986 for Hebrew) – suggest treating examples such as (2) or (19) below as instances of S-DE, i.e., as comparable to (4) and (16).

(19) a. *Rostou* ***nám*** *zoubky*! (Janda 1993: 92)
grow:PRES:3PL 1PL:DAT tooth:NOM:PL:M
'**We** have new teeth growing!'

b. {A: *ale, člověče, už tady skoro nic nemáš na stěnách.*}
B: *no, von* ***mi*** *to všechno sfouknul vítr*
yeah DM 1SG:DAT it all blow.off:PST:SG:M wind:NOM:SG:M
{*a já sem hrozně líná to tam ...*} [PMK]
{A: 'but, man, there's hardly anything left [hanging] on your walls here'}
B: 'yeah, it all got blown off by the wind **on me** {and I'm too lazy to ...}'

Granted, the example in (19a) is an exclamation expressing excitement over the state of a baby's mouth, and S-DEs often, though not necessarily, are exclamations as well (4). However, given the lexical content of this sentence, the speaker can be just as easily and naturally interpreted simply as an event participant (i.e., construed as the co-owner of the growing teeth),[5] without carrying any discourse-related function. Furthermore, the sentence would be equally well-formed with any other referent (*mu* '3SG:M:DAT', *vám* '2PL:DAT', *jim* '3PL:DAT', *dětem* 'children:DAT', etc.), unlike the sentences in (4) and (16). Moreover, the dative in (19a) cannot be left out without leaving the expression incomplete; that – by definition – is not the case with DEs. The imperative form as well as the context in (19b) make the semantic interpretation of the dative *mi* 'to me' even more plausible, casting the speaker as a beneficiary of the narrated event ('it got blown away – to my loss, as I now have bare walls as a result'), rather than as an outside observer commenting on the report of a particular event ('it got blown away – what a notion, if you ask me!'). Furthermore, the datives in (19) behave like true arguments in other ways: they can be questioned (e.g. *Komu*$_{\text{DAT}}$ *rostou zoubky*? 'Whose$_{\text{DAT}}$ teeth are growing?) or relativized, which certainly is not true of the datives in (4) and (16).

However, the crucial element in establishing S-DEs as distinct from DIs, is their textual status, which can only be identified when we consider the discourse context in which they are used. The interactional – rather than event-based – nature of S-DEs, as compared to the DIs, is easily corroborated by corpus evidence, which reveals two additional features that help identify S-DEs as a category *sui generis*. First, the S-DE is always a direct reaction to an immediately preceding utterance or turn; it is thus dialogical in nature. No such textual requirement is associated with DIs, nor should we expect such an association. This is particularly well illustrated in (19a), which can be very naturally used as a discourse-initial exclamation, in contrast to any of the S-DE examples. And second, just like with C-DE, the speaker's reaction is always an expression of surprise; something in the preceding discourse motivates the

5 Among its other functions, the 1st pers. plural is commonly used by parents or otherwise emotionally involved adults referring to their children's doings, as an inclusive 'we' in which the adult identifies with the child (cf. the observations in *Mluvnice češtiny* II, 1986: 370).

speaker's emotional reaction to an unexpected piece of news. In (4), the speaker is incredulous of certain developments to be random, in (16a) the reaction evaluates the unlikely grouping presented by the interlocutor, and the token in (16b) is taken from a dialog in which one interlocutor describes his escapades and the speaker of our token reacts with skepticism. In contrast, (19b) cannot be misidentified as an S-DE token, even though the text is a dialogical exchange: speaker B explicitly agrees with speaker A and only offers an explanation for A's assertion, as also confirmed by the follow-up; there is no surprise to react to.

Finally, notice that all the examples in (4) and (16) display the same syntactic pattern: the clause is introduced by a demonstrative pronoun in the invariant form of a singular neuter (*to*) and the non-verbal predicate is always a noun in the nominative (all other possibilities are excluded – adverbs, adjectives, prepositional phrases of location, and nouns in the instrumental). In contrast, 1st pers. sg. pronouns in other grammatical patterns, such as (19), seem to show at best an ambiguity between the discourse-based DE on the one hand and the DI event role on the other. The potential ambiguity between the event-role and a DE interpretation may, in certain contexts, make the distribution of the S-DE somewhat more fluid than is the case with C-DEs, but that is not the same as concluding that the DE use cannot be identified as a distinct entity. The proposed analysis simply acknowledges the complexity associated with the interaction between various grammatical configurations and with incorporating the discourse-deictic dimension of linguistic structure. The patterns observed in authentic usage lead to the conclusion that an indisputable S-DE status of the 1st pers. pronoun is conventionally associated only with a particular syntactic construction – copular existential statements, as in (4) or (16) – but a discourse-pragmatic interpretation may be invited also in specific morphosyntactic contexts (the dative must be 1st pers. clitic pronoun) and in particular communicative contexts, apparently facilitated by an exclamative, emotional nature of a given utterance, where it is layered over the primarily semantic status of the dative as an event participant (19).[6]

The S-DE is schematically captured in Diagram 2. Like in C-DE, the construction must specify the following: the pragmatic function of S-DE and the register, in the *prag* statement; the 1st pers. reference as the only possible one; and the second-position enclitic status of the pronoun (in this case obligatorily so – the corpus does not show any other placement). The rest is handled through inheritance relations: the dative form and semantics of interest is again shared with the DI construction, but unlike C-DEs, this pattern also inherits a particular type of a copular construction, which ensures the distribution of S-DE as restricted to this specific clause type; optionally it inherits also the Exclamation Construction, whose existence I simply assume here.

6 It may also be worth noting that, somewhat surprisingly, the spoken corpora within *CNK* seem to contain no example of S-DE. This finding will require further research.

Subjective DE
inherit "to...COP...NP$_{NOM}$" construction
inherit DI construction
(*inherit Exclamative construction*)

cat **pro$_{pers.}$**
prosody [**stress -** / **2P enclitic**]
prag [**discourse-frame** [**PRIVATE CONVERSATION** / **register informal**] / **involvement 'Speaker's subjective attitude to current discourse'**]
morphol. **[pers. 1]**

Diagram 2. Subjective DE construction

2.4. *Distancing DE*

The reflexive pattern illustrated in (5) and in (20) below is perhaps the most elusive of the three types, since the reflexive dative necessarily blurs the distinction between the semantic and discourse domains even more thoroughly than the 1st person referent in S-DE.

(20) a. {*mohl by to dělat v podstatě v pracovní době, jenže*}
von ***si*** *radši ten pracovník támle*
3SG:M:NOM RF:DAT more.happily that worker:NOM:SG:M there
bude ležet {*a neudělá navíc ani ň*} [PMK]
be:FUT:3SG lie.down:INF
'{he basically could get it done during regular work hours, but} he'll, the worker would rather be lying around, [**happily idling away**] {and won't lift a finger for any extra work}'

b. {*Našlapoval a funěl, funěl a našlapoval,*}
ale motor ***si*** *klidně stávkoval* [SYN]
but engine:NOM:SG:M RF:DAT calmly be.on.strike:PST:SG:M
'{He kept stepping [on the gas] and huffing, huffing and stepping [on the gas]} but the engine [**had the temerity**] to simply remain on strike.'

c. *Nechceš-li mě milovati, nechtěj* ***si***
want:PRES:2SG-if 1SG:ACC love:INF NEG:want:IMP:2SG RF:DAT
'If you don't want to give me your love, [**help yourself**] and go on not wanting, [**see if I care**] …' [Czech folk song]

Reflexive datives are a common sight in Czech; they express self-interest or self-directed affectedness and cover the entire semantic spectrum of dative-marked arguments, governed or ungoverned.[7] The corpus examples in (21)

7 The reflexive pronoun in Czech comes in the accusative (*se*) and dative (*si*) forms. The details of the use of the accusative form are also quite complex, but it generally has to do with marking self-affectedness and is independent of the dative form (Fried 2007 gives a full constructional treatment of the *se* variant).

below show a productive use of *si* in the DI role, parallel to the non-reflexive uses we saw in (17): the dative marks a participant in an action performed in the subject's own interest, without any pragmatically based overtones. Note that the subject can be any person and that in all cases, the reflexive pronoun simply signals co-reference between the subject and the DI nominal. As shown in the examples, the pronoun *si* could be replaced by any kind of non-reflexive NP, pronominal or otherwise, thus corroborating the argument status of *si*, with a link to a particular semantic relation.

(21) a. *{ježiš já tady mám půl roku tu špaldu,}*
tak ***si/mu/jim*** *ji uvařim* [oral2006]
so self/3SG:M/3PL:DAT 3SG:F:ACC cook:FUT:1SG
'{gee, I've had this spelt here for half a year,} so that's what I'll cook **for myself/him/them**'
b. *tak to* ***si/někomu/otci*** *přečteš* [oral2006]
so it:ACC self/someone/father:DAT read:FUT:2SG
'so you'll read it **for your/someone's/father's** information.'

The distribution of *si* in patterns like these presents an interesting and complicated topic in itself, involving a number of morphosyntactic and semantic nuances. For example, it is not always easy to classify a given instance of *si* as an argument or simply a particle that has become a fixed part of a lexical item (there is certainly a large number of dative-reflexive verbs in Czech that cannot be analyzed as phrasal). It is not necessary in the context of this paper to discuss all the intricacies involved in the patterns illustrated in (21), since those issues are not directly relevant to the problems at hand. It should be clear, however, that we cannot lump them together with the uses shown in (20); these two sets represent two semantically and functionally distinct kinds of *si*.

Specifically, the set of sentences in (5) and (20) cannot be treated simply in terms of co-reference between the subject and a dative NP nor simply in terms of the semantic notion of interest. They exemplify a more complex category that has features of both, further overlayed with a discourse-pragmatic meaning. It is, of course, true that the examples in (5) and (20) express the subject's self-interest; this follows from the presence of the reflexive form. For example, in (20a) the subject is lying around for his own enjoyment, in (5) the subject is given the opportunity to beat up somebody else in order to feel good, and so on. In this respect, the referent of *si* is necessarily a participant in the event. But there are, again, idiosyncratic restrictions that apply only to these examples and not to those in (21). Morphosyntactically, the utterances in (5) and (20) display the same behavior as the Contact and Subjective DEs. The reflexive DE can be used together with a thematic dative, as demonstrated by its co-occurrence with the recipient of the verb *dát* 'give' in (5), and, like the rest of DEs, it cannot be questioned or relativized. Also, only the clitic form *si* is possible, as shown by the ungrammaticality of the full-form dative reflexive *sobě* 'to/for self' in (22), as a modification of (20a).

(22) **von* ***sobě*** [...] *bude ležet* {*a neudělá navíc ani ň*}
3SG:M:NOM self:DAT be:FUT:3SG lie.down:INF

But most importantly, and in further contrast to the semantic uses of *si*, the presence of *si* in (5) and (20) also carries with it unique pragmatic features. These examples encode more than just the event of somebody doing something for his own benefit. They present a commentary of sorts, conveying the *speaker's* own evaluation of the reported state of affairs. These utterances typically involve elements of sarcasm, irony, defiance, or, less commonly, admiration, but whatever attitude is expressed, it always implies the speaker's disapproval of the reported event, or at least distancing himself from it. The subjective stance is sometimes strengthened by an additional commentary: the D-DE naturally collocates with the adverb *klidně* in the sense of 'without concern [for the rest of the world]', or the explicit disclaimer *mně je to jedno* 'I don't' care'.

Distancing DE
inherit Self-Interest construction

cat **pro$_{refl.}$**
prosody [**stress** - / **2P enclitic**]
prag [**discourse-frame** [**PRIVATE CONVERSATION** / **register informal**] / **involvement 'Speaker's distancing judgment of X's self-interest'**]

Diagram 4. Distancing DE construction

It follows from the pragmatic function of this pattern that it is also associated exclusively with spoken and informal discourse, thus sharing another important property with the rest of DEs. The reflexive pattern is just more complex than the other two. Overall, it is also speaker-centered, like S-DE, since it presents the speaker's attitude toward the grammatical subject's reported self-interest, but it also puts a distance between himself and the subject's activities. The speaker essentially rejects someone else's self-interest, recasting it as excessive self-indulgence.

The representation in Diagram 2 captures this particular variant of DE. The D-DE inherits properties associated with regular dative reflexives, which accounts for the details of the coreferential mapping, but it also adds features of its own, namely, the pragmatic function that is particular only to this use of the dative reflexive.

3. DEs in a constructional network

Several descriptive conclusions emerge from the analysis:

- The three dative patterns (contact, subjective, and distancing) have enough in common to be treated as instances of a single functional category. They all belong to the same register; serve a distinct pragmatic function within the domain of interpersonal relations; reinterpret the semantic notion of 'interest' as a discourse-deictic concept; and display the same morphosyntactic idiosyncracies (non-argument syntactic status, clitic form, and, canonically, clitic placement).
- They are distinct from semantic datives in that their integration into the sentence is limited: they all lack clear semantic status as event roles and therefore are not subject to the bi-uniqueness condition.
- At the same time, they are distinct from other devices that index speaker-hearer relations, in that they may be sensitive to a clause type and/or the referential identity of event participants of the sentence in which they appear.

The question is how this web of relationships can be best represented so that our generalizations reflect the speakers' conventional knowledge as accurately as possible.

To the extent that this issue has been addressed at all, the most insightful treatments have tended to focus predominantly on either the semantic (Janda 1993) or the pragmatic (King 1997) aspect of the problem, to the exclusion of most other features. Janda's 1993 cognitively based analysis concentrates on motivating the meaning of the DEs in relation to the meaning(s) of other Czech datives, specifically through the shared feature of affectedness. This is evidently an important part of the picture, but her analysis leaves aside the pragmatic and morphosyntactic dimensions of the DEs. King 1997 is more interested in the pragmatic aspects of Czech datives and thus notices the contextual and register-related features of DEs but overall does not go beyond speculating that in sentences with multiple datives, the referents should be all distinct. Nor does she consider any of their morphosyntactic properties.

Other possible analyses fall short as well. It is clearly not enough to assume a speech-act analysis proposed for 'ethical datives' in the spirit of generative approaches, in which the dative has been treated as an argument of a verb of saying, and therefore residing outside of the rest of the clause both structurally and semantically. Such a treatment would not address all the constraints associated with the Contact DE itself and it obviously could not apply to the remaining types. Nor can we reduce DEs to being a matter of contextual interpretation and inference, by positing them as inherently semantic entities (together with other 'semantic' datives, along the lines of Dąbrowska's 1997 proposal) and leave the rest to speakers' assumptions based on extralinguistic knowledge. While this is no doubt a valid view of the utterances in which both semantic and pragmatic interpretations are possible (recall the examples in 19), it fails to identify the cases in which the DE interpretation is unambiguous, grammatically constrained, and the only one available.

The empirical observations and the generalizations we can draw from them thus call for a multidimensional representation, as proposed in the preceding section for each DE. Once we identify the definitional features of each pattern we can also more easily map them onto the functional space they share and treat each of the three DEs as a particular linguistic strategy for expressing different kinds of speaker-hearer relations and particularly, in the domain that has to do with emotional involvement of discourse participants.

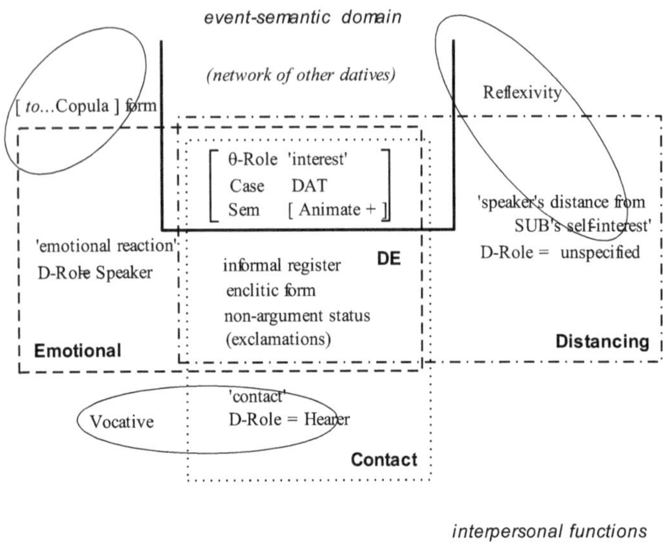

Diagram 5. DE network

I propose the following motivation for the relations that the DEs hold toward semantic datives on the one hand and other discourse-deictic devices on the other. The central notion that holds it all together is ‘indirect affectedness’. It motivates, in a robust and grammatically regular way, the use of the dative form in many semantic modulations: as we have seen, indirect affectedness plays a role in conceptualizing certain types of event participants as having an ‘interest’ in the narrated event. It is cognitively plausible to extend this notion from the event-semantic domain into the discourse domain, where indirect affectedness or interest can be easily reconceptualized in terms of interactional affectedness: as interest in (or affectedness by) hearing, assessing, or reacting to the content of the narrated event, rather than participating in it. Through this central connection between event-based affectedness and interactional affectedness, we can treat the dative form in the DE patterns as a motivated extension of a regular and highly salient linking relationship, rather than as a mere coincidence.

These connections can be organized in the form of a constructional map (in an application parallel to Fried 2007), such as in Diagram 4, that suggests an overlap between two domains: the event-semantic domain (the upper half of the diagram, schematically delimited by the thick line) and the interpersonal domain (the lower half). The central region of the map shows the event-semantic, dative-related features, overlapping with the common DE features listed below the dative. Each type of DE is then enclosed in its own rectangle in such a way that we can see exactly which features are shared either within the DE space or with other, non-DE constructions; the latter are indicated by the ovals extending to specific other patterns (the vocative, the reflexive, or copular structures).

Setting up the constructional map can be taken as a way to model grammatical categorization, with explicitly articulated continuities and discontinuities across the relevant patterns, based on specific, empirically grounded criteria. The crucial role given to the *semantic* motivation, revolving around a particular type of affectedness, is also important in that it defocuses the dative *form* as, potentially, the central motivating factor. The dative case may be associated with various semantic concepts in different languages, and it remains an empirical question to determine whether developments similar to the DEs in Czech can be established as a general property of datives. However, without relevant studies concerning other languages and other datives, the present analysis, captured in the configuration of the map, leads at least to the following hypothesis: interactional datives (such as DEs) should not be expected to develop in languages where the dative (or its apparent equivalent) is not associated with expressing the notion of indirect affectedness.

4. Summary and conclusions

We can conclude that all three DE types serve a discourse-deictic function, in three distinct configurations of speaker-hearer relations, and that certain other properties follow from this fact. For one thing, the deictic function motivates the restrictions on DE referents: in the hearer-directed Contact DE, the only referent can be the 2nd person; the speaker-oriented dative in the Subjective DE is limited to the 1st person; and the reflexive of the distancing DE draws attention, through the speaker's disapproving attitude, to another participant's self-indulgence perceived as excessive.

The decision to use a DE in a concrete utterance is, of course, the speaker's choice since the DE's presence does not depend on any grammatical requirements of the structure of a given sentence. In this respect, DIs are purely a speaker's choice as well. The point of the present investigation and analysis is to show that both the production and reception of DEs in actual discourse involves conventional expectations about their form, meaning, and function, on a par with any other piece of grammatical knowledge speakers must share in order to use and interpret these items with a native-like fluency. Put differently, DEs provide a conventionalized way of pointing to discourse participants for specific interactional purposes and speakers know what type of text and context allows them. Crucially, though, the use of these forms is also constrained by certain grammatical aspects of a given utterance. This further entails that DEs represent relatively complex grammatical objects with specific syntactic, semantic, and pragmatic requirements. At the same time, it is clear that there are some overlaps between DEs and other constructions.

The complexity of the data makes a case for a constructional analysis. Specifically, the material provides evidence that speakers must have access to abstract grammatical patterns, which, in turn, make connections to other patterns. DEs carry a relationship to semantic datives but they also include specific discourse-pragmatic conditions among the factors that affect their grammar. The multi-layered architecture of CxG offers a convenient model not only for representing complex grammatical objects, but also for mapping out the relationships that hold between them. The latter is especially valuable in that it allows us to represent grammar as a network organized around shared features, whether formal, semantic, or pragmatic.

Sources of data

Czech National Corpus (*ČNK*), accessible at http://www.korpus.cz:
ČNK - BMK. Ústav Českého národního korpusu FF UK, Praha 2002.
ČNK - ORAL2006. Ústav Českého národního korpusu FF UK, Praha 2006.
ČNK - ORAL2008. Ústav Českého národního korpusu FF UK, Praha 2008.
ČNK - PMK. Ústav Českého národního korpusu FF UK, Praha 2001.
ČNK – SYN2000. Ústav Českého národního korpusu FF UK, Praha 2000.
ČNK – SYN2005. Ústav Českého národního korpusu FF UK, Praha 2005.
ČNK – SYN2006PUB. Ústav Českého národního korpusu FF UK, Praha 2006.

References

Authier, J-Marc & Lisa Reed. 1992. Case Theory, Theta Theory, and the distribution of French affected datives. *West Coast Conference on Formal Linguistics* 10, 27-39. Stanford, CA: CSLI Publications.

Berman, Ruth A. 1982. Dative marking of the affectee role: data from Modern Hebrew. *Hebrew Annual Review* 6, 35-59.

Borer, Hagit & Yosef Grodzinsky. 1986. Syntactic cliticization and lexical cliticization: the case of Hebrew dative clitics. In H. Borer (ed.), *Syntax and Semantics* 19, 175-217. New York: Academic Press.

Croft, William. 1992. Voice: beyond control and affectedness. In B. Fox & P. Hopper (eds.), *Voice: form and function*, 89-117. Amsterdam & Philadelphia: John Benjamins.

Dąbrowska, Ewa. 1997. *Cognitive semantics and the Polish dative*. Cognitive Linguistics Research Vol. 9. Berlin and New York: Mouton de Gruyter.

Fillmore, Charles J. 1988. The mechanisms of 'Construction Grammar'. *BLS* 14, 35-55.

Fillmore, Charles J. 1989. Grammatical Construction theory and the familiar dichotomies. In R. Dietrich & C.F. Graumann (Eds.), *Language processing in social context*, 17-38. Amsterdam: North-Holland/Elsevier.

Fried, Mirjam. 1994. Second-position clitics in Czech: syntactic or phonological? *Lingua* 94: 155-175.

Fried, Mirjam. 1999a. From interest to ownership: a constructional view of external possessors. In D. Payne & I. Barshi (eds.), *External possession*: 473-504. Amsterdam: John Benjamins.

Fried, Mirjam. 1999b. The 'free' datives in Czech as a linking problem. In K. Dziwirek, H. Coats, & C. Vakareliyska (eds.), *Annual Workshop on Formal Approaches to Slavic Linguistics* 7: 145-166. Ann Arbor: Michigan Slavic Publications.

Fried, Mirjam. 2007. Constructing grammatical meaning: isomorphism and polysemy in Czech reflexivization. *Studies in Language* 31(4): 721-764.

Fried, Mirjam & Jan-Ola Östman. 2004. Construction Grammar: a thumbnail sketch. In M. Fried & J-O. Östman (Eds.), *Construction Grammar in a cross-language perspective*, 11-86. Amsterdam & Philadelphia: John Benjamins.

Grepl, Miroslav & Petr Karlík. 1986. *Skladba spisovné češtiny*. Praha: SPN.

Hopper, Paul. 1991. On some principles of grammaticization. In E. C. Traugott & B. Heine (Eds.), *Approaches to grammaticalization*, Vol. I, 17-35. Amsterdam & Philadelphia: John Benjamins.

Janda, Laura A. 1993. *A geography of case semantics: the Czech dative and the Russian instrument*. Berlin: Mouton de Gruyter.

King, Katerina P. 1997. *The Czech dative of interest: the hierarchical organization of possession in discourse and pragmatics*. Unpublished PhD Dissertation. Harvard University.

Kemmer, Suzanne. 1993. *The middle voice*. Amsterdam & Philadelphia: John Benjamins.

Leclere, Christian. 1976. Datif syntaxique et datif éthique. In J.-Cl. Chevalier & M. Gross (eds.), *Méthodes en grammaire francaise*, 73-96. Paris: Klincksieck.

Mluvnice češtiny, Vol. II & III. 1986. Praha: Academia.

Poldauf, Ivan. 1962. Místo dativu ve výstavbě věty. *Acta Universitatis Carolinae, Slavica Pragensia* IV, 335-345.

The Russian instrumental-of-comparison: constructional approach

Ekaterina Rakhilina, Russian Academy of Sciences, Moscow
Elena Tribushinina, University of Antwerp

1. Introduction

The Russian instrumental case is notoriously polysemous. It counts about twenty meanings including an instrumental meaning (e.g. *kopat' lopatoj* 'dig with a spade'), an agentive meaning (e.g. *sdelan Petrom* 'made by Peter'), a comitative meaning (e.g. *s čemodanom* 'with a suitcase'), a locative meaning (e.g. *idti polem* 'walk through the fields'), a temporal meaning (e.g. *večerami* 'in the evenings'), a comparative meaning (*vyt' volkom* 'howl like a wolf') and others. Several attempts have been made to provide a unified account of case in general and of the Russian instrumental in particular (e.g. Jakobson 1936; Janda 1993; Mrázek 1964; Panov 1999; Rakhilina 2000; Wierzbicka 1980). Most of them take the argument structure of the predicate as a starting point for description. An alternative, noun-centred approach to the Russian instrumental was put forward by Rakhilina (2000); see also Panov (1999).

This paper will reflect upon these two major approaches to the instrumental case and assess them by focusing on one specific type of instrumental – a so-called *instrumental of comparison*. We will then propose a new integrated approach to instrumentals inspired by construction grammar (Fillmore et al. 1988; Goldberg 1995, 2006; Lakoff 1987).

The paper begins with an overview of the verb-centric and noun-centric models of case (Section 2). In section 3, we argue that construction grammar provides a very fruitful way of analyzing the Russian instrumental. Section 4 gives a detailed description of the instrumental of comparison within the constructionist framework. In Section 5, we summarize the main findings and present the conclusions from this study.

2. Theoretical background

2.1. *Verb-centric approaches to case*

The first attempt towards a unified account of case was Jakobson (1936). Jacobson's primary objective was to distinguish the instrumental case from the other cases using three basic features. Within this paradigm, the invariant meaning of the instrumental is constituted by the features '+marginal',

'−quantified' and '−directional'. Marginality highlights the object's peripheral relation to the described action; quantification involves the degree of the object's participation in the action; and directionality concerns the direction of an action towards the object. Notice that all these features are verb-related. Jakobson's theory of case is therefore a prime example of a verb-centric view of case.

The approach advocated by Jakobson (1936) can elegantly distinguish between different cases, but is not able to say much about the variety of specific uses of the instrumental. Later studies have tried to fill this gap by focusing on various uses of the instrumental case. Mrázek (1964), for example, presents an elaborated analysis of nineteen (seemingly unrelated) meanings of the instrumental.

Another key publication about the Russian instrumental case is Wierzbicka (1980). Unlike Jakobson (1934), Wierzbicka pays a lot of attention to the particular senses of the instrumental case. She also refutes a feature-based approach and suggests using metalinguistic definitions of the following type:

> 'Something happened to Y
> because something happened to IN
> because X did something.' (p. 6)

Crucially, this definition is intrinsically verb-centred too. It enriches the verb schema with the semantic roles of an agent (X), a patient (Y) and an instrument (IN). However, just like Jakobson's theory, it does not predict which nouns can (or cannot) be used in the instrumental form with a particular meaning (instrument, time, location, etc.).

Janda (1993) presents a cognitive semantic view of the instrumental case. Her major goal is to provide a unified description of various meanings of the Russian instrumental. To this end, she uses four image-schemas with the "conduit instrumental" (instrumental of means) as a prototype. This is how Janda schematically presents the conduit meaning of the instrumental:

NOM ========➔ INSTR =======➔ ACC

This schema shows that the action is directed towards the instrument, although the instrument is not the main participant, but only a conduit between the agent and the patient. Generally, what such schemas do is capture a relation of various roles to the head predicate. In this sense, Janda's case schemas, like Wierzbicka's definitions, are essentially a reinterpretation of Jakobson's predicate-argument structure.

2.2. *Noun-centric hypothesis*

An alternative to the verb-centred view of case is the so-called *noun-centric hypothesis* introduced in Rakhilina (2000). This hypothesis posits that case cannot be reduced to or completely predicted by verb semantics (cf. Kuryłowicz 1949) and that lexical semantics of nouns should play a key role in the description of case (cf. Panov 1990, 1992, 1999).

Recall that the Russian instrumental has about twenty senses. If the argument structure of the verb was the only factor determining the contextually relevant meaning of an instrumental form, then every noun in the instrumental case taken in isolation would be ambiguous and have about twenty different interpretations ('means', 'time', 'location', 'agent', etc.). This prediction is, however, not confirmed by the natural language data, where nouns seem to be associated with one particular instrumental meaning. The instrumental form *palkoj* (stick-INS), for instance, is unambiguously interpreted as denoting an instrument. In contrast, *lesom* (forest-INS) is most likely to be the locative instrumental, whereas *bantikom* (bow-INS) is unequivocally interpreted as denoting similarity in shape. Although unexpected from the verb-centric point of view, this pattern is perfectly explicable. Every noun is associated with a particular taxonomic class, which determines the role of the described object in a situation and thereby shapes the situation itself (cf. Panov 1999:198). When processing an instrumental noun taken in isolation, we mentally reconstruct the predicate and interpret not the noun as such, but the whole construction. Which predicate will be reconstructed (and, thus, which interpretation will be assigned to the instrumental form) depends on the lexical semantics of the noun. An axe is a tool; therefore *toporom* (axe-INS) is straightforwardly assigned an instrumental meaning. A forest is a location; the form *lesom* (forest-INS) is then naturally interpreted as a locative instrumental. In the same vein, night is a temporal notion, and the instrumental form *nočami* (nights-INS) is unequivocally understood as a temporal instrumental.

Importantly, the range of predicates associated with a given noun in the instrumental case is very restricted. Due to the specific semantics of the noun, a whole panoply of predicates can in principle never be evoked by that noun. Take, for instance, the noun *palka* 'stick'. It is possible to say <*muravej polz*> *po palke* 'the ant crawled along the stick' just as it is possible to say <*mal'čik šel*> *po polju* 'the boy walked across the field'. Notice, however, that only the instrumental form *polem* (field-INS) is possible in this context, but not **palkoj* (stick-INS), since a field is a location and a stick is a tool.

This explanation may give the wrong impression that the lexical meaning of the noun is the only factor determining which meaning will be assigned to a particular instrumental form. The situation is, however, more complex. First, the semantics and the grammatical properties of the verb are certainly important too.

Some verbs, for example, obligatorily take nouns in the instrumental, irrespective of the nominal semantics (e.g. *vosxiščat'sja toporom / lesom / nočami* 'to admire an axe / a forest / nights'). Second, each kind of instrumental construction itself has a conventionalized meaning which is not equivalent to the sum of the constituent meanings. In the following section, we will briefly consider an example of such constructional meaning and propose a theoretical account which can neatly accommodate these phenomena.

3. Constructional meaning

Natural languages often witness a phenomenon which might be termed *multiple taxonomy*. The same noun may be used to profile different aspects of the object of conceptualization and may therefore be expected to function with various instrumental meanings. Take, for instance, the noun *poezd* 'train'. It can be conceptualized as a means of transport and at the same time a sort of location (for animate and inanimate objects). Since both these meanings can be expressed by means of the Russian instrumental case, we may expect the instrumental form *poezdom* to be ambiguous and felicitous in both transport and locative constructions, as in *exat' poezdom* 'travel by train' (means of transport) and *idti poezdom* 'walk through the train' (location). This is, however, not the case. The instrumental form *poezdom* is unambiguously interpreted as denoting a means of transport. The locative construction **idti poezdom* 'walk through the train' is unacceptable.

An important question to ask is why a noun which *does*, in fact, denote a location (*poezd* 'train') cannot be used in the locative instrumental. A closer scrutiny reveals that the answer should be sought in the nature of the instrumental constructions. We would like to claim that the construction which has been traditionally called *locative instrumental* is, in fact, much more specific in its meaning. Constructions such as *idti polem* 'walk through the field' and *exat' lesom* 'ride through the forest' denote a type of route rather than a location. Furthermore, it should be an established, recognized route through a larger space, such as a field or a forest. Together with a motion verb, the instrumental of route denotes a purposeful movement to a specific endpoint. Authentic examples from the Russian National Corpus (henceforth RNC) include, for instance, *exat' beregom* 'ride along the shore', *idti bolotami* 'walk through the swamps', *idti bul'varom* 'walk along the boulevard', *bežat' kraem lesa* 'run along the edge of the wood', *exat' lugom* 'ride through the meadows', *vezti morem* 'transport by sea'. Importantly, all these cases evoke a fixed (i.e. recognized) trajectory across a larger area (swamps, meadows, sea, etc.).

Notice that the combination ??*guljat' polem* 'stroll through the fields' is odd, since there is neither a fixed route nor a purposeful movement involved. In contrast, all motion verbs preceding *pole*-INS in the RNC denote purposeful movement along an established route: *bežat'* 'run', *vozvraščat'sja* 'return', *dvinut'* 'start off', *ezdit'* 'travel, ride', *idti* 'walk', *krast'sja* 'sneak', *pereletet'* 'fly over', *polzti* 'crawl', *pomčat'sja* 'dart', *ponestis'* 'rush along', *pustit'sja* 'set out', *taščit'* 'drag', *taščit'sja* 'drag oneself along', *sledovat'* 'follow'. Similarly, **plyl okeanom* (sailed ocean-INS) 'sail across the ocean' and **letel nebom/vozduxom/kosmosom* (flew sky/air/space-INS) 'fly through the sky/air/space' are infelicitous, since there is no fixed route across these spaces.

Further, the instrumental is not used in cases where a route is not made explicit, but is pre-determined by the topological properties of the described entity. For this reason, the noun *reka* 'river' is very rare in the route-instrumental construction: there are only 5 instances in the RNC (cf. 56 cases with *morem* 'sea-INS' and 51 cases with *polem* 'field-INS'). Furthermore, as is evident from examples (1)–(5), these uses are quite old: the most recent one is from 1913.

(1) *Kegress pokazyval novyj avtomobil', kotoryj proxodit po vsjakoj mestnosti.*
Kegress showed new automobile which passes on any area
Proexali dal'še i lesom spustilis' k našemu mestu u Dnepra.
Travelled further and forest –INS went.down to our place at Dnieper
Pokatalsja s Grabbe v dvojke. ***Vernulis' rekoju*** *k*
Rode with Grabe in carriage.and.two returned river-INS towards
6 čas. (Nicholas II. Diary, 1913–1916)
6 hours
'Kegress showed his new car that could drive on any type of road. We drove further through the forest and descended to our place by the Dnieper. I rode a carriage with Grabbe. We returned along the river at around 6 o'clock.'

(2) *Čuvaš možet vsegda pet' ėkspromtom:* ***edet lesom*** *— vospevaet*
Chuvash may always sing impromptu drives forest-INS sings.of
les, ***edet rekoju*** *– vospevaet reku i skladyvaet v pesni raznye*
forest drives river-INS sings.of river and puts.together in songs various
byli i nebylicy. (N.D. Telešov, 1892)
facts and tales
'A Chuvash can always sing impromptu. Travelling through the forest he sings of the forest, travelling along the river he sings of the river and puts together true stories and tall tales.'

(3) *On* ***exal rekoju Dvinoj*** *bez ostanovki do Sijskogo*
he travelled river-INS Dvina-INS without stop until Siysky
monastyrja. (Pavel Nikolaevskij, 1885)
monastery
'He travelled non-stop by the Dvina river to the Siysky Monastery.'

(4) *Ermak s obetom doblesti i celomudrija, pri zvuke trub*
Ermak with vow valour and chastity at sound trumpets

voinskix, 1 sentjabrja 1581 goda ***otplyl rekoju Čusovoju*** *k*
military 1 September 1581 year set.sailing river-INS Čusovaja-INS towards
goram Ural'skim, na podvig slavy, bez vsjakogo sodejstvija, daže
mountains Ural on feat glory without any assistance even
bez vedoma gosudareva. (N.M. Karamzin, 1816–1820)
without knowledge tsar
'On 1 September 1581, having made a vow of valour and chastity, Ermak set sailing to the Ural Mountains along the Chusovaya river. Accompanied by the sound of soldiers' trumpets, he left for his feat of glory, without any assistance and without even letting the tsar know of his departure.'

(5) *Mitropolit ot Vyšnego Voločka* ***plyl rekoju Mstoju*** *do*
metropolitan from High Volochok sailed river-INS Msta-INS as.far.as
Novagoroda, gde, ravno kak i vo Pskove, duxovenstvo i
Novgorod where equally like and in Pskov clergy and
graždanstvo iz''javilo userdnuju k nemu ljubov' darami i
civilians expressed zealous towards him love gifts and
piršestvami. (N.M. Karamzin, 1809–1820)
feasts
'The metropolitan sailed by the Msta river from the High Volochok to Novgorod where, just as in Pskov, the clergy and civilians expressed their great love to him with gifts and feasts.'

Bearing the above discussion in mind, it is not difficult to explain why **idti poezdom* 'walk through the train' is infelicitous. Even though the noun *poezd* has a locative meaning and the verb *idti* is a motion verb and even though it is possible to walk from the front to the back of the train (or the other way around), it is certainly not a fixed route through a larger space.

The above discussion clearly demonstrates that the nature of the verb+instrumental construction cannot be reduced to a grammatical rule for combining atomic syntactic units into a complex phrase. It is not the case that a combination of any locative noun with any motion verb will automatically result in a route instrumental. The route instrumental, just as the other instrumentals, is a conventionalized pairing of form and function (cf. Croft 2001; Fillmore et al. 1988; Fillmore and Kay 1993; Fried and Boas 2005; Goldberg 1995, 2006; Michaelis 2004; Michaelis and Lambrecht 1996). The constituents partly determine the meaning (or rather the domain) of the instrumental construction. For instance, the combination of *idti* 'walk' with the instrumental form *polem* 'field' evokes a locative domain, whereas the combination with another noun in the instrumental case (e.g. *nočami* 'nights') profiles a different domain of the instrumental case (temporal). However, the construction also contains information that is not found in the constituent parts and therefore imposes restrictions on verb-noun combinability. The route instrumental construction, for instance, denotes a purposeful movement along a fixed route through a larger space. This construction-specific semantics explains why *idti polem* 'walk through the field' is felicitous, whereas ??*guljat' polem* 'stroll through the field' and **idti poezdom* 'walk through the train' are odd. Similarly, if the instrumental case were purely compositional (i.e. words combined by rules), we would not be able to predict the restrictions of the type ??*plyt' rekoj* 'sail by river' and **letet' nebom* 'fly by sky'. Thus, the semantics of the instrumental construction largely relies on the low-level meanings of the specific nominal and verbal constituents, but is not directly predictable from the general rules of morphology, syntax and semantics, since the construction as a whole also has a meaning (e.g. purposeful movement along a fixed route through a larger space). The data in this study therefore support the idea put forward within the framework of *construction grammar* that there is no clear-cut borderline between grammar and lexicon; rather there is a syntax-lexicon continuum (Fillmore et al. 1988; Goldberg 1995, 2006; Jackendoff 2002). A similar approach to case may be found in Barðdal (2008, 2009a, 2009b).

In what follows, we will zoom in on one specific construction from the family of Russian instrumental constructions – the instrumental of comparison. We will compare this construction with the more general *kak*-similative in order to describe the construction-specific meaning of the instrumental of comparison.

4. The instrumental-of-comparison construction

4.1. *Form*

The instrumental-of-comparison (IoC) has two formal realizations: a stative nominal construction (implying the predicate verb *byt'* 'be') as in (6) and a verbal construction usually containing a dynamic predicate as in (7).

(6) *U tebja mogut byt' pyšnye volosy, kak u Bejonse, zadumčivye*
at you may be luxuriant hair like at Beyoncé thoughtful
glaza, kak u Ališi Kis, i ***gubki bantikom****, kak u*
eyes like at Alicia Keys and lips-DIM.NOM bow-DIM.INS like at
Andželiny Džoli. (*Tvoj kurs*, 2004.11.10)
Angelina Jolie
'You may have luxuriant hair like that of Beyoncé, thoughtful eyes like those of Alicia Keys and lips shaped as a bow like those of Angelina Jolie.'

(7) *Nikto iz nix ne* ***letel kamnem*** *– vse opuskalis' zamedlenno.*
none from them NEG flew stone-INS all went.down slowly
(L. Ulitskaya, 2000)
'None of them dropped like a stone; they all got down slowly.'

Like other comparative constructions (equatives and similatives), the IoC usually consists of four elements: a comparee (topic of comparison), a standard of comparison, a standard marker (pivot, marker of comparison) and a parameter (Cuzzolin and Lehmann 2004; Haspelmath and Buchholz 1998). A comparee can be expressed by the head-noun of the IoC (*gubki* 'lips' in example 6) or the subject of the sentence (*nikto* 'nobody' in example 7). A standard of comparison is expressed by the noun in the instrumental case: *bantik* 'bow' in (6) and *kamen'* 'stone' in (7). The standard marker in the IoC is the instrumental morpheme. A parameter is usually not expressed overtly in the IoC construction: in (6) the parameter is shape, and in (7) it is a manner of motion.

4.2. Content

4.2.1. The IoC and the kak-similative. Wierzbicka (1980:86) defines the semantics of the IoC construction in the following way:

> 'I say: imagine Y (IN),
> because I want you to be able to imagine X'

Applying this definition to the sentence *Kolesom za sini gory solnce krasnoe skatilos'* 'wheel-INS behind blue mountains sun red rolled down', we arrive at the following definition: 'I say: imagine a wheel, because I want you to be able to imagine the sun' (cf. Turovskij 1988:137; Zaliznjak 1996:172). Such definitions, however, cannot predict the semantic and combinatorial restrictions in the IoC construction. Nor can they account for the differences between the IoC construction and an apparently synonymous construction with *kak* 'like, as' (e.g. *kruglyj kak koleso* 'as round as a wheel'). In the remainder of this section we will present a constructional analysis of the IoC. We will argue that a constructionist approach has a significant predictive power and is able to solve the problems of the earlier approaches.

As mentioned above, the IoC is seemingly very much like the similative *kak*-construction. An initial look at the distributional properties of the two constructions gives an impression that that the *kak*-construction is very broad in meaning and can be used in all cases where the IoC is used. By way of illustration, see examples (8)-(11), where (a) sentences contain IoC constructions and (b) sentences their similative *kak*-counterparts:

(8a) *Udaril vypjativšijsja iz-za sosny mesjac v devič'e lico, i Petr*
struck protruded from pine.tree moon in maiden face and Petr
Kirilyč xorošo vidit slezinki vo vpalyx ščekax i ***grud' doskoj****,*
Kirilych well sees tears in hollow cheeks and breast-NOM board-INS
i sarafan takoj že goluboj i s kolokol'čikami po podolu
and sarafan such PCL light.blue and with bells on hem
i po grudi, tol'ko visit on slovno na palke.

and on breast only hangs 3.SG.M as.if on stick
(S.A. Klychkov, 1926)
'The moon protruded from a pine-tree and lit the girl's face. And then Petr Kirilych could clearly see the tears on her hollow cheeks and flat breast. Her sundress was also blue, with bells along the hem and the breast, but the dress was hanging like on a stick.'

(8b) *"**Grud', kak doska**, zadu i sovsem net", – podumala Katerina.*
breast like board bottom and entirely NEG thought Katerina
(P.S. Romanov, 1928)
'Her breast is as flat as a board, the bottom is altogether absent, Katerina was thinking.'

(9a) ***Letela streloj** Stal'naja, vovsju staralas'.*
flew arrow-INS Stalnaya as.hard.as.one.can tried
(I.S. Shmelev, 1927–1944)
'Stalnaya flew like an arrow, she did her utmost best.'

(9b) *Gody **leteli kak strela**. Dni tjanulis' kak smola.* (O.S. Minor, 1933)
years flew like arrow days dragged like pitch
'Years flew like an arrow. Days dragged like pitch. '

(10a) *I boli togda takie, čto gotov **revet' korovoj**.*
and pains then such that ready howl cow-INS
(R. Ungern, 1926–1938)
'And the pain is then so strong that you can howl like a cow.'

(10b) *Ja nalivaju sebe polstakana kon'jaku, vypivaju i, položiv golovu na*
I pour self half.glass cognac drink and having.put head on
*ruki, **revu, kak korova**.* (V. Voinovich, 1999)
arms howl like cow
'I pour some cognac for myself, drink it all, put my head on my arms and howl like a cow.'

(11a) *Smotrju, moj sidit, vse lico v gari, i rža v borode ot železa.*
look my sits all face in cinders and rust in beard from iron
***Gljadit volkom**.* (B.S. Zhitkov, 1924)
looks wolf-INS
'I see my husband sitting, his whole face is covered with cinders, his beard with iron rust. He is looking like a wolf.'

(11b) *Pritom že ėto členy Dumy – narod pokuda ešče dikij,*
besides PCL this members Duma people still yet wild
*nepriručennyj... **gljadjat, kak volki**, i, vidimo, sovsem ne ponimajut,*
untamed look like wolves and apparently totally NEG understand
čto značit – Car'! (M. Gorky, 1906)
what means tsar

'And in addition to that, the members of the State Duma are still wild and untamed. They look like wolves and seem not to understand what it means to be a tsar!'

The reverse is, however, not the case. It is only a limited number of *kak*-constructions that can be substituted by the IoC. Witness examples (12)–(14):

(12a) *Nočevali v izbuške egerja Dormidonta, predannogo baram s*
spent.night in hut huntsman Dormidont devoted masters from
malyx let, kogda sam on, Volodja Greve, prjatalsja na senovale,
small years when self he Volodya Greve hid on hayloft
droža, kak zajac. (L.S. Sobolev, 1932–1962)
trembling like hare
'They spent the night in the hut of huntsman Dormidont who had been devoted to his masters since his childhood when even Volodya Greve himself was hiding in the hayloft trembling like a rabbit.'

(12b) **Nočevali v izbuške egerja Dormidonta, predannogo baram s*
spent.night in hut huntsman Dormidont devoted masters from
malyx let, kogda sam on, Volodja Greve, prjatalsja na senovale,
small years when self he Volodya Greve hid on hayloft
droža zajcem.
trembling hare-INS

(13a) *On priletel v Moskvu. My* ***obnjalis' kak brat'ja.*** (N. Burlyayev, 2003)
he arrived in Moscow we embraced like brothers
'He arrived in Moscow. We embraced like brothers.'

(13b) *On priletel v Moskvu. *My* ***obnjalis' brat'jami.***
he arrived in Moscow we embraced brothers-INS

(14a) *Tol'ko kogda ix glaza vstretilis', oni pali drug drugu v ob''jatija*
only when their eyes met they fell each other in arms
i zarydali, ***kak malye deti.*** (M. Volkova, 2000)
and started.crying like small children
'Only when their eyes met, they fell into each other's arms and started crying like small children.'

(14b) **Tol'ko kogda ix glaza vstretilis', oni pali drug drugu v ob''jatija*
only when their eyes met they fell each other in arms
i zarydali, ***malymi det'mi.***
and started.crying small-INS children-INS

It is well-established that languages tend to avoid constructional synonymy. One of the basic principles of construction grammar – the *Principle of No Synonymy* – holds that "if two constructions are syntactically distinct, they must be semantically or pragmatically distinct" (Goldberg 1995:67). If two constructions become very close in function (semantics and pragmatics), one of them is likely to be eliminated from the language (see also Apresjan 1974, 2000; Kustova 2004; Padučeva 2004). This phenomenon has been convincingly demonstrated with respect to the loss of case morphology in Germanic languages by Barðdal (2009b). It is usually the low type frequency construction that is blocked. Given that the IoC is less frequently used to express comparison and is more specialised in its meaning than the *kak*-construction, we might expect the former to be eliminated. Thus, the following question arises: Why do we still use the instrumental case to express comparison, if the more frequent and more evenly distributed similative construction may replace it?

4.2.2. Prototypical core: quasi-comparison. Since the two constructions coexisted already in the 18th century and still coexist in present-day Russian, it is plausible to assume that the IoC has its own prototypical meaning that cannot be expressed by the similative construction. The analysis of extensive corpus material strongly suggests that this is indeed the case. There is one particular meaning which seems to be the "specialty" of the IoC. Only the IoC, but not the *kak*-construction, can express similarity in shape after position verbs (e.g. *stojat'* 'stand', *ležat'* 'lie', *raspolagat'sja* 'be situated'), causative locative verbs (e.g. *položit'* 'put', *sostavit'* 'put together') and motion verbs (e.g. *idti* 'walk', *letet'* 'fly'). By way of illustration, consider the following cases:

(15a) *Na stole* ***ležat stopkoj*** *10 odinakovyx knig.* (V. Lukashik, E. Ivanova, 2003)
on table lie pile-INS 10 identical books
'Ten identical books are piled on the table.'

(15b) **Na stole* ***ležat kak stopka*** *10 odinakovyx knig.*
on table lie like pile 10 identical books

(16a) ***Složil pis'mo treugol'nikom****, nadpisal: Dnepropetrovsk,*
folded letter triangle-INS wrote.above Dnepropetrovsk
kafedral'nyj sobor. (V. Timakov, 1998)
cathedral
'He folded the letter as a triangle, addressed it to the Dnepropetrovsk Cathedral.'

(16b) ****Složil pis'mo kak treugol'nik****, nadpisal: Dnepropetrovsk,*
folded letter like triangle wrote.above Dnepropetrovsk
kafedral'nyj sobor.
cathedral

(17a) *No u samogo kamnja dovol'no širokaja bolotnaja tropa* ***rasxodilas'***
but at very stone rather wide marsh path split
vilkoj: *odna, xorošaja, plotnaja tropa šla napravo, drugaja*
fork-INS one good firm path went right another
slaben'kaja, – prjamo. (M.M. Prishvin, 1945)

weak straight
'But by the stone itself a rather wide marsh path forked: a good firm path went to the right, the other path was weak and went straight ahead.'

(17b) **No u samogo kamnja dovol'no širokaja bolotnaja tropa* ***rasxodilas'***
but at very stone rather wide marsh path split
kak vilka*: odna, xorošaja, plotnaja tropa šla napravo, drugaja*
like fork one good firm path went right another
slaben'kaja, – prjamo.
weak straight

(18a) *My sročno sobrali vsex sotrudnic apparata CK, vsex*
we urgently gathered all employees apparat Central Committee all
rabotnic – bol'šinstvo azerbajdžanki, no byli i russkie, – i oni
workers majority Azerbaijanis but were and Russians and they
vyšli cep'ju*, vzjavšis' za ruki, i xorom*
went.out chain-INS having.grabbed behind hands and in.chorus
prokričali "Net!" (V. Morozov, 2004)
cried no
'We urgently gathered all female members of the Central Committee and all female workers most of whom were Azerbaijanis, but there were also a few Russians. They formed a chain, went out, joined their hands and all together cried "No!"'

(18b) **My sročno sobrali vsex sotrudnic apparata CK, vsex*
we urgently gathered all employees apparat Central Committee all
rabotnic – bol'šinstvo azerbajdžanki, no byli i russkie, – i oni
workers majority Azerbaijanis but were and Russians and they
vyšli kak cep'*, vzjavšis' za ruki, i xorom*
went.out like chain having.grabbed behind hands and in.chorus
prokričali "Net!"
cried no

The infelicity of (b)-sentences shows that the instrumental case in (a)-sentences does not really denote comparison. Rather, the instrumental forms in (15a)–(18a) *describe* the objects in terms of shape. Similative constructions can also characterize their referents, but only through comparison of two situations. Consider, for instance, (19):

(19) *Ivan poet kak ego otec.*
Ivan sings like his father
'Ivan sings like his father.'

In this example, two situations are compared: <Ivan sings> and <his father sings>. The similative construction suggests that the two situations are alike: they both sing well or they both sing badly. More complex similatives can compare a referential situation with a non-referential standard of comparison as in (20):

(20) *Ėta šuba greet kak pečka.*
this fur.coat keeps.warm like stove
'This fur-coat keeps me warm like a stove.'

Two situations are evoked by the similative construction in (20) – a referential situation <this fur-coat keeps me warm> and a non-referential situation <a stove can keep one warm>. Since any stove can keep one warm, and since this fur-coat is like a stove in this respect, the fur-coat can keep me warm. Thus, warmth as a property of the fur-coat is established by comparing two situations.

Now notice that the IoC examples in (15a)–(18a) are very different from the similative construction in (20). Unlike in similatives, there is only one situation involved, and this situation is further specified by means of the instrumental case. Obviously, no comparison takes place in (15a)–(18a). It is not the case that (15a) compares <books are on the table> with <piles are on the table>. Books and piles, in fact, become closely integrated within this single scene: books positioned one on top of the other *do* constitute a pile. Similarly, if we fold a piece of paper as a triangle (example 16), is *does* as a matter of fact become a triangle. In the same vein, a path that splits is, strictly speaking, not a single path, but a fork (example 17) and a set of connected people has the shape of a chain (example 18). These examples show that the cognitive distance between the trajector and the landmark in the IoC construction is much smaller than in the case of the *kak*-comparative, since the described objects (e.g. books, letters) are cognitively contingent (Kibrik et al. 2006) with their shapes. For one, shape does not exist without an object.

Since the prototypical core of the IoC construction involves no comparison, the whole construction should rather be called *quasi-comparative*. The quasi-comparative instrumental construction is syntactically and semantically similar to the *instrumental of additional characterization* illustrated in (21) (Wierzbicka 1980, cf. Bogusławski and Karolak 1970), whose semantics can be described against the background of the instrumental of main predication (example 22).

(21) *Ivan rabotal šaxterom.*
Ivan worked miner-INS
'Ivan was working as a miner.'

(22) *Ivan byl šaxterom.*
Ivan was miner-INS
'Ivan was a miner'

The instrumental of main predication, as in (22), is used to describe temporary situations that are presumably no longer applicable to the subject (Wade 1992: 101-103). Therefore, it is impossible to use this construction in the present tense with the omitted link verb. The instrumental of main predication cannot express full identity between the referent of the subject and the referent of the instrumental noun, because an instrumental predicative is only applicable to the past dimension of the subject's existence.[1] This idea is captured by the following metalinguistic definition of (22):

> 'I say this because I want you to be able to imagine
> X-at-that-time
> not because I want you to think of X as a Y' (Wierzbicka 1980:120)

The instrumental of additional characterization in (21) involves an even bigger cognitive distance than (22), because (21) "describes the way Ivan lives rather than Ivan as a person" (Wierzbicka 1980:121). In this sense, Wierzbicka maintains, the instrumental of additional characterization occupies an intermediate position between a predicative and a manner adverbial. Just as in the IoC cases in (15a)–(18a), the instrumental of additional characterization denotes neither comparison, nor full identity. Rather, both constructions are used to provide an additional description of the subject.

At first sight, the shape-describing IoCs in (15a)–(18a) seem very different from other types of IoCs. For example, the IoC constructions in (8a)–(11a) apparently involve two situations and compare a referential situation with a non-referential standard. These situations can also be described by means of a similative construction as shown in (8b)–(11b). However, taking the prototypical core of the IoC construction into account, we may further hypothesize that the IoC cases beyond the domain of shape are, in fact, also quasi-comparative in the sense that they profile a much tighter cognitive contingence between the comparee and the standard than in the case of the prototypical similative construction. In Section 4.3.1, we will demonstrate that this is indeed the case.

4.2.3. Why shape? At this point, we may ask ourselves what makes shape so suitable for the semantic prototype of the quasi-comparative instrumental construction. The answer might be sought in the domain of categorization. Shape is one of the most important visual cues to object categorization. Friedrich (1970), for instance, argues that shape is a semantic primitive, which is amply used by world languages to categorize and grammatically encode objects,

---

1 For full identity, the nominative predicative must be used, as in *Ivan – šaxter* 'Ivan is a miner'.

but also actions and locations. To mention just one example, Tarascan grammar makes use of three shape-based patterns – numeral classifiers, classificatory verbs and suffixes of locatives (see also Aikhenvald 2004; Heine 1982; Serzisko 1982). Relatedly, a number of studies have shown that geometric properties of common nouns play a crucial role in the semantics of prepositions (Brugman 1981; Herskovits 1986; Lakoff 1987: 416–461; Plungian and Rakhilina 2000; Talmy 2000; Vandeloise 1994).

Landau et al. (1988) have shown that both adults and children use similarity in shape much more often than similarity in size or texture to extend nouns to new objects. In a similar fashion, Bornstein (1985) suggests that shape has more functional significance than color in the categorization process, for it is mostly on the basis of shape that objects of different classes can be identified as such. For this reason, Bornstein (1985) maintains, children learn to use shape adjectives correctly before they fully master colour terms.

Since shape is a major visual cue to categorization, the role of the standard of comparison in the IoC constructions may be compared to that of numeral classifiers. The noun in the instrumental case is a kind of quasi-grammatical means of object (sub)categorization. This observation has two implications for our analysis. First, the comparee and the standard of comparison in the IoC constructions are to a greater extent cognitively contingent than the elements of the similative *kak*-construction. Second, we may hypothesize that the similative construction cannot be employed in cases which are reduced to purely quasi-grammatical (shape-based) similarity and therefore requires diversification of parameters. We will take this issue up in Section 4.3.1.

4.2.4. The proximity principle. Both cognitive linguistics and semiotics assume that language is largely iconic (Haiman 1980). Iconicity is defined as a conceived similarity between the form of a linguistic sign and its meaning. One of the manifestations of iconicity is the so-called *proximity principle*, which posits that conceptual distance is related to formal distance. Thus, if two conceptual entities are more related than others, their linguistic labels are likely to be placed closer to each other.

Now compare the constructions under analysis in terms of linguistic distance. In the *kak*-construction, the linguistic signs for the compared elements are separated by the conjunction *kak*; the two elements are juxtaposed without grammatical dependence. In contrast, the elements of the IoC construction are related by grammatical government, which is a much tighter formal relation. Smaller linguistic distance between the constituents of the IoC construction evidences smaller cognitive distance between the conceptual entities compared by means of this construction, which can be taken as evidence of cognitive contingence of the IoC-participants.

4.3. *Inheritance hierarchies in the instrumental network*

One of the major tenets of construction grammar is that language involves various degrees of generalization and that constructions occupy a specific position in a hierarchical construction network. In Goldberg's (2006:13–14) words: "Broad generalizations are captured by constructions that are inherited by many other constructions; subregularities are captured by positing constructions that are at various midpoints of the hierarchical network. Exceptional patterns are captured by low-level constructions" (see also Kay and Fillmore 1999; Langacker 1987). Taking this assumption as a starting point, we make two predictions. First, since similarity in shape constitutes the prototypical core of the IoC construction, various non-prototypical instances of the IoC may inherit some essential properties of the prototype. Second, since the IoC construction itself is a member of a larger family of instrumental constructions, some of its basic properties may be inherited from the instrumental prototype and/or other kinds of instrumental constructions. These two predictions will be discussed in turn.

4.3.1. Internal relations in the IoC family. In the previous sections we saw that similarity in shape (in its pure form) constitutes the prototypical core of the IoC construction. However, the IoC can also describe similarity qua manner of motion (example 23) and produced sound (example 24). Further, the IoC construction can be used after visual predicates to denote similarity in the manner of looking (example 25) or in overall outlook (example 26).

(23) *Nikto iz nix ne* ***letel kamnem*** *– vse opuskalis' zamedlenno.*
none from them NEG flew stone-INS all went.down slowly
(L. Ulitskaya, 2000)
'None of them dropped like a stone; they all got down slowly.'

(24) *Ja tri dnja* ***revela belugoj****, neskol'ko mesjacev na mužikov smotret'*
I three days cried beluga-INS several months on men look
ne mogla, a sejčas dumaju SLAVA BOGU.
NEG could CONJ now think glory god
(Web-forum *Ženščina + mužčina*, 2004)
'I was bellowing like a whale for three days, could not look at men for several months, and now I think – Thank God.'

(25) *Spiros sidel razvaljas', kovyrjal v zubax,* ***smotrel barinom****.* (D. Karalis, 2001)
Spiros sat sprawling picked in teeth looked master-INS
'Spiros was sitting sprawled, picking his teeth and looking on like a lord.'

(26) *Mužčina v kostjume* ***smotritsja Apollonom,*** *daže esli emu ot prirody*
man in suit looks Apollo-INS even if him from nature
dostalas' jajceobraznaja figura. (L. Stotskaya, 2004)

received egg.shaped figure
'A man in suit looks like Apollo even if he has an egg-shaped body.'

It is plausible to assume that the less prototypical uses exemplified above inherit some of the essential characteristics of the prototypical, shaped-related IoC. The analysis of corpus examples suggests that there are at least three such properties.

First, all shape-related instances of the IoC construction are intrinsically simple, i.e. they describe objects in terms of shape and nothing more than that. We will refer to this property as *monotonicity* of the IoC construction. The *kak*-construction, in contrast, is quite diverse and can denote comparison along a number of dimensions. Crucially, monotonicity is relevant not only in shape-based instances of the IoC construction, but also in less prototypical cases. Also beyond the domain of shape, the semantics of IoCs is limited to only one parameter, such as spatial configuration, manner of looking or motion. Compare, for instance, the IoC constructions in (27) and (28) with the *kak*-similatives in (29) and (30):

(27) *Konečno, on i teper'* ***smotrel volkom****, kosil na storonu*
certainly he and now looked wolf-INS looked.asquint on side
i kak budto kogo-to s''est' sobiralsja. (F.M. Dostoyevsky, 1846)
and as if someone.PCL eat gathered
'Of course, now he was also looking daggers out of the corner of his eyes and it seemed like he was going to eat someone up.'

(28) *V kameru on vsegda* ***smotrel volkom****, govoril korotko i zlo, no*
in camera he always looked wolf-INS spoke briefly and angrily but
kak-to tak, čto emu xotelos' verit', zagljadyvat' v glaza i
somehow.PCL so that him wanted believe look.into in eyes and
deržat'sja rajdom. (T. Ustinova, 2003)
keep.oneself near
'He always looked daggers in the camera, spoke briefly and angrily, but in a way that you wanted to believe him, look into his eyes, stay near him.'

(29) *No emu ne vezlo v svatovstve; delo v tom, čto Kuz'ma*
but him NEG had.luck in marriage.proposals affair in that what Kuzma *byl fenomenal'no bezobrazen: on byl sutulyj, neukljužij, imel ryžie* was phenomenally ugly he was round.shoulered awkward had red
volosy, grubyj golos, ***smotrel, kak volk, ispodlob'ja****, a ego*
hair hoarse voice looked like wolf from.under.forehead CONJ his
gromadnyj, dlinnyj nos i širokij, ljagušačij rot s tolstymi gubami
huge long nose and wide frog.like mouth with thick lips
i krivymi želtymi klykami vnušali nevol'nyj strax i otvraščenie.
and crooked yellow canines filled involuntary fear and disgust
(S.G. Petrov, 1900)
'But he was not lucky in marriage proposals. The thing is that Kuzma was extremely ugly. He was round-shouldered, awkward, red-haired, he had a hoarse voice and

peered like a wolf from under his eyebrows. And his huge, long nose and his wide frog-like mouth with crooked yellow canines involuntarily filled you with fear and disgust.'

(30) *Zdes' vse čužoe, vse vyloženo kamnjami, net ni odnoj*
here everything strange everything paved stones no NEG one
travki. Steny ***smotrjat, kak volki na jagnenka***. (E.L. Schwarz, 1934)
grass walls look like wolves on lamb
'Everything is strange here, everything is paved with stones, not a blade of grass. The walls look like wolves at a lamb.'

Since the IoCs are confined to only one parameter, they do not have to be further specified by adjuncts. The IoC constructions in (27) and (28) are unambiguously associated with one particular meaning – a frowning manner of looking. Since the similative *kak*-construction is semantically broader than the IoC, one specific aspect of looking as a wolf must be explicitly highlighted for comparison – looking from under the brows in (29) and looking terrifyingly as at a victim in (30). The same difference holds for the IoC construction in (31) and its *kak* counterparts in (32) and (33):

(31) *Rb7. Prixoditsja lad'e* ***metat'sja zajcem***. (S. Shipov, 2004)
Rb7. has.to rook dash hare-INS
'Rb7. The rook has to dash like a hare.'

(32) *Ljudi, zaxvačennye poxot'ju,* ***mečutsja, kak zajac v zapadne***.
people captured lust dash like hare in snare
(L.N. Tolstoy, 1910)
'People captured by lust dash like a hare in a snare.'

(33) *I kuda ja ni mečus',* ***kak zajac na ugonkax*** *...vse to že,*
and where I NEG dash like hare on hunting all that PCL
to že! (I.S. Turgenev, 1878–1882)
that PCL
'And wherever I run like a hare chased after, it's all the same, it's all the same.'

As evidenced by the adjuncts *v zapadne* 'in a snare' and *na ugonkax* 'chased after' the similative construction requires parameter specification due its diversity. The monotonous IoC, on the contrary, maps onto a single parameter (a visual image of a dashing hare) and does not require further specification.

A second property of the IoC outside the domain of shape that might be inherited from the prototypical IoC construction is cognitive contingence of the comparee and the standard of comparison. Earlier in this paper, we saw that the cognitive distance between the trajector and the landmark in the shape-related IoC construction is much smaller than in the similative *kak*-construction, because shape and its possessor are cognitively contingent upon each other. It is noteworthy that this property of the IoC is also relevant in cases which do not describe shape. Compare examples (34) and (35):

(34) *Mnogie iz mužčin zastavili nas smejat'sja. Inoj voobražaet sebja puškoju*
many from men made us laugh some imagines self canon
i besprestanno palit rtom svoim; drugoj ***revet medvedem*** *i xodit*
and incessantly fires mouth own other roars bear-INS and walks
na četveren'kax. (N.M. Karamzin, 1793)
on on.all.fours
'Many of the men made us laugh. One thought he was a canon and was constantly firing with his mouth. Another one was roaring like a bear and walking on all fours.'

(35) *Čego* ***reveš kak medved'****? Ubegaj otsjudova podal'še.* (Ju. Mamleev, 1997)
why roar like bear run.away from.here farther
'Why are you roaring like a bear? Get away!'

The psychiatric patient in (34) thinks he is a bear and therefore produces bear-like sounds, which significantly reduces the cognitive distance between the trajector (patient) and the landmark (bear). In contrast, (35) simply compares two sounds – that of a bear and that of a person.[2]

Third, an important property of the IoCs in the domain of shape is the *visualness* of the described parameter; both the comparee and the standard in the IoC are directly observable by the human eye. Compare examples (36) and (37):

(36) *Stixija razygralas' protiv včerašnego, volny razbivalis' o granitnye*
element played against yesterday's waves broke about granite
naberežnye, ***vstavaja stenoj*** *bryzg.* (A. Bitov, 1980–1990).
quays rising wall-INS splashes
'The storm broke as against yesterday. The waves were breaking against the granite quays and rising in a wall of splashes.'

(37) *Vspomni Kavkaz. Xrebty* ***stojat, kak steny****.*
recall Caucasus mountain.ranges stand like walls
(K. Serafomov, 1978–1996)
'Recall the Caucasus. The mountain ranges stand like walls.'

The conceptualizer of (36) sees the storming waves as a wall of splashes, whereas the mountain range in (37) is construed as a mountain range that only resembles a wall. It is important to note that there is no objective ontological difference between cases such as (36) and (37). It is not the case that waves are

2 It should be mentioned, however, that example (34) is marginally acceptable in present-day Russian, which is an indication of the ongoing semantic change. The historical development of the IoC construction, though interesting and worth pursuing, is beyond the scope of this paper. We will leave this point for future investigation.

more similar to walls than mountains are. It is a matter of dynamic construal (Langacker 1987; Verhagen 2007) in the naïve worldview (Apresjan 1995, 2006; Zaliznjak 2006) of a human conceptualizer (Kustova 2004; Padučeva 1996, 2000). The conceptualizer sees the configuration of waves in (36) as a shape that forms an insurmountable obstacle; this interpretation is exclusively visual. In contrast, the conceptualizer in (37) focuses on internal properties of walls such as firmness and stability.

Further, we claim that the IoCs outside the domain of shape also tend to construe perceptually observable situations, simple events directly accessible to the human eye: e.g. *bit' ključom* 'spout', *vzletet' pticej* 'take wing', *gljadet' sokolom* 'look like a falcon', *sijat' zolotom* 'shine like gold'. Observability determines the choice of predicates which can be used in the IoC construction. Position verbs (e.g. *stojat'* 'stand', *sidet'* 'sit', *ležat'* 'lie', *raspolagat'sja* 'be situated'), causative locative verbs (e.g. *položit'* 'put', *sostavit'* 'put together', *svernut'* 'roll up', *složit'* 'lay') and motion verbs (e.g. *xodit'* 'walk', *letet'* 'fly', *polzti* 'crawl', *izvivat'sja* 'coil', *prygat'* 'jump') are acceptable in this construction due to the intrinsic observability of the situation construed by them. Observability also motivates the use of numerous visual predicates in the IoC (e.g. *smotret'* 'look, watch', *vygljadet'* 'look', *kazat'sja* 'seem', *prikidyvat'sja* 'pretend').

A critical reader might object by indicating that events denoted by auditory predicates (e.g. *vyt' volkom* 'howl like a wolf', *pet' solov'em* 'sing like a nightingale') are not visually observable. It should, however, be taken into account that auditory events are robustly related to vision across human languages (Geeraerts 1988; Ullmann 1972; Viberg 1984; Williams 1976). For example, in languages such as Walbiri (West Australia), Djaru (West Australia) and Lesghian (East Caucasus) the word for *hear* is derived from the verb *see*. It is therefore not surprising that the IoC has also been extended to the auditory domain.[3]

The observability criterion explains why we cannot use the IoC construction in cases like **greet pečkoj* (gives.out.warmth stove-INS), **taet l'dom* (melts ice-INS), **plavaet ryboj* (swims fish-INS), **utonul toporom* (drowned axe-INS) –

3 Interestingly enough, the 18th century corpus contains only one instance of the IoC after a sound verb (*revet' medvedem* 'roar like a bear'). The rest of the occurrences of the construction describe visually perceivable events (e.g. *lit' vedrom* 'bucket, rain cats and dogs', *katit'sja gradom* 'roll down', *polzat' čerepaxoj* 'crawl like a turtle', *nos kljapom* 'gag-shaped nose'). In early 19th century, we observe a significant extension of the IoC construction into the domain of sound (e.g. *vyt' volkom* 'howl like a wolf', *kričat' kukuškoj* 'cry like a cuckoo', *vorkovat' golubkom* 'coo like a pigeon'). However, this outcome should be taken as suggestive only, because the 18th century corpus contains about 2 million words, whereas the corpus of 19th century texts counts about 25 million words.

all these situations are not directly observable. Expressions such as **begal ugoreloj koškoj* (ran mad-INS cat-INS) and **drožal zajcem* (trembled hare-INS) are unacceptable because they depict internal states and traits – disorderliness and cowardice, respectively. All unobservable cases are too complex for the IoC and are therefore taken over by the *kak*-construction: *greet kak pečka* 'keeps warm like a stove', *taet kak led* 'melts like ice', *plavaet kak ryba* 'swims like fish', *utonul kak topor* 'sank like an axe', *begaet kak ugorelaja koška* 'runs like a mad cat', *drožit kak zajac* 'trembles like a hare'.

Compare in this respect also the following pair: *gljadet' Napoleonom* (look Napoleon-INS) vs. *voevat' kak Napoleon* (fight like Napoleon), but **voevat' Napoleonom* (fight Napoleon-INS). Only the observable situation – external properties of a prototypical winner (e.g. barrel-chested, eagle eye) – can be described by means of the IoC. Fighting as Napoleon, in contrast, involves an array of internal properties that cannot be captured by the perceptually "simple" construction with the instrumental case.

The observability criterion is crucial not only to the selection of predicates participating in the IoC construction, it also plays an important part in determining which nouns are acceptable and/or frequent in the IoCs. More specifically, these nouns have to denote standards of comparison with an established visual image within the Russian worldview. The more established a visual image is, the more frequently the noun will be used in the IoC and the more idiomatic the specific construction will be.

For example, the commonality of the expression *nos kartoškoj* 'bulbous nose' is motivated by the fact that the potato has been a vital product in Russia for centuries.[4] In contrast, the visual image of a water-melon or a banana is much less prominent in the Russian culture. This explains why the RNC contains seventeen occurrences of the noun *kartoška* 'potato' in the IoC construction (all in combination with *nos*) and only one instance of *arbuz* 'water-melon' (in *život arbuzom* 'melon-shaped belly'). Likewise, the nouns *banan* 'banana' and *jabloko* 'apple' are not used in the IoC construction at all despite the fact that they both have a very distinct shape.

In the same vein, the noun *kalačik* is remarkably frequent in the IoC construction (e.g. *svernut'sja kalačikom* 'roll up like a kalatch', *spat' kalačikom* 'sleep in kalatch-shape', *leč' kalačikom* 'lie in kalatch-shape', *podžat' nogi kalačikom* 'sit with one's legs tucked up'). In 269 out of 381 (70%) occurrences in the RNC, this noun is used to denote a standard in the IoC. A *kalatch* – a kind

---

4 It is interesting to mention that visual images associated with particular standards of comparison are culturally-specific. For instance, the Russian IoC *nos kartoškoj* denotes a bulbous (fat and round) nose, whereas the English noun phrase *potato nose* and the corresponding Dutch compound *aardappelneus* are used to refer only to irregular, reddish-purple noses.

of fancy loaf – is a salient visual image in the Russian culture, a lot more salient than, say, a croissant. Therefore, there are no instances of the noun *kruassan* as part of the IoC construction in the RNC.

Frequent IoC phrases, based on culturally significant visual images, gradually become increasingly idiomatic and acquire additional meanings. For instance, the IoC *grud' kolesom* (lit. chest wheel-INS) 'with chest well out' currently denotes not only a particular shape, but also a show of bravery, whereas *gubki bantikom* (lit. lips bow-DIM.INS) 'bow-shaped lips' evokes the image of a girl with her nose in the air.

The inherent observability of the IoC has several theoretical implications. To begin with, it shows that neither a verb-centric nor a purely noun-centric hypothesis can adequately account for the combinatorial patterns within the IoC construction. The semantics of the IoCs is not fully predictable by the verb, nor can it be fully predicted by the semantics of the noun. Rather, the construction itself has a meaning, observability being an important part of it. And this construction-specific meaning imposes restrictions on verbs and nouns that may participate in this IoC. As shown earlier in this section, only specific types of predicates can be used in the IoC construction; and only nouns denoting culturally significant standards of comparison are completely acceptable in the IoC. Relatedly, as shown by Desjatova et al. (2008), there are also low-level semantic restrictions on the types of nouns which can be employed to denote a comparee. For example, shape IoCs frequently describe human and animal body parts (e.g. *ruki* 'hands, arms', *grud'* 'chest', *nogi* 'feet, legs', *golova* 'head'; *xvost* 'tail', *xobot* 'trunk', *kljuv* 'beak'), and (more rarely) names of clothing (e.g. *jubka* 'skirt', *plat'e* 'dress', *rubaxa* 'shirt'), as well as paths (e.g. tropa 'path', *reka* 'river', *doroga* 'road').

The principle of observability also justifies the frequently made claims about the similarity between the IoC and metaphor (e.g. Arkad'ev et al. 2008:86; Timberlake 2004:335; Zaliznjak 1996:173), since as demonstrated by Arutjunova (1983: 7, 1990:28) observability is an inherent property of metaphor.

Finally, the idea of observability is related to the primacy of shape in the conceptual sphere of the IoC. As noticed in Tribushinina (2005), shape predicates are abundantly used in genres where visualization of the described situation is of paramount importance. For instance, popular-scientific texts that are meant to make very complex scientific matter accessible (more visible as it were) to the general public contain a large number of adjectives of the type *heart-shaped*. Notice that these adjectives, just as the IoCs, compare two objects in terms of shape in order to visualize a complex situation. Recall also that shape is one of the crucial visual cues to categorization (see Section 4.2.3). We may therefore conclude that descriptions of shape constitute the semantic core of the IoC due to being inherently observable.

4.3.2. The IoC in the instrumental family. The IoC construction is but one type of instrumental constructions and not the prototypical one. As shown in Janda (1993), the conduit instrumental constitutes the prototype of the instrumental family. More precisely, the prototypical meaning of the instrumental case, as the term suggests, is one of an instrument (Wierzbicka 1980:4). Given that constructions occupy a specific position in a hierarchical network of related constructions (Kay and Fillmore 1999; Goldberg 2006), the IoC construction may be expected to inherit some of its essential properties from the prototypical instrumental construction.

The analysis of the empirical data suggests that observability is one such property. Obervability is intrinsic to the instrumental of instrument (e.g. *rubit' toporom* 'cut with an axe'), since using an instrument is an essentially observable situation. The observability requirement accounts for the fact that nouns in the instrumental case cannot be combined with abstract verbs. Therefore, even when the same objective situation is meant (assassination in revenge), **otomstil toporom* (avenged axe-INS) is infelicitous, whereas *ubil toporom* (killed axe-INS) is perfectly acceptable. Observability of the prototypical instrumental is inherited by various instrumental constructions including instrumentals of route, agent and comparison. All these constructions lay out the specifics of the described scene, thereby *visualizing* the object of conceptualization.

A further property of the IoC that might be inherited from other instrumental constructions is the cognitive distance between a trajector and a landmark. As mentioned in Section 4.2.2, there is a difference between nominative and instrumental constructions in terms of cognitive distance. The instrumental constructions, unlike the nominative ones, can never denote full identity. Various instrumental constructions may be claimed to occupy transitional positions on the following continuum of increasing cognitive distance: instrumental of main predication > instrumental of additional characterization > instrumental of shape > instrumental of quasi-comparison. We would like to point out that this range is a continuum without strict borderlines, which is supported by the observation that Wierzbicka's (1980:121) example of the instrumental of additional characterization in (38) can also be seen as an example of the instrumental of quasi-comparison:

(38) *On xodil za nej ten'ju.*
he walked behind her shadow-INS
'He followed her like a shadow.'

Notice that what is traditionally called the *instrumental of comparison* has been now split into two sub-constructions – the instrumental of shape (e.g. *stojat' ovalom* 'be arranged in the oval shape') and the quasi-comparative instrumental construction (e.g. *upast' kamnem* 'drop like a stone'). This distinction is motivated by the finding that shape instrumentals do not really denote comparison (see Section 4.2.2); rather this construction is used to describe objects in terms of shape and is in this respect very similar to the instrumental of additional characterization. Shape and its possessor are cognitively contingent just as a person and his/her profession are (cf. examples 21 and 22). In contrast, the instrumental of quasi-comparison does involve comparison of two distinct entities. Therefore, the cognitive distance between the trajector (compare) and the landmark (standard) is this case bigger than in the non-comparative shape instrumental. Further increase in cognitive distance between the comparee and the standard of comparison brings us to the realm of the similative *kak*-construction (see Section 4.3.1).

The distinction between purely shape-describing instrumentals and cases of quasi-comparison is further supported by the distribution of similatives in case-free languages such as English and Dutch. Critically, these languages construe cases such as *stojat' ovalom, složit' ruki krestom na grudi* and *letet' klinom* only non-comparatively (Eng. *be arranged in an oval shape / *be arranged like an oval, cross one's arms / *put arms like a cross on one's breast, fly in V-shape / *fly like a wedge(V)*; Dut. *in de vorm van een ovaal staan / *als een ovaal staan, armen kruizen / *armen als een kruis op de borst leggen, in V formatie vliegen / *als een wig(V) vliegen*). Recall that in these cases Russian does not allow the similative construction either (**stojat' kak oval; *složit' ruki kak krest na grudi; *letet' kak klin*). In cases like these, English and Dutch often use denominal verbs as in *cross one's arms / armen kruizen* or compound nouns as in *pruimmondje trekken* 'purse one's lips' (lit. pull a plum mouth, cf. *složit' gubki bantikom*). The fact that the IoC construction often corresponds to compound nouns and denominal verbs in Germanic languages is very instructive, since these forms iconically represent a smaller cognitive distance between the two objects compared in this manner (see Section 4.2.4).[5] This confirms our claim that an important difference between the IoC construction and the *kak*-comparative is the degree of cognitive contingence between a comparee and a standard of comparison.

In contrast, quasi-comparative situations such as *letet' streloj, poletet' kamnem vniz, vyt' volkom* and *sijat' zolotom* are construed as similatives in

---

5 In this connection, it is important to note that expressions such as 'cross one's arms' and 'oval-shaped' also exist in Russian (*skrestit' ruki, oval'nyj*). In other words, Russian has a number of quasi-synonymous constructions for expressing similarity in shape. This observation suggests an immediate direction for further research. It is important to investigate the functional differences between various shape-comparing constructions (IoC, compound adjectives, denominal vebs) in order to establish the exact correspondences between their form and function, and to be able to place these constructions in a larger network of comparative constructions.

English and Dutch: Eng. *fly like an arrow, drop like a stone, howl like a wolf, shine like gold*; Dut. *als een pijl (uit de boog) vliegen, als baksteen naar beneden vallen, huilen als een wolf, als goud schijnen*. Importantly, in all these cases Russian also allows the similative *kak*-construction: *letet' kak strela, poletet' kak kamen' vniz, vyt' kak volk, sijat' kak zoloto*. This outcome confirms our claim that quasi-comparative instrumentals such as *letet' streloj* occupy an intermediate position between instrumentals of additional (shape) characterization and similatives. Therefore, languages which have no instrumental construction adapt such cases to similatives. Due to having both the instrumental and the similative construction, Russian is then better able to manipulate the cognitive distance between the comparee and the standard of comparison.[6]

4.4. *Vantage configuration*

In the foregoing sections, we have been analysing the distribution of the IoC construction vis-à-vis the *kak*-similative. The analysis has shown that these constructions are not synonymous. Rather the semantic relation between them is one of co-extension, a type of semantic relation first described by MacLaury (1995, 1997) within the framework of *vantage theory*. Co-extensive categories largely overlap, but still retain their own prototypical core. They do not have an equal status though. One of the categories is dominant; it is more frequent and has an even distribution. The other category – called *recessive vantage* – has a narrow range and a very specific prototype located towards the periphery of the co-extensive category. The dominant category usually covers a great deal of the recessive vantage.

If we apply vantage theory to the analysis of the two comparative constructions, we see that the *kak*-comparative is the dominant category applicable to a wide range of comparative situations. In contrast, the IoC has a very narrow range of application and a very skewed distribution: it is confined to quasi-comparative observable situations and descriptions of shape, manner of motion, looking and sound. The IoC vantage is located towards the periphery, since it is restricted to situations where the cognitive distance between a comparee and a standard of comparison is small. The dominant comparative construction and the recessive vantage significantly overlap, in the sense that the *kak*-comparative is applicable in most cases where the IoC can also be used, with the exception of the cases of pure (quasi-grammatical) shape descriptions

6 This idea is consistent with Timberlake's claim that the instrumental manipulates two situations, it "both differentiates them and also connects them as part of a larger picture" (Timberlake 2004:337).

(e.g. *svernut' treugol'nikom* 'fold in the shape of a triangle') which constitute the prototypical core of the IoC.

5. Conclusions

One of the most significant findings of this investigation is that the semantics of the IoC cannot be adequately accounted for by either a verb-centric or an entirely noun-centric approach. The preceding sections provide ample evidence that the meaning of the instrumental construction is not entirely predictable in terms of the argument structure of the predicate; nor is it completely determined by the semantics of the nominal element(s). The construction as such has its own semantics; essential properties of its content are observability, monotonicity and a small cognitive distance between the trajector and the landmark.

The data in this study also suggest that construction traditionally called *instrumental of comparison* is rather quasi-comparative in nature. The prototypical core of this construction covers descriptions of shape, where the relation between shape and its possessor is cognitively contingent, but obviously non-comparative. These cases are semantically very close to the instrumental of additional characterization. Less prototypical cases of the quasi-comparative construction describe manner of motion, sound and outlook; these cases occupy an intermediate position between the instrumental of additional characterization and the similative *kak*-construction. Like the other instrumentals, quasi-comparatives denote a smaller cognitive distance between the trajector and the landmark. However, unlike the instrumental of additional (shape) characterization, this construction construes two distinct situations and is in this respect similar to the comparative *kak*-construction. The similative construction is preferred in less observable and more diversified cases, as well as in the cases with a larger cognitive distance between the comparee and the standard. The instrumental construction and the *kak*-similative were shown to constitute a co-extensive range, with the similative construction as a dominant vantage and the instrumental as a recessive vantage.

A major result of interest is that not only "exotic" languages, such as African, Austronesian and Mayan languages, but also a Slavic language – Russian – can grammatically encode shape. We have argued that the IoC construction is a quasi-grammatical means of object (sub)categorization. This finding suggests two immediate directions for future study. First, it would be interesting to know whether other Slavic languages can also grammatically encode shape the way Russian does. Second, many questions remain about the cross-linguistic realizations of categorization on the basis of shape. We have seen that English and Dutch also distinguish between pure similarity and additional shape characterizations and use derivational morphology to encode the latter. It will be a matter for future research to establish how other Indo-European languages encode similarity (or additional characterization) qua shape and how this typology is related to the cognitive salience of shape in the categorization process.

References

Aikhenvald, Alexandra Y. 2004. Nominal classification: Towards a comprehensive typology. *Sprachtypologie und Universalienforschung* 57(2/3). 105–116.

Apresjan, Jurij D. 1974. *Leksičeskaja semantika: sinonimičeskie sredstva jazyka* [Lexical semantics: Synonymic means in language]. Moskva: Nauka.

Apresjan, Yurij D. 1995. Obraz čeloveka po dannym jazyka: Popytka sistemnogo opisanija [The image of a human being through language: An attempt towards a systematic description]. *Voprosy jazykoznanija* 1. 37–67.

Apresjan, Jurij D. 2000. *Systematic Lexicography*. Oxford: Oxford University Press.

Apresjan, Jurij D. (ed.). 2006. *Jazykovaja kartina mira i sistemnaja leksikografija* [The linguistic worldview and systematic lexicography]. Moskva: Jazyki skavjanskix kul'tur.

Arkad'ev, Petr M, Grigorij E. Krejdlin & Aleksandr B. Letučij. 2008. Semiotičeskaja konceptualizacija tela i ego častej. 1. Priznak "forma" [Semiotic conceptualization of body parts. 1. Feature "shape"]. *Voprosy jazykoznanija* 6. 78–97.

Arutjunova, Nina D. 1983. Toždestvo ili podobie? [Equality or similarity?] In Aleksej A. Leont'ev (ed.), *Problemy strukturnoj lingvistiki* [Issues in structural linguistics], 3–22. Moskva: Nauka.

Arutjunova, Nina D. 1990. Metafora i diskurs [Metaphor and discourse]. In Nina D. Arutjunova (ed.), *Teorija metafory* [The theory of metaphor], 5–32. Moskva: Nauka.

Barðdal, Jóhanna. 2008. *Productivity: Evidence from case and argument structure*. Amsterdam: John Benjamins.

Barðdal, Jóhanna. 2009a. Lexical vs. structural case: A false dichotomy. To appear in Cathryn Donohue & Jóhanna Barðdal (eds.), *Empirical approaches to morphological case*.

Barðdal, Jóhanna. 2009b. The development of case in Germanic. To appear in Jóhanna Barðdal & Shobhana Chelliah (eds.), *The role of semantic, pragmatic and discourse factors in the development of case*. Amsterdam: John Benjamins.

Bogusławski, Andrzej & Stanisław Karolak. 1970. *Gramatyka rosyjska* [Russian grammar]. Warszawa: Wiedza Powszechna.

Bornstein, Marc H. 1985. Color naming versus shape-name learning in young children. *Journal of Child Language* 12. 387–393.

Brugman, Claudia. 1981. *Story of over*. Berkley: University of California MA thesis.

Croft, William. 2001. *Radical construction grammar: Syntactic theory in typological perspective*. Oxford: Oxford University Press.

Cuzzolin, Pierluigi & Christian Lehmann. 2004. Comparison and gradation. In Geert Booij, Joachim Mugdan, Stavros Skopeteas & Christian Lehmann (eds.), *Morphology: An international handbook on inflection and word-formation. Vol. 2*, 1212–1220. Berlin & New York: Mouton de Gruyter.

Desjatova, Anna V., Olga Lashevskaja & Alexandra A. Makhova. 2008. Konstrukcija s tvoritel'nym formy "X Y-om" [The instrumental of shape construction X Y-om]. *Proceedings of the Annual International Conference "Dialogue"* 7(14). 133–139.

Fillmore, Charles J. & Paul Kay. 1993. *Construction grammar coursebook*. Berkley: University of California manuscript.
Fillmore, Charles J., Paul Kay & Mary Catherine O'Connor. 1988. Regularity and idiomaticity in grammatical constructions: The case of *let alone*. *Language* 64. 501–538.
Fried, Mirjam & Hans C. Boas. (eds.). 2005. *Grammatical constructions: Back to the roots*. Amsterdam and Philadelphia: John Benjamins.
Friedrich, Paul. 1970. Shape in grammar. *Language* 46(2). 379–407.
Geeraerts, Dirk. 1988. Where does prototypicality come from? In Brygida Rudzka-Ostyn (ed.), *Topics in cognitive linguistics*, 207–229. Amsterdam: John Benjamins.
Goldberg, Adele. 1995. *Constructions: A Construction Grammar approach to argument structure*. Chicago: University of Chicago Press.
Goldberg, Adele. 2006. *Constructions at work: The nature of generalization in language*. Oxford: Oxford University Press.
Haiman, John. 1980. The iconicity of grammar: Isomorphism and motivation. *Language* 56(3). 515–540.
Haspelmath, Martin & Oda Buchholz. 1998. Equative and similative constructions in the languages of Europe. In Johan van der Auwera (ed.), *Adverbial constructions in the languages of Europe*, 277–334. Berlin & New York: Mouton de Gruyter.
Heine, Bernd. 1982. African noun class systems. In Hansjakob Seiler & Christian Lehmann (eds.), *Apprehension: Das sprachliche Erfassen von Gegenständen. V.I: Bereich und Ordnung der Phänomene*, 189–216. Tübingen: Narr.
Herskovits, Annette. 1986. *Language and spatial cognition: An interdisciplinary study of the prepositions in English*. Cambridge: Cambridge University Press.
Jackendoff, Ray. 2002. *Foundations of language: Brain, meaning, grammar, evolution*. Oxford: Oxford University Press.
Jakobson, Roman O. 1971 [1936]. Beitrag zur allgemeinen Kasuslehre. In Roman O. Jakobson *Selected Writings, Vol. II*, 23–71. The Hague: Mouton.
Janda, Laura. 1993. *A geography of case semantics: The Czech dative and the Russian instrumental*. Berlin: Mouton de Gruyter.
Kay, Paul & Charles J. Fillmore. 1999. Grammatical constructions and linguistic generalizations: The *What's X doing Y?* construction. *Language* 75. 1–33.
Kibrik, Aleksandr E., Marija M. Brykina, Aleksej P. Leont'ev & Andrej N. Xitrov. 2006. Russkie posessivnye konstrukcii v svete korpusno-statističeskogo issledovanija. *Voprosy jazkoznanija* 1. 16–45.
Kuryłowicz, Jerzy. 1949. Le probléme du classement des cas. In Jerzy Kuryłowicz. 1960. *Esquisses linguistiques*, 131–154. Wrocław & Kraków: Polska Akademia Nauk.
Kustova, Galina I. 2004. *Tipy proizvodnyx značenij i mexanizmy jazykovogo rasširenija* [Types of derivative meanings and mechanisms of linguistic extension]. Moskva: Jazyki slavjanskoj kul'tury.
Lakoff, George. 1987. *Women, fire, and dangerous things*. Chicago: The University of Chicago Press.
Landau, Barbara, Linda B. Smith & Susan S. Jones. 1988. The importance of shape in early lexical learning. *Cognitive Development* 3. 299–321.
Langacker, Ronald W. 1987. *Foundations of Cognitive Grammar*, *Vol.1*, *Theoretical prerequisites*. Stanford, CA: Stanford University Press.
MacLaury, Robert E. 1995. Vantage theory. In John R. Taylor & Robert E. MacLaury (eds.), *Language and the cognitive construal of the world* (Trends in Linguistics. Studies and Monographs 82), 231–275. Berlin & New York: Mouton de Gruyter.

MacLaury, Robert E. 1997. *Colour and cognition in Mesoamerica: Constructing categories as vantages*. Austin: University of Texas Press.

Michaelis, Laura A. 2004. Type shifting in construction grammar: An integrated approach to aspectual coercion. *Cognitive Linguistics* 15. 1–67.

Michaelis, Laura A. & Knud Lambrecht. 1996. Toward a construction-based model of language function: The case of nominal extraposition. *Language* 72. 215–247.

Mrázek, Roman. 1964. *Sintaksis russkogo tvoritel'nogo* [The syntax of the Russian instrumental]. Praha.

Padučeva, Elena V. 1996. *Semantičeskie issledovanija* [Semantic investigations]. Moskva: Jazyki russkoj kul'tury.

Padučeva, Elena V. 2000. Nabljudatel' kak Eksperient za kadrom [The observer as an exprerience behind the scenes]. In Leonid L. Iomdin & Leonid P. Krysin (eds.), *Slovo v tekste i v slovare.* [A word in text and in dictionary], 185–201. Moskva: Jazyki russkoj kul'tury.

Padučeva, Elena V. 2004. *Dinamičeskie modeli v semantike leksiki* [Dynamic models in lexical semantics]. Moskva: Jazyki slavjanskoj kul'tury.

Panov, Mixail V. 1990. O pozicionnom čeredovanii grammatičeskix značenij [On positional alternation of grammatical meanings]. In Viktor S. Khrakovskij (ed.), *Tipologija i grammatika* [Typology and grammar], 82–90. Moskva: Nauka.

Panov, Mixail V. 1992. Predskazuemost' allomorfa [The predictability of an allomorph]. In *Rusistika segodnja. Funkcionirovanie jazyka: leksika i grammatika* [Russian linguistics today. Functions of language: lexis and grammar], 30–36. Moskva: Nauka.

Panov, Mixail V. 1999. *Pozicionnaja morfologija russkogo jazyka* [Positional morphology of the Russian language]. Moskva: Nauka.

Plungian, Vladimir A. & Ekaterina, V. Rakhilina. 2000. Po povodu "lokalistskoj" koncepcii značenija: Predlog *pod* [On the localist theory of meaning: Preposition *pod*]. In Denis Paillard & Olga N. Seliverstova (ed.), *Issledovanija po semantike predlogov* [Papers on the semantics of prepositions], 115–133. Moskva: Russkie slovari.

Rakhilina, Ekaterina V. 2000. *Kognitivnyj analiz predmetnyx imen: semantika i sočetaemost'* [A cognitive analysis of concrete nouns: Semantics and combinability]. Moskva: Russkie slovari.

Serzisko, Fritz. 1982. Gender, noun class and numeral classification. In René Dirven & Günter Radden (eds.), *Issues in the theory of the universal grammar*, 95–123. Tübingen: Narr.

Talmy Leonard. 2000. *Toward a cognitive semantics. Vol. I.* Cambridge, MA: MIT Press.

Taylor, John R. 2003. Near synonyms as co-extensive categories: 'High' and 'tall' revisited. *Language Sciences* 25(3). 263–284.

Timberlake, Alan. 2004. *A reference grammar of Russian*. Cambridge: Cambridge University Press.

Tribushinina, Elena. 2005. *Models of denotative shape adjectives in English*. Amsterdam: VU University manuscript.

Turovskij, Vladimir V. 1988. *Kak poxož, napominat'*, tvoritel'nyj sravnenija: Tolkovanija dlja gruppy kvazisinonimov [*Kak poxož, napominat'*, instrumental of comparison: Definitions for quasisynonyms]. In Nina D. Arutjunova (ed.), *Logičeskij analiz jazyka: Referencija i problemy tekstoobrazovanija* [Logical analysis of language: Reference and text formation], 30–36. Moskva: Nauka.

Ullmann, Stephen. 1972. *Grundzüge der Semantik: Die Bedeutung in sprachwissenschaftlicher Sicht*. Berlin & New York: Walter de Gruyter.

Vandeloise, Claude. 1994. Methodology and analyses of the preposition *in*. *Cognitive Linguistics* 5(2). 157–184.
Verhagen, Arie. 2007. Construal and perspectivisation. In Dirk Geeraerts and Hubert Cuyckens (eds.), *Handbook of cognitive linguistics*, 48–81. Oxford: Oxford University Press.
Viberg, Ake. 1984. The verbs of perception: a typological study. *Linguistics* 21(1). 123–162.
Wade, Terence. 1992. *A comprehensive Russian grammar.* Oxford, UK and Cambridge, USA: Blackwell.
Wierzbicka, Anna. 1980. *The case for surface case*. Ann Arbor: Karoma.
Williams, Joseph M. 1976. Synaesthetic adjectives: a possible law of semantic change. *Language* 52(2). 461–478.
Zaliznjak, Anna A. 1996. The case for instrumental case reopened, ili ešče raz o značenii russkogo tvoritel'nogo padeža. *Die Welt der Slaven* 41. 167–184.
Zaliznjak, Anna A. 2006. *Mnogoznačnost' v jazyke i sposoby ee predstavlenija* [Polsysemy in language and means of expressing it]. Moskva: Jazyki slavjanskix kul'tur.

Constructional realizations of affirmation in Slavic[1]

Marcin Grygiel, University of Rzeszów

1. Introduction

The aim of this article is to account for a number of linguistic manifestations of affirmation in Slavic languages – mainly in Serbian – in terms of Construction Grammar (Goldberg 1995, Goldberg 2006). Affirmation is not a part of traditionally understood "core grammar" nor does it hold an important place in linguistic investigations and its documentation across languages is still very poor (Liu and Soehn 2009). This default, mostly covert and formally unmarked category has long been neglected and considered to be uninteresting. Construction Grammar, on the other hand, claims that fundamental insights into the nature of grammar and linguistic coding can be gained from examining marginal, problematic, fuzzy and non-core idiosyncrasies. Another property which makes Construction Grammar particularly suitable for the study of affirmation is the claim that there is no strict division between the lexicon and syntax, thus elevating constructions to the position of basic units of language. Consequently, a Construction Grammar approach to a functionally defined category such as affirmation seems to be especially propitious because of this theoretical framework's long lasting interest in the semantics of grammar and the relation between the nature of verb meaning and sentential meaning.

Affirmation is primarily a semantic function whose expression relies on the cooperation among various subsystems of the language. In Slavic, affirmation is lexically marked with the existential BE forms. This meaning is transferred onto syntactic constructions with BE used as an auxiliary verb and copula. The process of affirmation marking is further extended on enclitic constructions where BE becomes optional and gets dropped. Thus, the study of affirmation shows that grammatical constructions develop out of lexical constructions whereas abstract syntactic constructions arise as a result of de-lexicalization and inheritance of the properties of semantically transparent syntactic combinations. Consequently, grammar may be conceived as a process of meaning schematization and semantic coding leading to the rise of more and more abstract formal representations.

---

1 I would like to thank Laura Janda for all her comments and suggestions which helped me to reformulate many of the ideas in this paper. Thanks are also due to Diana Prodanović-Stankić, who checked the Serbian examples for accuracy. Obviously, I bear sole responsibility for any remaining errors.

2. The concept of affirmation

Unlike negation, the study of affirmation has never been high on the linguistic agenda. The problem hardly ever gets even a brief mention in grammar reference books and the lemma is hard to find in most linguistic encyclopedias. Because of the fact that negative sentences use more uniform syntactic marking in comparison to their affirmative counterparts, the status of negation as a coherent, homogenous and universal category with linguistic validity has been generally recognized while that of affirmation is still questioned, widely avoided or confused with other categories such as assertion. The only laudable exception here is the generative tradition where affirmation is studied in the context of polarity and distribution of polarity sensitive items (Progovac 1994, Tovena 1998). Nevertheless, a visible disproportion and a bias toward negation are still present.

Contrary to the generative tradition, I believe that affirmation is primarily a semantic category whose formal marking is of secondary importance as it tends to be not fully coherent or even absent. Consequently, affirmation may be described as a functional and onomasiological class whose homogeneity resides merely in the conceptual plan. In a similar vein, it seems reasonable to argue that linguistic affirmation is closely related to a universal human cognitive domain associated with counter-reaction or disagreement to negation. Thus, by affirmation I mean the modification of an utterance by specifying the positive attitude of the speaker with regard to the content of the sentence. In the case of affirmation the speakers transform their utterances by stressing or attributing a positive value as an additional component to the semantic structure of a proposition. This type of affirmative polarization is triggered in opposition to negation or hypothetically negative contexts.

Affirmation is primarily a semantic category because it is a linguistic and sometimes non linguistic manifestation of dynamic mental operations which I will refer to as positive polarization. A given sentence is considered to be affirmative only in the context of being potentially negative. As such, it is semantically marked due to the cognitive on-line processes accompanying it while formally it may be marked only by the absence of a negator. Affirmation thus pertains not only to purely linguistic structures, but also to the communicative contexts within which these structures are embedded.

One of the consequences of the context dependence and the semantic nature of affirmation is the fact that it is not an absolute but rather a graded category. The weakest forms of affirmation may be signaled only by the "proto-grammar" devices (Givón 2001:34) such as intonation rules, spacing rules, sequence rules and quantity rules. Strong or emphatic affirmation is syntactically realized, for example, by means of verb reduplication (Martins 2006), full forms of auxiliary verbs (Grygiel 2010), affirmative particle reinforcement strategy used in Spanish e.g.: *Luis sí fue a Roma* 'Luis did go to Rome' or European Portuguese e.g.: *O João não comprou o carro, pois não? Comprou sim* 'John didn't buy the car, did he? Yes, he certainly did'.[2] Affirmation may also be represented or intensified with a number of lexical items: particles, numerals, modifiers and discourse markers (Piper 2009).

3. The mechanism of de-lexicalization and inheritance

Construction Grammar defines constructions as form and meaning pairings which appear at all levels of grammatical analysis and include morphemes, words, idioms and phrasal patterns. Goldberg (2006:5) claims that any linguistic pattern is recognized as a construction provided no element of its form or meaning is strictly predictable from the properties of its component parts or other constructions. Another important property is that constructions have meaning independently of verbs – generally regarded to be crucial for the expression of sentence meaning. As a consequence, Goldberg (1995:224) believes that "an entirely lexically based approach to grammar is inadequate, and that lexically unfilled constructions must be recognized to exist independently of the particular lexical items which instantiate them".

One of the examples given by Goldberg (2006) to illustrate this point is the dative subject construction used in South Slavic languages to imply an attitudinal stance where the meaning cannot be attributed to or inferred from any specific word or morpheme in the construction:

(1) Srb. *Pilo mi se pivo* (Goldberg 2006:7)
[Drink.3SG.PAST I.DAT REF beer]
'I felt like drinking beer'

Notice that the construction is used to express a desire or craving which is hard to control and which may be caused by the functioning of the human body. It is typically employed in the context of expressing physiological need, dispositional or involuntary state. Observe that only in two of the below given situations a parallel construction may be encountered in Polish:

(2a) Srb. *Zeva mi se*
[Yawn.3SG.PRES I.DAT REF] 'I feel like yawning/I can't help yawning'
(2b) Pol. *Ziewa mi się*
[Yawn.3SG.PRES I.DAT REF] 'I feel like yawning/I can't help yawning'

---

2 All the examples are mine unless stated otherwise. They come from a variety of sources: TV programs, magazines, websites, informants. In most cases they have been shortened or reformulated. Names of sources are provided only for longer sentences which have not been modified in any way.

(3a) Srb. *Povraća mi se*
[Vomit.3SG.PRES I.DAT REF] 'I feel like vomiting/I can't help vomiting'
(3b) Pol. *Rzyga mi się* [3]
[Vomit.3SG.PRES I.DAT REF] 'I feel like vomiting/I can't help vomiting'

(4a) Srb. *Spava mi se*
[Sleep.3sg.PRES I.DAT REF] 'I feel like sleeping'
(4b) Pol. *Chce mi się spać*
[Want.3SG.PRES I.DAT REF sleep.INF] 'I feel like sleeping'

(5a) Srb. *Jede mi se čokolada*
[Eat.3SG.PRES I.DAT REF chocolate] 'I feel like eating chocolate'
(5b) Pol. *Chce mi się jeść czekoladę*
[Want.3SG.PRES I.DAT REF eat.INF chocolate.ACC] 'I feel like eating chocolate'

(6a) Srb. *Danas mi se ne kuva ručak*
[Today I.DAT REF NEG Cook.3SG dinner.ACC]'Today I don't feel like cooking dinner'
(6b) Pol. *Dzisiaj nie chce mi się gotować obiadu*
[Today NEG want.3SG I.DAT REF cook.INF dinner.GEN] 'Today I don't feel like cooking dinner'

The examples show that in Polish the construction which implies an attitudinal stance is very similar to that used in South Slavic languages except for the fact that it includes a lexical element *chce*/3sg WANT – crucial for the semantic interpretation of the whole construction. In the South Slavic dative subject construction the meaning of the lexical element is incorporated into the grammatical structure of the construction and the lexical element becomes redundant. The dative subject construction with the verb *hoće*/3sg WANT is still preserved in Serbian (7), but its occurrence is quite limited as it has been generally replaced with its de-lexicalized equivalent. The de-lexicalized construction is so widely used that it may appear in contexts (8) where in Polish the dative subject construction with 3sg WANT may seem dubious in affirmative sentences (?*Chce mi się oglądać ten film* 'I feel like watching this film'), but quite correct in negative utterances (*Nie chce mi się oglądać ten film* 'I don't feel like watching this film'). In Polish the de-lexicalized – without 3sg WANT – dative subject construction is restricted only to contexts referring to very basic physiological needs, but even then the 3sg WANT dative subject construction may be used instead (9, 10). A language where both constructions are permitted is, for example, Slovene (11):

3 The dative subject construction followed by instrumental case may also be used in very colloquial Polish in the sense 'to be fed up with', e.g.: *Rzyga mi się tą robotą* 'Im fed up with this work'.

(7) Srb. *Njemu se hoće lako živeti i vladati* (Kryzia 2005:27)
[He.DAT REF want.3SG easily live and rule] 'He would like to live and rule easily'

(8) Srb. *Hajdemo u bioskop, gleda mi se taj film*
[Go.1PL.IMPER to cinema, see.3SG I.DAT REF this film]
'Let's go to the cinema, I feel like watching this film'

(9a) Pol. *Chce mi się ziewać*
[Want.3SG.PRES I.DAT REF yawn.INF] 'I feel like yawning/I can't help yawning'

(9b) Pol. *Ziewa mi się*
[Yawn.3SG.PRES I.DAT REF] 'I feel like yawning/I can't help yawning'

(10a) Pol. *Chce mi się rzygać*
[Want.3SG.PRES I.DAT REF vomit.INF] 'I feel like vomiting/I can't help vomiting'

(10b) Pol. *Rzyga mi się*
[Vomit.3SG.PRES I.DAT REF] 'I feel like vomiting/I can't help vomiting'

(11a) Slo. *Janezu se hoče plesati*
[Janez.DAT REF want.3SG dance.INF] 'Janez feels like dancing'

(11b) Slo. *Janezu se pleše*
[Janez.DAT REF dance.3SG] 'Janez feels like dancing'

The assumption that construction with overt *hoće*/3sg WANT must be prior to dative subject construction is further supported by the fact that the acquisition of dative subject construction expressing attitudinal stance is more complicated for child-learners and it is mastered later than the equivalent WANT construction (Stojanović 2004). Stojanović (2004:217) claims that "not all attempts at producing a 'feel-like dative' in child Serbian are successful" and provides the following example:

(12) context: looking at the picture of Mickey Mouse yawning (Stojanović 2004:217)
Child: *ovaj Miki hoće da spava*
'this Mickey wants to sleep'
Mother: *Spava mu se?*
[Sleep.3SG he.DAT REF]
'He feels like sleeping?'
Child: **ovoj se spava*
[This.DAT.FEM REF sleep.3SG]
**se spava Miki Maus*
[REF sleep.3SG M. M.]
**ovaj se spava*
[This.NOM.MASC REF sleep.3SG]
Same context, one day later:
Child: **ovaj se meni spava, a nema krevet*
[this.NOM.MASC REF I.DAT sleep.3SG] (but doesn't have a bed)

The above discussion seems to suggest that the South Slavic dative subject construction expressing attitudinal stance may lack an element denoting the meaning associated with the construction, but this meaning is indirectly motivated by an element which has been removed. Still, the semantic structure and some syntactic properties of the construction – impersonal 3sg form – have been incorporated into the de-lexicalized construction. The situation where a given construction inherits its structure from other constructions in the language and becomes motivated in this way is not rare in grammar (Lakoff 1987). The relations of inheritance between constructions are also amply presented by Goldberg (1995). In the next sections I will try to show how a similar mechanism of de-lexicalization and inheritance resulted in the formation of affirmation constructions.

4. The verb BE as a marker of affirmation

The verb BE in its full, most lexical and primary use is associated with the expression of EXISTENCE and this quality predisposes it for the function of affirmation marker. Notice that one of the conceptual metaphors on which the semantics of affirmation is based may be phrased as AFFIRMATION IS EXISTENCE. By stressing that something exists, language users attach a positive value to it and contrast it with an opposite situation which may be metaphorically presented as NEGATION IS NONEXISTENCE. In East Slavic languages all the present forms of BE have disappeared except for 3sg BE which is used in the sense of EXISTENCE both for singular and plural mostly in the sentence initial position:

(13) Rus. *Est` dalekaja strana, vy ne znaete ee* (Lubocha-Kruglik 2001:52)
[Be.3SG faraway country, you NEG know.2PL she.GEN]
'There is (exists) a faraway country, you don't know it'

(14) Rus. *Est` kartiny, kotorye vlijanie okazyvajut na sud`bu* (Lubocha-Kruglik 2001:38)
[Be.3SG pictures which influence show.3PL on fate.ACC]
'There are (exist) pictures that influence fate'

Having a relatively limited occurrence in Russian, BE in its present form – *est'* – has clearly acquired an affirmative function and stands in sharp contrast to its negative counterpart *net*. Notice that *est'* cannot be used with the usual negator *ne* and become negated in the same way as other verbs in Russian. In order to express the negative meaning of NONEXISTENCE, speakers of Russian have to resort to the negative particle *net* equivalent to English *no*, for example:

(15) Rus. *Est' tam dom?* *Est'/Net*
[Be.3SG there house Be.3SG/No (NEG.be.3SG)]
'Is there a house there? Yes, there is. /No, there isn't'

Similarly, in Ukrainian and Polish the verb BE in the existential sense cannot be negated and the meaning of non-existence is expressed by the negative form of the verb HAVE:

(16) Ukr. *Je tam dom?* *Je/Nemaje*
[Be.3SG there house Be.3SG/NEG.have.3SG]
'Is there a house there? Yes, there is / No, there isn't'

(17) Pol. *Jest tam dom?* *Jest/Nie ma*
[Be.3SG there house Be.3SG/NEG have.3SG]
'Is there a house there? Yes, there is / No, there isn't'

The meaning of EXISTENCE and the function of affirmation marking in existential constructions is further metaphorically extended to lexicalizations of the present form of BE in 3rd person singular where it is used as an affirmative particle with a contextually limited occurrence. For example, Russian *Est'!* used by soldiers or subordinates expresses agreement in situations when a command has been given. Similarly, Polish *Tak jest!,* which is semantically equivalent to English *Yes, sir!*, performs a function of confirming that an order has been accepted and understood by a soldier in the army.

Treated as auxiliary verbs – the present forms of BE – are widely used in Serbian as markers of affirmation. The distinction between short and long forms makes it possible to differentiate weak from strong/emphatic affirmative meanings. The long forms (*jesam*, *jesi*, *jeste*, *jesmo*, *jesu*) may be treated as expressions of affirmation *par excellence* due to the fact that they cannot be negated in any way and are often used on their own as affirmative rejoinders or even affirmative particles – the occurrence of *jeste* is so widespread that it may be compared only to that of the main affirmative particle *da.*

Table1. Present imperfective BE forms in Serbian

	strong affirmative forms	weak affirmative forms	negative forms
1.SG	*jesam*	*sam*	*nisam*
2.SG	*jesi*	*si*	*nisi*
3.SG	*jeste*	*je*	*nije*
1.PL	*jesmo*	*smo*	*nismo*
2.PL	*jeste*	*ste*	*niste*
3.PL	*jesu*	*su*	*nisu*

Unlike the long forms, the short forms of BE (*sam*, *si*, *je*, *smo*, *ste*, *su*) can be contrasted with their negative counterparts (*nisam*, *nisi*, *nije*, *nismo*, *niste*, *nisu*).

These shorter positive forms, however, cannot be negated in the same fashion as other Serbian verbs where the negator *ne* must be obligatorily used (*govorim – ne govorim*). As a consequence, short forms may also be treated as markers of affirmation as shown in the examples given below:

(18a) Srb. *Ona je moja*
[She be.3SG my] 'She's mine'
(18b) Srb. *Ona nije moja*
[She NEG.be3SG my] She's not mine'

(19a) Srb. *Mi smo se vratili*
[We be.1PL REF return.PERF] 'We have returned'
(19b) Srb. *Mi se nismo vratili*
[We REF NEG.be.1PL return.PERF] 'We haven't returned'

Notice that in Serbian and Slovene the perfect tense always requires the use of the present form of the auxiliary verb BE even in 3rd person. Although in both languages the present tense forms are negated with the negative particle *ne*, in the case of the present forms of BE only contracted negative forms may be used and their phonetic forms substantially vary from their positive counterparts. For example in Slovene 3sg *je* contrasts with *ni* 'is not':

(20a) Slo. *Peter je bil v knjižnici* (Herrity 2000:181)
[Peter be.3SG be.PERF.MASC in library] 'Peter was in the library'
(20b) Slo. *Petra ni bilo v knjižnici*
[Peter.GEN NEG.be.3SG be.PERF.NEUT in library] 'Peter wasn't in the library'

(21a) Slo. *Ona je videla psa* (Herrity 2000:181)
[She be.3SG see.PERF.FEM dog.ACC] 'She saw a dog'
(21b) Slo. *Ona ni videla psa*
[She NEG.be.3SG see.PERF.FEM dog.ACC] 'She didn't see a dog'

Except for West South Slavic languages (Slovene, Serbian, Croatian, Bosnian, Montenegrin) and Sorbian, in all the other Slavic languages that use complex forms, the 3rd person perfect/past tense forms do not require the auxiliary verb BE. Also in some languages, like Czech and Slovak, all the remaining past forms cannot be constructed without the present forms of BE (*volal jsem/som* 'I called', *volal jsi/si* 'you called', *volal* 's/he called', *volali jsme/sme* 'we called', *volali jste/ste* 'you called', *volali* 'they called'). Furthermore, the Czech and Slovak BE forms cannot carry negation in the past tense – negation can only appear on the *l*-form as in the case of all Polish past tense forms (*nie wołałem* 'I didn't call', *nie wołałeś* 'you didn't call', *nie wołał* 's/he didn't call', *nie wołaliśmy* 'we didn't call', *nie wołaliście* 'you didn't call', *nie wołali* 'they didn't call').

Table 2. Serbian and Czech perfect/past forms

	Serbian affirmative	Serbian negative	Czech affirmative	Czech negative
1.SG	*zvao sam*	*nisam zvao*	*volal jsem*	*nevolal jsem*
2.SG	*zvao si*	*nisi zvao*	*volal jsi*	*nevolal jsi*
3.SG	*zvao je*	*nije zvao*	*volal*	*nevolal*
1.PL	*zvali smo*	*nismo zvali*	*volali jsme*	*nevolali jsme*
2.PL	*zvali ste*	*niste zvali*	*volali jste*	*nevolali jste*
3.PL	*zvali su*	*nisu zvali*	*volali*	*nevolali*

In Bulgarian, on the other hand, the 3sg auxiliary form of BE – *e* – is optional in the perfect tense and it is generally used to mark strong/emphatic affirmation. When this form is omitted, the sentence expresses weak affirmation:

(22a) Blg. *Tja sledvala v Sofija* (Koseska-Toszewa 2007:286)
[She study.PERF in Sofia] 'She (probably) studied in Sofia'
(22b) Blg. *Tja e sledvala v Sofija*
[She be.3SG study.PERF in Sofia] '(I'm sure) she studied in Sofia/She did study in Sofia'

(23a) Blg. *Toj pročel tazi kniga* (Koseska-Toszewa 2000:51)
[He read.PERF this book] 'He (probably) read this book'
(23b) Blg. *Toj e pročel tazi kniga*
[He be.3SG read.PERF this book] '(I'm sure) he read this book/He did read this book'

These examples show that the 3sg present form of BE seems particularly important for the expression of affirmation. In Serbian, for example, the long 3sg BE – *jeste* – is very close to acquiring the function of general affirmative particle used regardless of person, number and tense (24-27) – some of its uses, however, are more typical and readily acceptable (26-27) than others (24-25). Its shorter variant – *je* – is also used in sentence initial (24b, 25b, 26b) or final position (27) as an affirmative part of the interrogative particle – *je li* (*jel'*) – almost irrespective of person, number and tense. The semantics of this compositional interrogative particle may be interpreted as checking or confirming affirmation – yes with a question mark. The 3sg *je* form may be used instead of the more frequent in this role affirmative particle *da*. However, the particle *da li* is limited to only sentence initial position and a more formal register:

(24a) Srb. *Da li stvarno tamo ideš?* *? Jeste*
[AFF INTER really there go.2SG] [Be.3SG.LONG (AFF)] 'Are you really going there? Yes'
(24b) Srb. *Je li stvarno tamo ideš?* *? Jeste*
[Be.3SG INTER there go.2SG] [Be.3SG.LONG (AFF)] 'Are you really going there? Yes'

(25a) Srb. *Da li žurite?* *? Jeste*
[AFF INTER hurry.2PL] [Be.3SG.LONG (AFF)] 'Are you in a hurry? Yes'

(25b) Srb. *Je li žurite?* ? *Jeste*
[Be.3SG INTER hurry.2PL 2PL] [Be.3SG.LONG (AFF)] 'Are you in a hurry? Yes'

(26a) Srb. *Da li je ovo banka?* *Jeste*
[AFF INTER be.3SG this bank] [Be.3SG.LONG (AFF)] 'Is this a bank? Yes'
(26b) Srb. *Je li (je) ovo banka?* *Jeste*
[Be.3SG INTER (be.3SG) this bank] [Be.3SG.LONG (AFF)] 'Is this a bank? Yes'
(27) Srb. *Vi ste došli iz Beograda, jel`?* *Jeste*
[You.PL be.2PL come.PAST.PL from Belgrade be.3SG INTER] [be.3SG.LONG (AFF)]
'You have come from Belgrade, haven't you? Yes, we have'

This short discussion should suffice for the conclusion that the verb BE plays a crucial role in affirmation marking. The occurrence of 3sg BE as a marker of affirmation is not limited to Slavic languages exclusively. Quite the contrary, for example in European Portuguese 3sg BE present – *é* – and past – *foi* – are used both as elements in question tags and short affirmative answers to questions with conjugated verbs of all persons e.g.: *Vocês gostam desta música, não é? É, é* 'You like this music, don't you? Yes, we do', *Tu tiveste sorte, não foi? Foi, foi* 'You were lucky, wasn't that so? Yes, yes'.

5. Affirmation constructions resulting from inheritance

It may seem that in Serbian and other West South Slavic languages, sentences in the present tense with ordinary conjugated verbs do not include any kind of syntactic or morphological formal marker of affirmation. However, the abundance of periphrastic constructions with the verb BE makes it possible for a considerable number of present tense sentences to contain markers of affirmation similar to those encountered in the past tense. The widespread use of the dative subject, passive and copula BE constructions with meanings equivalent to those of the major verbs explains why 3sg *je* is almost ubiquitous in these languages allowing for formal marking of affirmation. It also appears that affirmation marking is the most important function performed by this verb. Here are a few examples:

UNDERSTAND
(28) Srb. *Jasno mi je zašto on to radi*
[Clear I.DAT be.3SG why he this do.3SG] 'It's clear to me/I understand why he's doing it'

NEED
(28) Srb. *Potreban joj je ključ za kola*
[Needed She.DAT be.3SG key for car] 'She needs a key for the car'

FEEL

(29) Srb. *Žao mi je tog čoveka*
[Sorry I.DAT be.3SG this.GEN man.GEN] 'I feel sorry for that man'

(30) Srb. *Milo/drago mi je*
[Nice/glad I.DAT be.3SG] 'I feel glad'

(31) Srb. *Dosadno mi je*
[Bored I.DAT be.3SG] 'I feel bored'

(32) Srb. *Hladno mi je*
[Cold I.DAT be.3SG] 'I feel cold'

(33) Slo. *Mraz mi je bilo* (Herrity 2000:253)
[Cold I.DAT be.3SG be.PERF] 'I felt cold'

LIKE

(34) Slo. *Darilo jim je všeč* (Herrity 2000:252)
[Present they.DAT be.3SG likeable] 'They like the present'

WANT

(35) Srb. *Želja mu je da ode u inostranstvo*
[Wish he.DAT be.3SG to go.3SG abroad] 'His wish is/he wants to go abroad'

(36) Srb. *Ona je za još jedno piće*
[She be.3SG for another drink] 'She wants to have another drink'

KNOW

(37) Srb. *To mi je vrlo dobro poznato*
[It I.DAT be.3SG very well known] 'It's very well known to me/I know it very well'

MISS

(38) Slo. *Zelo mi je dolgčas po njej* (Herrity 2000:252)
[Very I.DAT be.3SG long.time after her] 'I miss her very much'

DOUBT

(39) Srb. *Ovo je meni sumnjivo*
[This be.3SG I.DAT doubtful] 'It seems doubtful to me/I doubt it'

CAN/BE ABLE TO

(40) Srb. *On je u prilici da ti pomogne*
[He be.3SG in opportunity.DAT to you.DAT help.2SG] 'He's able to/he can help you'

(41) Srb. *On je sposoban da ti pomogne*
[He be.3SG able to you.DAT help.2SG] 'He's able to help you'

(42) Srb. *On je u stanju da ti pomogne*
[He be.3SG in state to you.DAT help.2SG] 'He's able to help you'

SHOULD

(43) Srb. *Znam šta mi je sada činiti*
[Know.1SG what I.DAT be.3SG now do] 'I know what I should/I am to do now'

HAVE

(44) Srb. *Sutra mi je slobodan dan*
[Tomorrow I.DAT be.3SG free day] 'Tomorrow I'm having a day off'

(45) Srb. *U laži su kratke noge*
[At lie.DAT be.3PL short legs] 'A lie has short legs'

ALRIGHT/OK

(46) Srb. *U redu je*
[At order be.3SG] 'It's alright'

(47) Slo. *Meni je prav, če je tebi prav* (Herrity 2000:252)
[I.DAT be.3SG alright if be.3SG you.DAT alright]
'It's alright with me if it's alright with you'

(48) Srb *U pravu je*
[At right be.3SG] 'S/He's right'

The Serbian 3sg form *je* acts as a marker of affirmation not only because it is formally contrasted with its negative counterpart *nije*, but also because of the asymmetrical distribution of the two markers. Except for its occurrence as a part of interrogative particle, *je* is always used as an enclitic – an element which appears last in clitic clusters – whereas Slavic negative markers are generally distributed as proclitics:

(49a) Srb. *Tako je*
[Like that be.3SG] 'That's right'

(49b) Srb. *Nije tako*
[NEG.be.3SG like that] 'It's not like that/That's wrong'

(50a) Srb. *Kupila je*
[Buy.PERF be.3SG] 'She bought'

(50b) Srb. *Nije kupila*
[NEG.be.3SG buy.PERF] 'She didn't buy'

Although all Serbian short BE forms are enclitics and markers of affirmation, the 3sg form *je* presents some particularities. It obeys distinct ordering constraints in clitic clusters – it always comes last, whereas the other forms precede all pronominal clitics (e.g: *zvao si ga* 'you called him' vs. *zvao ga je* 'he called him') and it is the only short BE form used on its own as a stressed head of a clitic cluster in questions (25b, 26b). The 3sg *je* also stands out because all the other clitic BE forms are identical to the full forms without their initial *je* syllable whereas 3sg *je* appears to be the stem rather than an ending (Table 1). As a consequence, the 3sg form seems to be better equipped for the function of affirmation marking in comparison to the other short BE forms and the last position in clitic clusters occupied by it may be considered crucial for the expression of affirmation in Serbian declarative sentences.

Positive and negative clitic clusters also show considerable differences in their distribution within a clause. As I have already mentioned, negative markers tend to occupy preverbal position whereas affirmative markers normally follow the verb or other host elements. Consequently, in Serbian enclitic BE forms may always occur after a finite verb, regardless of that verb's position in the sentence or clause. However, there is a tendency to mark sentences affirmatively in the most initial position possible. In the case of enclitic markers of affirmation this position is the so called Wackernagel or second position in the clause (Franks and King 2000). Enclitic markers of affirmation in Serbian tend to move to Wackernagel position in any clausal configuration despite the fact that in doing so they often break up a syntactic unit such as adjective + noun or other constituents. The examples (51a-f) show that the 3sg BE form is transferred from post-verbal position (51a) and placed after the first word or first phrase in the clause. Even if additional elements are being added clause-initially, the 3sg form moves accordingly to mark affirmation in the most forward clause position possible :

(51a) Srb. *Veliku ulogu odigralo je pismo upućeno predsedniku vlade*
[Big.ACC part.ACC play.PERF be.3SG letter addressed prime minister.DAT]

(51b) Srb. *Veliku ulogu je odigralo pismo upućeno predsedniku vlade*
[Big.ACC part.ACC be.3SG play.PERF letter addressed prime minister.DAT]

(51c) Srb. *Veliku je ulogu odigralo pismo upućeno predsedniku vlade*
[Big.ACC be.3SG part.ACC play.PERF letter addressed prime minister.DAT]

(51d) Srb. ?*U celoj je priči veliku ulogu odigralo pismo upućeno predsedniku vlade*
[In whole.DAT be.3SG story.DAT big.ACC part.ACC play.PERF letter addressed prime minister.DAT]

(51e) Srb. ? *U ovoj je celoj priči, pismo upućeno predsedniku vlade, odigralo veliku ulogu*
[In this.DAT be.3SG whole.DAT story.DAT letter addressed prime minister.DAT play.PERF big.ACC part.ACC]

(51f) Srb. *Zaista je, u ovoj celoj priči, pismo upućeno predsedniku vlade, odigralo veliku ulogu*
[Really be.3SG in this.DAT whole.DAT story.DAT letter addressed prime minister.DAT play.PERF big.ACC part.ACC]
'The letter addressed to the prime minister has really played a big part in this whole story'

According to Diesing et al. (2009), Serbian allows both the first word and first phrase placement options for Wackernagel clitics, but in Croatian the sentential position for enclitics is after the first word. As a consequence, the examples (51d-e) would be judged as perfectly OK in Croatian while being problematic in

Serbian. Also, cases such as (52), where an initial phrase constituting a proper name is broken up with the BE 3sg form, are rare in modern Serbian whereas in Croatian and in 19th century Serbian (the language of Vuk Karadžić) the placement of Wackernagel clitics after the first word is the norm:

(52) Srb. *Jovo je Protić znao govoriti turski i grčki* (Stanojčić 1996:133)
[Jovo be.3SG Protić know.PERF.MASC speak Turkish and Greek]
'Jovo Protić could speak Turkish and Greek'

Diesing et al. (2009) designed a series of experiments to elicit judgments from native speakers about the role of semantic/pragmatic factors in clitic placement. The experiments seem to prove that first phrase placement has unmarked intonation, and is acceptable in "out-of-the-blue" contexts. The first word placement, on the other hand, has a marked contour, and is only acceptable in contexts with contrastive emphasis.[4] Here are some of their conclusions::

> "Our claim is that not only do we not find optionality unrestricted by grammar, but that clitic placement is dependent on both syntactic and pragmatic factors. (...) The least studied facet of clitic positioning is the impact of information structure, yet intuitively we feel that this is where the action is. In particular, differences between the four cases we have identified may well be correlated with different discourse conditions between the first word and first phrase positions within each category" (Diesing et al. 2009:73).

It seems very probable to me that the syntactic/pragmatic factors mentioned by Diesing et al. (2009) have also to do with affirmation marking. Notice that the first word position is used for more emphatic marking of affirmation as the enclitic affirmation marker placed in this position cannot fall within the scope of the negator. In the linear structure of an utterance the information about the positive polarization is expected to appear as early as possible. This kind of polarization is more general than negative polarization which is usually limited to the verb or particular lexical items. Similarly, in the case of dependant clauses, enclitic markers of affirmation are placed directly after the complementizer or relative pronoun (53a). This position cannot be reached by the negator (53b, 53c) unless it is immediately followed by the verb, but even then a sentence with a negative BE form in the Wackernagel position will sound rather awkward (53d):

(53a) Srb. *On veruje da je u ovoj celoj priči, pismo upućeno predsedniku vlade, odigralo veliku ulogu*
[He believe.3SG that be.3SG in this.DAT whole.DAT story.DAT letter addressed prime minister.DAT play.PERF big.ACC part.ACC]

---

4 The 'first word and the first phrase position' refers here to 'peninititial position' i.e. 'the position after the first word or after the first phrase in the clause'.

'He believes that in the whole story a letter addressed to the prime minister has played a big part'

(53b) Srb. **On veruje da nije u ovoj celoj priči, pismo upućeno predsedniku vlade, odigralo veliku ulogu*
[He believe.3SG that NEG.be.3SG in this.DAT whole.DAT story.DAT letter addressed prime minister.DAT play.PERF big.ACC part.ACC]

(53c) Srb. *On veruje da u ovoj celoj priči, pismo upućeno predsedniku vlade, nije odigralo veliku ulogu*
[He believe.3SG that in this.DAT whole.DAT story.DAT letter addressed prime minister.DAT NEG.be.3SG play.PERF big.ACC part.ACC]
'He believes that in the whole story a letter addressed to the prime minister hasn't played a big part'

(53d) Srb. ?*On veruje da nije odigralo veliku ulogu pismo upućeno predsedniku vlade u ovoj celoj priči*
[He believe.3SG that NEG.be.3SG play.PERF big.ACC part.ACC letter addressed prime minister.DAT in this.DAT whole.DAT story.DAT]
'He believes that a letter addressed to the prime minister hasn't played a big part in the whole story'

The aim of the discussion so far has been to show that Serbian has developed a 3sg BE affirmation construction. It is highly restricted in the sequence of its elements and the positioning pattern makes part and parcel of the construction itself. Now I will try to demonstrate that the original lexical properties of BE – the expression of EXISTENCE – which must have been decisive in the choice of this verb for the function of affirmation marking, can be incorporated into the structure of the whole enclitic construction.[5] Consequently, the verb BE disappears in the process and its properties are inherited by other elements in the construction as has happened in the case of the verb WANT in the dative subject construction expressing an attitudinal stance.

The 3sg BE form *je* may usually be omitted in a dependent clause when another *je* already occurs in the main clause. This shows that the role of affirmation marking, and not that of linking a subject with its predicate, is the most important function performed by 3sg BE. The affirmation marker *je* may be omitted, once a given sentence has been positively marked, even in situations when the subjects in constituent clauses are different. This is exactly what is happening in the following example (54) with *je* used only once although it should be repeated in the subsequent clause (*Rekao je da je...*):

(54) Srb. *Rekao je da u ovoj celoj priči, pismo upućeno predsedniku vlade, odigralo veliku ulogu*
[Say.PERF be.3SG that in this.DAT whole.DAT story.DAT letter addressed prime minister.DAT play.PERF big.ACC part.ACC]
'He said that in the whole story the letter addressed to the prime minister played a big part'

5 For more information about the lexical properties of Slavic BE see Clancy (2010).

Notice that in Serbian 3sg BE *je* is usually absent in sentences which cannot be classified as expressing positive or negative polarization according to the definition I have given. Affirmative polarization entails a dynamic mental operation changing the modality of an utterance. In this process the speaker construes mentally a given state of affairs or action as having a polarized value and uses adequate linguistic resources to express it. This kind of cognitive operation, where assumed affirmative meaning of an utterance might be hypothetically questioned, is excluded in the situation of giving facts whose truthfulness cannot be denied as in the case of newspaper headlines, biographical data or in proverbs. This is a situation where Serbian uses the so called incomplete perfect tense "krnji perfekat" without an auxiliary verb:

(55) Srb. *Predsednik Tadić razgovarao sa šefom poljske diplomatije* (RTS 2009, 5, 20)
[President Tadić talk.PERF with head.INST Polish.GEN diplomacy.GEN]
'President Tadić talked to the head of Polish diplomacy'

(56) Srb.*1879 - Rođen sovjetski diktator poreklom iz Gruzije Josif Visarionovič Džugašvili Staljin, 1940 - Umro američki pisac Frensis Skot Ficdžerald* (B92 2008, 12, 30)
[1879 – born Soviet dictator origin.INST from Georgia.GEN Joseph Visarionovich Jugashvili Stalin, 1940 – die.PERF American writer Francis Scott Fitzgerald]
'1879 –Soviet dictator of Georgian origin Joseph Visarionovich Jugashvili Stalin was born 1940 –American writer Francis Scott Fitzgerald died'

(57) Srb. *Brada narasla, a pamet ne donela* (Stevanović 1980:405)
[Beard grow.PERF and intelligence NEG bring.PERF]
'The beard has grown, but intelligence hasn't arrived'

The incomplete perfect tense represents a situation where the BE affirmative construction is not used at all. There are cases, however, where the 3sg BE marker of affirmation *je* gets dropped, but the affirmative meaning is expressed by enclitic constructions. This type of *je* omission occurs most often when the reflexive *se* is already used. In colloquial rather than formal register also some pronouns *me*, *te, je* may have a similar effect and cause the disappearance of *je*. Notice that all of these pronominal enclitic elements resemble formally verbal *je* – all of them end in *–e*, include two phonemes and occupy exactly the same positions as *je* both within the sequence structure of an enclitic cluster and clause. A similar type of replacement cannot take place, however, in the case of other BE forms:

(58) Srb. *Probudio se (je) u osam sati*
[Wake up.PERF.MASC REF (be.3SG) at eight o'clock]
'He woke up at eight o'clock'

(59) Srb. *Milan vam se (je) predstavio*
[Milan you.PL.DAT REF (be.3SG) introduce.PERF.MASC]
'Milan introduced himself to you'

(60a) Srb. *Privukao ju je sebi* [6]
[Pull.PERF.MASC she.ACC be.3SG onself.DAT]
'He pulled her to himself'

(60b) Srb. *Privukao je sebi, osećajući njeno čvrsto telo tik uz sebe* (LVR, 18)
[Pull.PERF.MASC she.ACC onself.DAT]
'He pulled her to himself, feeling her strong body right next to him'

(61) Srb. *Pitala me (je) kako živim*
[Ask.PERF.FEM I.ACC (be.3SG) how live.1SG]
'She asked me how I lived'

Although the process of 3sg BE omission is quite limited in Serbian – it is even less frequent in Croatian and very rare in Slovene – its use is far more common and less restricted in Polish whereas in Russian it led to the almost complete disappearance of present BE forms. In Polish 3sg BE – *jest* – is optionally omitted in, for example, dative subject constructions:

(62a) Pol. *Przykro mi/ci/mu/jej/nam (jest)*
[Sorry I/you/he/she/we.DAT (be.3SG)] 'I/you/he/she/we feel sorry'

(62b) Pol. *Smutno mi/ci/mu/jej/nam (jest)*
[Sad I/you/he/she/we.DAT (be.3SG)] 'I/you/he/she/we feel sad'

(62c) Pol. *Słabo mi/ci/mu/jej/nam (jest)*
[Weak I/you/he/she/we.DAT (be.3SG)] 'I/you/he/she/we feel weak'

Unlike in Serbian where only genitive/accusative pronouns with the closest phonetic resemblance to *je* may be chosen for a particular affirmative enclitic construction replacing the 3sg BE form, in Polish both genitive/accusative and dative pronouns may serve this function irrespective of person and number. One should not forget, however, that it is not the dative subject construction, but the affirmative enclitic construction which is responsible for the omission of *jest*. It seems that the 3sg BE form is dropped because affirmation may be expressed by other enclitic elements and their sequence. Notice that *jest* cannot be dropped when a full – non-clitic – form of a pronoun (63) or non-pronominal (64a-b) dative subject is used or when there is no dative subject used at all (64c-d):

(63a) Pol. **Przykro tobie/jemu*
[Sad you/he.DAT]
(63b) Pol. *Przykro tobie/jemu jest*
[Sad you/he.DAT be.3SG] 'You/he feels sad'
(63c) Pol. **Jemu/tobie przykro*
[He/you.DAT sad]

6 In Serbian the sequence *je* + *je* (she.ACC + be.3SG) dissimilates into *ju* + *je*.

(63d) Pol. *Jemu/Tobie jest przykro*
[He/you.DAT be.3SG sad] 'You/he feels sad'

(64a) Pol. ?*Adamowi smutno*
[Adam.DAT sad]
(64b) Pol. *Adamowi jest smutno*
[Adam.DAT be.3SG sad] 'Adam feels sad'
(64c) Pol. **Smutno*
[Sad]
(64d) Pol. *Smutno jest/Jest smutno*
[Sad be.3SG/ be.3SG sad] 'It feels sad'

In the examples provided above there are no enclitic constructions. As a consequence, 3sg BE is the only marker of affirmation here and it cannot be deleted as in (62). Similarly, in Polish 3sg BE *jest* may become redundant in situations when affirmation is marked not only by an enclitic construction, but also by some other lexical or syntactic marker of affirmation. In the following examples 3sg BE may be optionally deleted because affirmation is already linguistically marked by *już* 'already' (65c-d, 67a), *ale* 'how' (66b-c, 67b), *strasznie* 'awfully' (66f-g, 67c), *sobie* 'to onself' (68b) and *raz* 'once' (68c). If these alternative affirmation constructions are not used, then 3sg BE is obligatory (65a-b, 66a-c, 68d). This explains why (64a) is weird – no affirmation construction used – and (64b, 67a-c) sound correct:

(65a) Pol. **Późno*
[Late]
(65b) Pol. *Późno jest/Jest późno*
[Late be.3SG/be.3SG late] 'It's late'
(65c) Pol. *Już późno*
[Already late] 'It's already late'
(65d) Pol. *Późno już*
[Late already] 'It's already late'

(66a) Pol. **Tutaj ciemno*
[Here dark]
(66b) Pol. *Ale tutaj ciemno!*
[How here dark] 'How dark it is here!'
(66c) Pol. *Ale tutaj jest ciemno!*
[How here be.3SG dark] 'How dark it is here!'
(66d) Pol. **Ale tutaj nie jest ciemno*!
[How here NEG be.3SG dark]
(66e) Pol. **Strasznie tutaj nie jest ciemno*
[Awfully here NEG be.3SG dark]
(66f) Pol. *Strasznie tutaj ciemno*
[Awfully here dark] 'It's awfully dark here'
(66g) Pol. *Strasznie tutaj jest ciemno*
[Awfully here be.3SG dark] 'It's awfully dark here'

(67a) Pol. *Adamowi już smutno*
[Adam.DAT already sad] 'Adam is already sad'
(67b) Pol. *Ale Adamowi smutno!*
[How Adam.DAT sad] 'How sad Adam is'
(67c) Pol. *Strasznie Adamowi smutno*
[Awfully Adam.DAT sad] 'Adam is awfully sad'

(68a) Pol. **Był król*
[Be.PERF king]
(68b) Pol. *Był sobie król*
[Be.PERF REF king] 'There was once a king'
(68c) Pol. *Był raz król*
[Be.PERF once king] 'There was once a king'
(68d) Slo. *Bil je kralj* (Herrity 2000:334)
[Be.PERF be.3SG king] 'There was once a king'

In these examples lexical items *sobie* 'to onself', *raz* 'once', *już* 'already', *strasznie* 'awfully' and the exclamation construction *ale* 'how' mark affirmation by referring to conceptual metaphors such as AFFIRMATION IS EXISTENCE, AFFIRMATION IS COMPLETION and AFFIRMATION IS INTENSIFICATION. Notice that the sentences cannot be negated (66d-e), unless the lexical items change their meaning, for example from *już* 'already' to *już nie* 'no longer'.

An interesting situation can be observed in (68a) where the Polish 3sg BE form was lost in the historical process of synthetic past tense formation. The meaning, however, is clearly existential and the Polish sentence in order to be correct needs an additional lexical item associated with the conceptual domain of EXISTENCE[7] such as *sobie* 'to onself' (68b) or *raz* 'once' (68c). In Slovene (68d) and Serbian, on the other hand, the same function is performed by the 3sg BE *je* form and the addition of lexical items such as Slo. *enkrat* 'once' or Srb. *jednom* 'once' is optional. In Serbian, however, if a lexical existential marker of affirmation is used, such as *jedan* 'one', the 3sg *je* may be omitted:

(69) Srb. *Bio jedan car* (Stanojčić and Popović 2000:275)
Be.PERF.MASC one czar 'There was (once) a king/czar'

(70) Srb. *Bio jednom jedan mali princ koji je živeo na jednoj planeti* (Mali Princ, 20
[Be.PERF.MASC once one little prince who be.3SG live.PERF.MASC on a.DAT planet.DAT] 'There was once a little prince who lived on a planet'

(71) Srb. *Bila jednom jedna zemlja*
[Be.PERF.FEM once one country] 'There was once a country'

---

7 The term 'conceptual domain' is used here in the sense of Grygiel (2004) and Grygiel (2008).

(72) Srb. *Jednog dana otputovao taj čovek nedge na ceo dan* (Emanuil, 80)
[One.GEN day.GEN travel.PERF.MASC this man somewhere for whole day]
'One day the man traveled away somewhere for the whole day'

The appearance of the numeral *jedno* 'one' in the idiomatic expression for a feeling of indifference can cause the same effect of 3sg BE dropping in Polish. Here we can also talk about affirmation construction similar to enclitic constructions described in the foregoing.[8] Compare, for instance, these examples from Polish with their Serbian equivalents:

(73) Pol. *Wszystko mi jedno*
[Everything I.DAT one] 'It's all the same to me'

(74a) Pol. *Wszystko mi jest znane*
Everything I.DAT be.3SG known 'Everything is known to me/I know everything'

(74b) Pol. **Wszystko mi znane*
[Everything I.DAT known]

(75) Srb. *Sve mi je jedno*
[Everything I.DAT be.3SG one] 'It's all the same to me'

(76) Srb. *Sve mi je poznato*
[Everything I.DAT be.3SG known] 'Everything is known to me/I know everything'

Summing up the discussion on affirmation constructions resulting from inheritance, it is possible to conclude that the 3sg BE form is a marker of affirmation placed in the last position within an enclitic cluster. In Serbian this form may be lost and its function inherited by the enclitic construction it would otherwise appear in. The same affirmative properties may be extended onto other enclitic constructions such as reflexives and middle passives where 3sg BE is never used. Notice that reflexive *se*, very much like 3sg *je*, is generally placed in Wackernagel position and comes last within an enclitic cluster if *je* is absent. Thus, the reflexive *se* tends to move to the second position in the clause where it indicates affirmative meaning whereas in negation it falls within the scope of the negator:

---

8 Notice that *jedno* 'one' may be treated as an existential marker of affirmation. For example, the Polish sentence *Mam jedno pytanie* 'I have one question' cannot be negated in a straightforward manner **Nie mam jedno pytanie*, instead two negators must be used, one for the verb and the other for *jedno* plus the accusative case is changed into genitive which is very typical of negation in Slavic: *Nie mam ani jednego pytania* 'I don't have even one question'. Similarly in Serbian *Imam jedno pitanje* is negated as *Nemam nijedno pitanje*.

(77) Srb. *Tvoja se struka veoma traži*
[Your REF profession very search.3SG] 'Your profession is very much in demand'

(78) Srb. *Noćima se u internatu, krijući se od svih, sastajali u Fazilovom gornjem krevetu* (Sneg, 120)
[Night.PL.INS REF in students' dorm.DAT, hiding REF from ererybody.PL.GEN, meet.PAST.PL.MASC in Fazil.POSS.DAT upper bed.LOC]
'At nights in the students' dorm they met in Fazil's upper bed hiding from everybody'

Notice that the reflexive *se*, apart from being phonetically similar to 3sg BE *je*, may be considered its rival in 3sg perfect tense forms. Also, in Serbian passive or pseudo-passive constructions with *se* are preferred to those with BE forms which are definitely rarer. If in one sentence there is a sequence of affirmative clauses with verbs in perfect tense and the first verb includes the reflexive clitic *se*, then all the verbs in the subsequent clauses may drop the auxiliary *je*. Thus, affirmation marking is transferred from the first *se* enclitic construction to all the following clauses:

(79) Srb. *Prodavačica se okrenula, skinula sa police jedan od tri plava automobila sa sijalicom, pružila ga Ivanu i kratko rekla* (Auto sa sijalicom, 35)
[Shop assistant REF turn around.PERF, take down.PERF from shelf.GEN one from three blue.GEN car.PL.GEN with flashing light.INST, hand.PERF it.ACC Ivan.DAT and shortly say.PERF]
'The shop assistant turned around, took down from a shelf one of three blue cars with a flashing light, handed it to Ivan and briefly said'

(80) Srb. *Taj čovek dočepao se nečijeg blaga, sve pretvorio u hiljadarke, pa u brizi kako to da sačuva metne novac u slamu* (Emanuil, 80)
[This man grab.PERF REF somebody.GEN treasure.GEN, everything change.PERF in one-thousand-dinar bills.ACC, and in care.LOC how it to save.3SG put.3SG money in straw.ACC] 'This man came into the possession of somebody's treasure, he exchanged it all for one-thousand-dinar bills, and in order to save it he puts the money into the straw'

The evolution of the two enclitic forms – *je* and *se* – also shows many similarities in moving along a cline from lexical meaning to grammatical function and ending in being merely a formal signal which marks, for example, affirmation only positionally. Mitkovska (1998:47) observes that:

> "The particle *se*, which derives from the direct object clitic of the reflexive pronoun exhibits a gradual increase in grammaticalization along a continuum of mutually related meanings. Grammaticalization reaches the highest degree in those constructions which express agent defocusing situations. There, the coreferential meaning of *se* has completely faded and it is interpreted as a signal for demotion of the initiating participant" (Mitkovska 1998:47).

Analogous processes can be found in other Slavic languages. Slovene serves as yet another very good example of how affirmation function is inherited and passed onto pronominal and reflexive forms. In Slovene the clitic forms that participate in the affirmation construction may be stressed and used on their own as affirmative rejoinders or affirmative particles as shown in the following examples from Greenberg (2007) and Piper (2009):

(81) Slo. *Si ga videla? Ga* (Greenberg 2007:132)
[Be.2SG he.ACC see.PAST.FEM] [He.ACC] 'Did you see him? Yes, (I did)'

(82) Slo. *Ali se umiva? Se* (Greenberg 2007:132)
[INTER REF wash.3SG] [REF] 'Is s/he washing up? Yes, (s/he is)'

(83) Slo. *Ali se to sliši? Se* (Piper 2009:369)
[INTER REF it hear.3SG] [REF] 'Can you hear it? Yes'

(84) Slo. *Ali je prišla? Je* (Piper 2009:351)
[INTER be.3SG come.PERF.FEM] [Be.3SG] 'Has she come? Yes'

The only logical explanation why Slovene makes use of short pronominal forms (*ga*, *se*) to mark affirmation instead of longer and supposedly more suitable stressed pronouns (*njega*, *sebe*) seems that affirmation is expressed by the enclitic construction (*videla si ga*, *umiva se*) and this meaning is inherited by pronominal and reflexive clitics standing for a predicate in affirmative rejoinders. The examples show that even very formal patterns such as ways of packaging sentence elements are associated with and motivated by functions such as that of affirmation.

6. Conclusions

The processes of de-lexicalization and inheritance which can be observed in the rise of Slavic affirmative constructions demonstrate that symbolic units differ along the parameters of complexity and specificity. The lexical items employed as markers of affirmation are the minimal and simplest constructions used to express or reinforce affirmation while syntactic constructions such as affirmative enclitic constructions are much more complex as they involve not only individual clitics, but also their distributional patterns within the clause. The semantics of these higher level constructions, however, is associated with particular clausal patterns and not individual elements because these have already been highly de-lexicalized, grammaticalized or dropped. As a consequence, symbolic units form a continuum from the highly specific to maximally schematic. The most schematic representation of affirmation is the unmarked construction where affirmative function of a sentence is indicated by a lack of the negator.

In the foregoing I have tried to show a bottom up approach to affirmation, going from the most specific forms to the most schematic de-lexicalized functions. According to Goldberg (1995), generalizations across constructions and links between semantic and grammatical relations can be captured by placing generalizations about them at the highest node in an inheritance hierarchy of constructions. In this way, generalizations are inherited through dominated constructions. In the case at hand, we should be looking at the semantics of affirmation as the highest and unifying node for all affirmation constructions. Although they considerably vary in form, all of them are semantically motivated.

References

Clancy, Steven J. 2010. *The Chain of BEING and HAVING in Slavic*. Amsterdam/Philadelphia: John Benjamins Publishing Company.

Diesing, Molly *et al.* 2009. Clitic placement in Serbian: Corpus and experimental evidence. In Sam Featherston and Susanne Winkler (eds.) *The Fruits of Empirical Linguistics, vol. 2*, 59-74. Berlin/New York: Mouton de Gruyter.

Franks, Steven and Tracy H. King. 2000. *A Handbook of Slavic Clitics*. Oxford: Oxford University Press.

Givón, Talmy. 2001. *Syntax: An Introduction, vol. 1*. Amsterdam/Philadelphia: John Benjamins.

Goldberg, Adele. 1995. *Constructions: A Construction Grammar Approach to Argument Structure*. Chicago: University of Chicago Press.

Goldberg, Adele. 2006. *Constructions at work: The nature of generalization in language*. Oxford: Oxford University Press.

Greenberg, Marc L. 2007. *A Short Reference Grammar of Slovene*. Munich: Lincom GmbH.

Grygiel, Marcin. 2004. Semantic change as a process of conceptual blending. In Francisco J. Ruiz de Mendoza (ed.) *Annual Review of Cognitive Linguistics 2*, 285-304. Amsterdam/Philadelphia: John Benjamins Publishing Company.

Grygiel, Marcin. 2008. *From Semantic Change to Conceptual Blending: Semantic Development of English Historical Near-synonyms of MAN/MALE HUMAN BEING*. Rzeszów: Wydawnictwo UR.

Grygiel, Marcin. 2010. Wyrażanie afirmacji za pomocą czasowników posiłkowych w języku serbskim i angielskim [Expressing affirmation by means of auxiliary verbs in Serbian and English]. In Artur Czapiga and Zofia Czapiga (eds.) *Słowo i tekst w opisie porównawczym*, 56-64. Rzeszów: Wydawnictwo UR.

Herrity, Peter. 2000. *Slovene: A Comprehensive Grammar*. London/New York: Routledge.

Koseska-Toszewa, Violetta (ed.). 2007. *Polsko-bułgarska gramatyka konfrontatywna* [Polish-Bulgarian contrastive grammar]. Warszawa: Dialog.

Koseska-Toszewa, Violetta. 2000. O bułgarskich czasach przeszłych w gwarach i języku literackim [On Bulgarian past tenses in dialects and literary language]. In *Bulletin de la Société Polonaise de Linguistique*, fasc. LVI, 49-58.

Kryzia, Władysław. 2005. *Polskie i słoweńskie predykaty modalne o znaczeniu 'chcieć', 'móc', 'musieć', 'powinien' na poszerzonym tle słowiańskim* [Polish and Slovene modal predicates meaning 'want', 'can', 'must', 'should' on an extended Slavic material]. Katowice: Wydawnictwo UŚ.

Lakoff, George. 1987. *Women, Fire and Dangerous Things: What Categories Reveal about the Mind.* Chicago: University of Chicago Press.

Liu, Mingya and Jan-Philipp Soehn. 2009. An empirical perspective on positive polarity items in German. In Sam Featherston and Susanne Winkler (eds.) *The Fruits of Empirical Linguistics, vol. 2*, 197-216. Berlin/New York: Mouton de Gruyter.

Lubocha-Kruglik, Jolanta. 2001. *Rosyjskie zdania egzystencjalne w konfrontacji z polskimi* [Russian existential sentences in comparison with Polish ones]. Katowice: Wydawnictwo UŚ.

Martins, A. M. 2006. Emphatic affirmation and polarity: contrasting European Portuguese with Brazilian Portuguese, Spanish, Catalan and Galician. In J. Doetjes and P. Gonzalez (eds.) *Romance Languages and Linguistic Theory*, 197-223. Amsterdam/Philadelphia: John Benjamins.

Mitkovska, Liljana. 1998. The *se* passive in Macedonian compared to the neighboring Slavic languages. In *Studia linguistica Polono-Meridiano-slavica*, vol. 9, 47-61. Toruń: Wydawnictwo UMK.

Piper, Predrag (ed.). 2009. *Južnoslovenski jezici: gramatičke strukture i funkcije* [South Slavic languages: grammatical structures and functions]. Beograd: Beogradska knjiga.

Progovac, Liljana. 1994. *Negative and Positive Polarity: A Binding Approach*. Cambridge: Cambridge University Press.

Radovanović, Milorad (ed.). 1996. *Srpski jezik* [Serbian language]. Opole: Uniwersytet Opolski.

Stanojčić, Živojin. 1996. Morfologija, sintaksa i frazeologija. In Milorad Radovanović (ed.) *Srpski jezik* [Serbian language]. Opole: Uniwersytet Opolski.

Stanojčić, Živojin and Ljubomir Popović 2000. *Gramatika srpskoga jezika* [Grammar of the Serbian Language]. Beograd: Zavod za udžbenike.

Stevanović, Mihailo. 1980. *Gramatika srpskohrvatskog jezika* [Grammar of Serbo-Croat] Cetinje: Obod.

Stojanović, Danijela. 2004. Children's interpretation of 'feel-like' constructions. In Sophie Burelle and Stanca Somesfalean (eds.) *Proceedings of the 2003 Annual Canadian Linguistics Association Meeting*, 214-233. Montréal: Université du Québec.

Tovena, M. Lucia. 1998. *The Fine Structure of Polarity Sensitivity*. New York/London: Garland Publishing.

Sources

Auto sa sijalicom, Geza Gulka i Vera Niderman, Novi Sad, 1985.
Emanuil, Vladika Nikolaj Velimirović, Beograd, 2002.
Ljubavni vikend roman (LVR), broj 368, Beograd 1991.
Mali Princ, Antoine de Saint-Exupéry prevela Bojna Vukšić, Beograd 1994.
Sneg, Orhan Pamuk, preveo Ivan Panović, Beograd, 2007.
Kurir www.kurir-info.rs
B92 www.b92.rs
RTS www.rts.rs

Possessive locative constructions in Macedonian

Liljana Mitkovska, FON University, Skopje

1. Introduction

This paper examines three types of external possessive constructions in Macedonian which code the possessum (PM) in a locative prepositional phrase, exemplified in (1-3).

(1) *Toj ja udri* ***po rakata.***
he DO.CL -3SG.F hit on arm- DEF
'He hit her **on the arm**.'

(2) *Gledaj gi kakvi se crveni* ***vo kosite.***
look- IMP.SG DO.CL -3PL how are-3PL red- PL in hair-DEF.PL
'Look at them, how strange they look with **their red hair**.'

(3) *Odnesi ja topkata* ***kaj čičko Milan vo dvorot.***
take - IMP.SG DO.CL -3SG.F ball-DEF.SG.F at uncle Milan in yard- DEF.SG.M
'Take the ball into **uncle Milan's yard**.'

(4) ***Mu*** *ja ukrale* ***kolata.***
IO.CL-3SG.M DO.CL -3SG.F stole -3PL car-DEF.SG.F
'They/some people have stolen **his car**.'

The term *external possession* is used, relying on Payne & Barshi (1999), for syntactic constructions in which the possessive relation is not expressed within the NP, but is implied between two participants in a clause. Unlike the internal possessive construction (IPC), the external possessive constructions (EPCs) "appear to impose a requirement that hearers consider the consequences of the event's outcome for the possessor" (O'Connor 2007: 578). In Macedonian there are several such constructions. Apart from those illustrated in examples (1) - (3), the so-called Possessive Dative construction is also commonly used in a similar way as in other Slavic languages (example 4).[1]

In this paper, I will examine the syntactic, semantic and pragmatic features of the three Possessive Locative constructions pointed out above, trying to pinpoint the factors conditioning their use and how each is differentiated in comparison with the IPC. At the end, they will be compared with the Possessive Dative. I will try to show that the two EPCs differ in the way they contribute to the interpretation of the utterance.

---

1 For more detailed information see Mitkovska (2000, 2005, 2007).

Unlike the Possessive Dative construction, which is quite widespread in Macedonian, the Possessive Locative constructions have a much more restricted range of use.[2] They represent types of EPCs which, according to Haspelmath (1999:119), are not common for the European Linguistic Area. The constructions exemplified in (1) and (2) are often referred to as possessor raising or possessor ascension, because the PR takes up the position which the PM occupies in the IPC with corresponding meaning, Direct Object (DO) in example (1) and Subject in example (2). In this respect, Heine (1997) argues that, considering their function and origin, the Possessive Locative constructions are better described in terms of the pragmatic strategy of possessor specification, while coding of the possessor (PR) in the dative case (i.e. indirect object) is a result of possessum specification. The analysis will show that the specific usage properties of the discussed constructions hinge on the process of PR specification.

2. Description of the three locative possessive constructions

2.1 *Contact*

In the constructions of the type illustrated in example (1) (*ja udri po rakata*) the PR occupies the DO position while the PM is expressed in a prepositional phrase with locative semantics (most commonly *vo* 'in', *po* 'on' and *za* 'at'). The PR is typically a person or an animal; plants and other inanimate objects, while possible, are rarely encountered and require special circumstances, such as the one in example (5). The PM is limited to inalienably possessed entities, typically body parts, but clothes and accessories worn in the given situation are also permitted (see example 6).

(5) *Nekoj ja izgrebal* ***kolata po vratata****.*
someone DO.CL -3SG.F scratched-3SG car -DEF.SG.F on door-DEF.SG.F
'Someone has scratched **the door of our car**.'

(6) *Vojnikot* ***go*** *fati* ***za remenot****.*
soldier-DEF.SG.M DO.CL -3SG.M seized-3SG by belt-DEF.SG.M
'The soldier seized **him by the belt**.'

These constructions occur with predicates expressing contact; those with negative effect prevail (hit, pull, strike, poke, press, squeeze, bite, scratch, pierce, stab, pierce, shoot etc.), but neutral predicates are also common (hold, touch, catch, etc) as well as those with positive effect (stroke, pat, kiss etc.). Metaphorically, some

2 See Mitkovska (2005).

verbs of mental contact are also possible candidates, especially verbs of perception (see, glance, observe, look at) (example 7) and a few others (e.g. recognize).

(7) *Taa vednaš* ***go*** *pogledna* ***vo čevlite***.
she immediately DO.CL -3SG.M looked-3SG in shoes-DEF.PL
'She immediately looked at **his shoes**.'

The possessive relation is thus inferred from the constellation of the elements involved. It relies on the part-whole image schema underlying the relation between the person and his/her body parts and the assumption that a contact with a body part is automatically a contact with the whole.

The directly involved PM is presented as the specification of the place in question. (Herslund & Baron 2001:19; Heine 1997) In fact, the contact point is not presented as an individual participant in the event, but only as part of the scene. The main actor that suffers the consequences is the PR whose part was affected and, even without the prepositional phrase, the meaning would basically remain the same. For instance, in the situation of example (1), just saying *Toj ja udri*. 'He hit her.', without specifying where, would convey the core information.

The PM cannot be made more prominent in this syntactic structure, even if it is highly conspicuous. The possessive pronoun is not allowed in the PM NP, as example (8a) suggests. It is possible to add it if the PM has some specific property that would render it particularly noticeable, such as the sentence in (8b). However, such examples are rarely encountered because, even without the possessive pronoun, the body part stands out by virtue of its specific state and the sentence *Nekoj go udri po bolnata noga*. 'Someone hit him on the sore leg.' would have the same effect. No matter if the possessive pronoun is used or not, the holistic effect of the construction remains. If the situation requires individualization of the part, we have to express the PM as a directly affected participant occupying the syntactic position of DO with the PR as its modifier. Such distribution is appropriate in example (9), where it is important that by hitting the woman's shoulder, the egg ruined her silk blouse.

(8) a. *Nekoj go udri po *negovata noga.*
someone DO.CL -3SG.M hit-3SG on his-DEF.SG.F leg-F
'Someone hit him on his leg.'

b. *Nekoj go udri po negovata bolna noga.*
someone DO.CL -3SG.M hit-3SG on his-DEF.SG.F sore-F leg-F
'Someone hit him on his sore leg.'

(9) *Vtoroto jajce go pogodi nejzinoto ramo,*
second-DEF egg-N DO.CL -3SG.N hit-3SG her-DEF.SG.N shoulder-N
(t.e. bluzata od čista svila).
'The second egg hit her shoulder, i.e. her pure silk blouse.'

2.2. *Personal feature*

In the second EP construction type discussed here (example 2) the PR is coded in subject position and the PM in a prepositional phrase, mainly with the preposition *vo* – 'in', indicating location.[3] They employ copular predicates with an adjectival subject complement which expresses some kind of physical (or sometimes personality) feature of the PR. Apart from the copula *sum* 'be', other verbs denoting change of state, such as the one in example (10), may occur. The PR is restricted to persons, but garments are also possible possessors in a limited range of use (see example 11).

(10) *Taa se zacrvene vo liceto.*
she REF reddened- 3SG in face- DEF.SG.N
'She got red in the face.'

(11) *Košulata beše izlitena vo rakavite.*
shirt DEF.SG.F was-3SG faded- PARTIC.F in sleeves- DEF.PL
'The shirt was worn out in the sleeves.'

As in the Contact subgroup of EPCs, the part-whole image scheme is also at work here, ensuring a possessive relation to be inferred between the referent of the subject and the referent of the complement in the prepositional phrase. In fact, this inference is to a large extent contextually conditioned, as only body parts (and parts of garments) are permitted as PMs, and in particular such body parts usually associated with a particular feature that marks the person's physical or mental state. Clothing items, even when worn, are not permitted. Often the PM is conventionally implied in the attribute, and it is easily invoked even if it is not mentioned (*bušav* 'scruffy' describes hair, *bled* 'pale' implies face). The most common are, however, those attributes which may refer to several body parts, but have one or two most typically associated with them. For instance, while it is natural that a person's face, neck, ears or eyes may be, or may turn, red for various reasons, nonetheless, the first interpretation in example (12) *Toj pocrvene* 'He got red' would be the face. If the context does not make it sufficiently clear, the locative phrase would be necessary for clarification, as for example in (12), where we may feel a need to add *vo ušite* 'in the ears', *vo vratot* 'in the neck' or *vo nosot* 'in the nose'.

---

3 Another type of constructions which code the possessor in subject position are those with reflexive predicates, such as the sentences in (a) – (d) below. They code different types of events from those discussed in this paper.
(a) *Pejdž mi se dopagja vo kosata.* 'I like Paige's hair.'
(b) *Se krena na prsti.* 'He stood on his toes.'
(c) *Se previtka vo polovinata.* 'She bent over.'
(d) *Sredi se vo kosata.* 'Fix your hair.'

(12) *Toj pocrvene vo ušite (vratot / nosot).*
he reddened- 3SG in ears- DEF.PL neck- DEF.SG.M nose- DEF.SG.M
'His ears (the neck/the nose) got red.'

It is, thus, obvious that the PM in the locative adjunct does not have a critical contribution to the meaning of the utterance, as it serves only for additional specification and can easily be omitted. This provides evidence for the low level of grammaticalization of these constructions. According to Haspelmath (1999: 119) this type of EPCs are rare in the European linguistic area and, where present, are not used for properties attributed to the locative PM phrase if they cannot be attributed with the same meaning to the whole person. This is generally true for the analysed Macedonian constructions, but there are situations when this condition does not hold. Examples (13) and (14) illustrate this. In (13) *bev šarena* 'I was stained' would imply the whole body was affected and in (14) *Paltoto mi e dolgo* 'The coat is too long for me' would imply the main part of the coat is too long. Thus the locative PM phrase is necessary here.[4] In the same vein, we sometimes encounter examples which present totally unconventional combinations of attributes and body parts. Examples given in (15) confirm that this syntactic-semantic configuration is employed creatively by the speakers.

(13) *Cela bev šarena posle vo liceto.*
all was-1SG colourful afterwards in face-DEF.SG.N
'My face was all stained in various colours after that.'

(14) *Paltoto mi e dolgo vo rakavite.*
coat- DEF.SG.N IO.CL-1SG is long-SG.N in sleeves- DEF.PL
'The coat is too long in the sleeves for me.'

(15) a. *Dlaboko emotiven vo glasot, (toj ja preživuva pesnata od prviot do posledniot zbor).*
deeply emotional in voice-DEF.SG.M
'With his deeply emotional voice (he experiences the song from the first to the last word).'

b. *Vakvi decentralizirani vo umot (izleguvaat od pod sekoj kamen).*
such decentralized in mind- DEF.SG.M
'Such decentralised in their mind people (spring up from under every rock)'.

What is their relationship to the implicit possessive construction? First, they do not allow insertion of the possessive pronoun for explicit marking of possession, as shown in (16). It is felt as totally superfluous, redundant and unnecessary. If the same event is presented by using an IPC, the body part occupies the subject

4 Native speakers have also accepted *Toj e dolg vo nozete/ racete* Lit. 'He is long in the legs/arms.' although without the adjunct the predicate does not imply the same.

position, which shifts the focus to the PM and presents the situation from a different angle, as in example (17).

(16) *Što si mi nekako crven vo *tvoite oči/ očite?*
what are-2SG IO.CL-1SG somehow red in your- 2SG eyes eyes- DEF.PL
'Why are your eyes red?'

(17) *Tvoite oči se nekako crveni (i dalecni).*
your-2SG eyes are-3PL somehow red-PL and distant-PL
'Your eyes seem red and distant.'

2.3. *Possessor of location*

The constructions of the type *kaj čičko Milan vo dvorot* 'in uncle Milan's yard' (example 3) are syntactically and semantically different from the two types discussed so far. Both the PR and the PM are expressed in prepositional phrases that follow one another, the PR is the complement of the preposition *kaj* 'at' and the PM of *vo* 'in'. It seems that the order of the two phrases is not fixed, but the one with the PR preceding the PM sounds more basic. The preposition *kaj* has its origin in the Old Slavic place adverb and adverbial pronoun *kəde/konde.* This preposition is commonly used to express location of a state or direction of movement without any possessive implication (example 18). The phrase *kaj* + a noun denoting a person implies being at or going to a place that the person owns or normally occupies, so the additional mention of the place in a *vo*-construction may, if the place is of a suitable type, result in a possessive implication. In such contexts the possessive interpretation is only a conversational implicature. The example in (19) is ambiguous: it could mean either the room where my mother was or in my mother's room, or both.

(18) *Bev kaj brat mi vo London.*
was-1SG at brother IO.CL-1SG in London
'I was at my brother's in London.'

(19) *(Stanav i gledam nema nikoj.)*
Vlegov ***kaj majka mi vo soba****, gledam spie.*
enter-1SG at mother IO.CL-1SG in room see-1SG sleep-3SG
'(I got up and saw there was no one.) I got in my mother's room, saw she was sleeping. '

It becomes clearer that the possessive interpretation is becoming attached to the construction when the context indicates that the PR is not present at the designated place at the time of the event, as example (20) clearly demonstrates. The implication is strengthened when the place mentioned is not such where one would expect to find the person and/or the predicate does not express canonical direction or location, as in example (21). Example (22) demonstrates how the context can induce reinterpretation. Here both prepositional phrases have personal referents, but a person cannot be at two places at the same time, so the possessive implication is invited. This process, known as context-induced reinterpretation (Heine 2003:580), is the first step towards grammaticalization.

(20) *Otidovme **kaj Ognen vo stanot**, (no toj ne beše doma).*
went-1 PL at Ognen in flat- DEF.SG.M
'We went to Ognen's flat, (but he wasn't at home).'

(21) *Košnicata ja zedov od **kaj baba mi vo podrumot**.*
basket- DEF.SG.F DO.CL -3SG.F took-1SG from at grandma IO.CL-1SG in cellar- DEF.SG.M
'I took the basket from my grandma's cellar.'

(22) *Ednaš se zaboraviv igrajḱi **kaj drugarka mi vo komšiite**.*
once REF forgot-1SG playing at friend IO.CL-1SG in neighbours-DEF.PL
'Once I lost track of the time playing at my friend's neighbours.'

The Possessive Locative constructions with *kaj* are clearly an instance of possessor specification. The first and most important information they convey is that the person indicated, which is coded as the complement of *kaj,* is the more salient participant and that it is closely connected to some place. This place coded as complement of *vo* 'in' is only supplementary information, which can, in most cases, be omitted without radical consequences. In contexts where the possessive implication is stronger, the omission may cause misunderstanding, as for instance in example (22).

The obvious question that arises here is the relation of these constructions with those of explicit possession. The possessive pronoun can sometimes be used as modifier in the locational NP, explicitly stating the possessive relation which is otherwise implied. This is rare and usually occurs if the speaker wants to make sure that the relation is understood, or wants simply to emphasize it (as in example 23). The brother, nevertheless, remains the more salient participant. The IPC, as illustrated in example (24), puts the place, i.e. the PM, into focus, while the importance of the brother in the described situation is reduced.

(23) *Rabotam **kaj brat mi vo negovata producentska kuḱa**.*
work-1SG at brother IO.CL-1SG in his- DEF.SG.F production house
'I am working at my brother's production company. / I am working at my brother's in his production company.'

(24) *Rabotam vo **producentskata kuḱa na brat mi**.*
work-1SG at production- DEF.SG.F house of brother IO.CL-1SG
'I am working at my brother's production company.'

3. Comparison with the Possessive Dative

In all the situations described so far, the Possessive Dative construction is also an option. It may be revealing to examine the specific conditions that would favour one over the other construction. Let us first consider what they have in common. Both the Possessive Dative and the Possessive Locative constructions present a type of external possession, so they do not code the possessive relation explicitly, rather it is implied on the basis of certain extra-linguistic inferences that rely on the whole-part and/or reference point conceptual schema (Langacker 1993). The PR is presented as more important than the PM in the described situation. In some cases this is manifested in the PR being presented as the affected entity. However, the syntactic and pragmatic factors that help put the PR in the focus and leave the PM in the background are quite different.

The Possessive Locative constructions present the PR in higher syntactic roles, DO in those expressing contact, Subject in those expressing personal features, prepositional object in prepositional *kaj*-constructions, whereas the PM is coded as a locative adjunct. Coupled with the semantic superiority of the PR as a person, i.e. whole, this feature renders the PR as the more important participant. As these constructions have not been highly grammaticalized in Macedonian, they still reveal the fact that the PM is additional information and can easily be omitted. Only rarely can this lead to misunderstanding.

The Possessive Dative construction, however, presents the PR in the indirect object, which is often an external syntactic role, while the PM is coded as a core role, either as DO or Subject. Syntactically, it is the PR that presents additional information, therefore there must be other factors responsible for the pragmatic salience of the PR. Velasquez-Castillo (1999:95) argues that the Dative case that accommodates the PR in Spanish (which is parallel to that in Macedonian) is responsible for that.[5] It brings in the second human referent, the receiver of the consequences of the event, in the predication and introduces a "subjective construal of the body-part – PR relation" (Velasquez-Castillo 1999:95). The speaker presents the event from his/her point of view empathizing with the PR.

The differences pointed out are reflected in the different level of affectedness of the PR expressed in the Locative and in the Dative constructions and the speaker's implied relation to the PR. In the Possessive Locative constructions of contact, the focus on the PR affectedness is more emphasized than in the corresponding Possessive Dative construction (compare example 25a

---

5 First, it provides a means for preserving the whole-part asymmetry and coding a human PR who is higher on the empathy hierarchy, and second, "the general meaning of an indirect object includes two crucial aspects relevant to a subjective construal of the body-PR relation: a sense of affectedness that captures the empathic link between speaker and PR and a sense of possession which obviates the need for an explicit coding of the possessive relation ..." (Velasquez-Castillo 1999:95).

and 25b). In the Possessive Dative (example 25b) the PM is more pronounced and bears more importance in the situation than is the case in the locative construction (example 25a). In (25b), the belt is an important participant as the sentence is about some event affecting it. The PR is seen as affected only through his relation to the belt. PR affectedness is, thus, a subjective evaluation of the situation by the speaker. King (1998) (among others) has argued convincingly that the distinct function of the dative of possession is to express Empathy with the possessor.[6] Such a subjective view is not present in the Possessive Locative construction. It presents the PR as objectively affected.

(25) a. *Vojnikot go fati za remenot.*
soldier-DEF.SG.M DO.CL -3SG.M seized-3SG for belt-DEF.SG.M
'The soldier seized him on the belt.'

b. *Vojnikot mu go fati remenot.*
soldier-DEF.SG.M IO.CL-3SG.M DO.CL -3SG.M seized-3SG belt-DEF.SG.M
'The soldier seized his belt (lit.to him).'

We can see a similar distinction between the Possessive Locative expressing personal features and the corresponding Possessive Dative (compare examples 26a and 26b). In (26a) the feature is directly attributed to the person and the specification of the body part is marginal information. In (26b) the feature is attributed to the body part and, by introducing the PR in the indirect object, the speaker expresses his/her view that this concerns the PR. Thus, if a mother uttered (26a) upon seeing the child in such condition, it would imply the placing of responsibility on the child. She may be expected to add *Što si pravel?* 'What have you been doing?'. If she uttered (26b) it would express concern about what has happened to the child, the child being seen as exposed to some event. The more appropriate follow up question would be *Što ti se slučilo?* 'What has happened to you?'.

(26) a. *Što si taka crven vo liceto?*
what are-2SG so red -M in face-DEF.SG.N
'Why are you so red in the face?'

b. *Što ti e taka crveno liceto?*
what IO.CL-2SG is so red -N face-DEF.SG.N
'Why is your face so red?'

---

6 "When the speaker chooses the Dative over any other available construction, he expresses a certain kind of knowledge. He shows that he is aware of a possessive relationship between two referents, one of which is animate, and most likely human. The speaker also knows that the referent, [A], is physically or psychologically affected, either positively or negatively, by something that happened to or with the other referent [B]" (King 1998: 250-151).

The difference is even more pronounced in constructions expressing possession of location. In example (19) above, repeated in (27a), the possessor of the location is seen as involved in the situation without any particular additional implications, but if the same event is expressed with Possessive Dative, the PR is presented as being subjected to something unwanted and unpleasant. Thus in (27b) there is an inference that the room was entered against the PR's will or unexpectedly and created a negative experience for her.

(27) a. *Vlegov kaj majka mi vo soba.*
enter-1SG at mother IO.CL-1SG in room
'I got in my mother's room.'

b. *Í vlegov na majka mi vo soba.*
IO.CL-3SG.F entered-1SG to mother IO.CL-1SG in room
'I got in my mother's room.'

4. Conclusions

Possessive Locative constructions are an alternative way of expressing possession in Macedonian, whereby the PR is put into focus. They have limited range of use, each one being restricted to a particular type of possession and invoking a particular type of additional implications. The differences between the Possessive Dative and the Possessive Locative constructions described above can be explained by their different origin: the Possessive Dative is a result of the pragmatic function of PM specification, so the PM has a more important role there than in the Possessive Locatives which are product of PR specification, where the PM is additional information. The topicality of the PR in the Possessive Dative comes from its higher position on the empathy hierarchy (Velasquez-Castillo 1999:95) which manifests itself in the speaker's attitude towards the effects of the event on the PR.

References

Baron, Irene, Michael Herslund & Finn Sorensen (eds.). 2001. *Dimensions of Possession.* Amsterdam: John Benjamins Publishing Company.

Haspelmath, Martin. 1999. External Possession in European Areal Perspective. In Doris L. Payne & Immanuel Barshi (eds.), 109-136.

Heine, Bernd. 1997. *Possession: Cognitive sources, forces and grammaticalization.* Cambridge: Cambridge University Press.

Heine, Bernd. 2003. Grammaticalization. In Brian D. Joseph and Richard D. Janda (eds.), *A Handbook of Historical Linguistics*, 575-601. Wiley - Blackwell Publishing Ltd.

Herslund, Michael & Irene Baron. 2001. Introduction: Dimensions of possession. In Irene Baron, Michael Herslund & Finn Sorensen (eds.), 1-26.

King, Katerina P. 1998. *The Czech Dative of Interest: The Hierarchical Organization of Possession in Discourse and Pragmatics*. Cambridge, Massachusetts: Harvard University dissertation.

Langacker, Roland W. 1993. Reference -point constructions. *Cognitive Linguistics 4-1*. 1-38.

Mitkovska, Liljana. 2000. On the possessive interpretation of the indirect object in Macedonian. *Linguistica Silesiana* 21. 85-101.

Mitkovska, Liljana. 2005. *Izrazuvanje posesivnost na nivo na imenskata sintagma vo makedonskiot i vo angliskiot jazik.* [Expressing Possession on the Noun Phrase Level in Macedonian and in English]. Skopje: University "Ss Cyril and Methodius" dissertation.

Mitkovska, Liljana. 2007. Why Double Marking in Macedonian Dativus Sympatheticus. In Dagmar Divjak & Agata Kochańska (eds.), *Cognitive Paths into the Slavic Domain*, 55-79. Berlin, New York: Mouton de Gruyter

O'Connor, Mary Catherine. 2007. External possession and utterance interpretation: a crosslinguistic exploration. *Linguistics* 45-3. 577-613.

Payne, Doris L. & Immanuel Barshi (eds.). 1999. *External Possession*. Amsterdam/Philadelphia: John Benjamins Publishing Company.

Velasquez-Castillo, Maura. 1999. Body-Part EP Constructions: A Cognitive/Functional Analysis. In Payne L. Doris & Immanuel Barshi (eds.), 77-108.

Abbreviations

EPC – External Possessive Construction
IPC – Internal Possessive Construction
PM – Possessum
PR – Possessor
DO – Direct Object
NP – Noun Phrase

Abbreviations for Glosses

CL – clitic
DEF – definite article
DO – direct object
F – feminine
IO – indirect object
M – masculine
N – neuter
PARTIC – participle
PL – plural
REF – reflexive
SG – singular

II.
Cognitive semantics

Not seeing trees for wood: a case study of metonymy-induced polysemy in Germanic and Slavic languages

Mario Brdar, University of Osijek
Rita Brdar-Szabó, Eötvös Loránd University, Budapest

1. Introduction

Although the phenomenon of polysemy is ubiquitous in all natural languages – and relatively easy to recognize – its precise characterization is fraught with many difficulties. As far as the cognitive linguistic approach to polysemy is concerned, we note that in addition to studying fundamental questions that were inherited from traditional approaches, the research is now beginning to focus on polysemy as a problem of categorization as well as on the motivation for linking various meanings (cf. Geeraerts 1993; Tuggy 1993; Ibarretxe Antuñano 1999; Cuyckens and Zawada 2001; Ravin & Leacock 2002; Nerlich, Todd, Vimala and Clarke 2003).

It is well-known that one of the core assumptions in cognitive linguistics is that large portions of language are motivated by the facts of human embodiment (physical, cognitive and social), i.e. by how these are reflected in cognitive structures, primarily through mechanisms such as conceptual metaphors and metonymies. This also applies to research on polysemy. In particular, cognitive linguistics attempts to show how meanings of polysemous lexical units are linked by metaphor and metonymy.

If much of what we consider to be the central facts of human embodiment is shared by humans, and therefore universal, we should expect human languages to be, if not the same, then at least extremely similar. Cognitive linguists have "naturally" exhibited more interest in demonstrating cross-linguistic similarities. In actual fact, the cognitive linguistic success in uncovering all that "hidden" systematicity and universality was long (and still is) advertised as one of its major comparative advantages over other approaches. As far as the extension of meaning is concerned that leads to polysemy, we might expect that similar patterns of extension and meaning shift will be attested in a number of languages. In other words, on the basis of the complex systems of conceptual metaphors and metonymies (often in interaction with each other), linguists assemble a cognitively plausible picture.

It seems that unconditionally stressing the universal aspects of language, often based on conscious introspection and decontextualized data, which appears to have been necessary while the cognitive linguistic movement was establishing and profiling itself against the background of the formal-generative framework from which it emerged on the one hand, and emancipating itself from the then prevalent objectivist philosophical atmosphere in and around linguistics, has now gradually come to be felt to be a potential obstacle to the development of cognitive linguistics. The attempts to redress this situation and make cognitive linguistics more realistic with respect to variation were heralded by the rise of usage-based models as envisaged by Langacker (1987).

The range of types and forms of mental structures that can be assumed to characterize the linguistic knowledge of native speakers must be limited to those that are warranted by the actual language data that can be observed in one way or another, and for which convincing psychological, acquisitional, synchronic and diachronic accounts can be offered. In other words, the evidence that cognitive linguistics should rely on is not only introspective and based on the embodiment hypothesis, but should include more empirical evidence such as the results of psycholinguistic tests, language acquisition data, diachronic and syntactic data (the either elicited from native speakers, or corpus data). Ideally, various types of evidence should point in the same direction. A hypothesis built on the basis of the facts of human embodiment should thus be corroborated by actual usage data, possibly by more than one type.

A possible problem is that we may be lacking certain types of evidence, but this can be relatively easily remedied, either by data supplied subsequently, or by looking at analogous existing data (i.e. considering the evidence available for a related linguistic phenomenon). Of course, the lack of diachronic evidence may in some cases be an irreparable problem.

However, it is possible to envisage an even worse scenario: when data are apparently in conflict with an otherwise conceptually plausible hypothesis based on embodiment, and perhaps even in conflict with some other set of data. The central methodological issue in the present paper is precisely of this type. In what follows we will demonstrate this in a case study on an instance of metonymy-induced polysemy in which diachronic and (anthropological) typological evidence appear to be in conflict with what seem to be an otherwise conceptually plausible hypothesis. The case study has to do with the polysemy of terms for 'tree' and 'wood' in a variety of languages, but especially in Germanic and Slavic languages.

The case study is organized as follows. After sketching a plausible hypothesis based on metonymy as one of the cognitive processes lying at the core of the embodiment hypothesis, and comparing trees and wood with other analogous cases, such as the phenomenon of so-called "animal grinding" with particular reference to Slavic languages, we review some empirical data presented in literature that does not square with our intuitively satisfying cognitive account. Specifically, we consider diachronic, anthropological/typological data, and critically examine the path of semantic change that Brown and Witkowski propose. In the final part of the case study, we return to the issue of the relative conceptual saliency of 'tree' and 'wood.'

What is at stake is of course not only the "correct", more plausible or realistic account of the polysemy in case, but also a "higher" methodological lesson. What we advocate is a step by step, or cyclic examination of both data and the hypothesis. First, we propose to consider a wider range of data in as meticulous a way as possible, and then if necessary, refine and upgrade the initial hypothesis so that it fits the data. The theoretical construct resulting from this process may of course be occasionally less elegant than the initial one, but this is of course unproblematic in a usage-based model of language.

The term was coined and first used by Langacker (1987:494). It was described there as giving substantial importance to:

> … the actual use of the linguistic system and a speaker's knowledge of this use; the grammar is held responsible for a speaker's knowledge of the full range of linguistic conventions, regardless of whether these conventions can be subsumed under more general statements.

Characterizing cognitive grammar as a usage-based model language structure, Langacker (1988:131) states that it differs from the generative tradition by being:

- maximalist
- non-reductive, and
- bottom-up oriented.

A maximalist view means here that language is seen as a massive, highly redundant inventory of conventional units. The units themselves can be anything from very general (or abstract) to utterly special and idiosyncratic. A usage-based model is an attempt to approach language in a nonreductive way by employing "fully articulated schematic networks and emphasizes the importance of low-level schemas." (ibid.)

The bottom-up approach means that cognitive grammar has a vested interest in investigating specific instances of patterns and not only high-level generalizations. In fact, the latter can only be extracted on the basis of a large amount of the former.

However, we can also interpret these three as applying in a more general sense. A most realistic description of linguistic phenomena need not always be the most elegant one, but can occasionally be quite redundant. Non-reductiveness can also be seen not only as the property of descriptive constructs but also as a property of the description itself in the sense that data cannot be arbitrarily excluded just because they would complicate the description, i.e. make it redundant. Finally, the description is bottom-up oriented in the sense that explanations should be made to agree with the data, and not vice versa.

2. Wood and trees: The case study on metonymy-induced polysemy in Germanic and Slavic languages

As a starting point in this case study, we may take the observations by Koch (2001: 1154), who sums up the possibilities for the polysemy in the frame FOREST/WOODS – TREE – WOOD as follows:

Table 1. Lexicalization patterns in the frame FOREST/WOODS - TREE - WOOD

		1: FOREST/WOODS	3: TREE	2: WOOD
Type A	3 different lexemes	Lat. *silva*	Lat. *lignum*	Lat. *arbor*
Type B	polysemy 1 – 3	Fr. *bois*	Fr. *bois*	Fr. *arbre*
Type C	polysemy 2 – 3	Russ. *l'es*	Russ. *d'er'evo*	Russ. *d'er'evo*

Languages belonging to Type A, like Latin, German (*Wald* 'woods, forest' – *Baum* 'tree' – *Holz* 'wood'), or Turkish (*koru* 'woods' - *odun* 'tree' – *ağaç* 'wood'), avoid polysemy altogether. Type B languages, like French, have a metonymic polysemy linking concepts 1 and 3. Among languages of Type C, which have a metonymic polysemy relating concept 2 and concept 3, in addition to Russian, we find Croatian (*šuma* 'forest' – *drvo* 'tree/wood') and Hungarian (*erdő* 'forest' – *fa* 'tree/wood'). A fourth type, with a triple metonymic polysemy, which is according to Koch (2001:1154) extremely rare may be instantiated by the Old Irish marginal lexeme *fid*.

Though English apparently has three different lexemes and thus belongs to type A, it must be noted, however that English in addition to keep the three concepts distinct by referring to FOREST by a distinct, borrowed lexeme, i.e. *forest*, probably coming from *forestis silva*, it also uses the lexeme *wood* marked as a plural. This means that we have a fifth type of polysemy (relating concepts 1 and 2), a type not noted by Koch.

Similarly, Slovene also has three different lexemes: *gozd* 'woods', *drevo* 'tree', and *les* 'wood', but *les* can also be used in the sense of 'woods' as well, though it is then often in the plural (e.g. *mračni lesi* 'dark woods', *iti po drva v les* 'go into the woods for wood' (cf. http://bos.zrc-sazu.si/). This means that Slovene is split between type A and a type not listed by Koch, where the metonymic polysemy obtains between concepts 1 and 2.

In this case study we focus on languages that belong to type C, i.e. those that exhibit the polysemy pattern 2 – 3 in using one and the same lexical item to refer to TREE and to WOOD. As a cognitively plausible story of how this type of polysemy arises, we can suggest a universally valid metonymic mapping of the type WHOLE FOR PART, whereby the whole plant, i.e. tree, can also be used to refer to one of its salient "parts", i.e. wood as the material gained from trees.

Witkowski, Brown and Chase (1981) claim in a series of studies that approximately two thirds of the languages of the world (their sample contains 66 languages all over the world) polysemously designate WOOD and TREE by a single lexical item and thus belong to type C. What is more, contrary to what we have just presented as a cognitively plausible scenario of the rise of metonymy-induced polysemy, they suggest that the concept TREE is more recent in the world's languages (although less so than other general plant classes), and that "languages either did not encode 'tree' in past millennia or if they did, it was typically encoded only as a low salience category by widening referential application of a term for 'wood'." In other words, both meanings were united for some time under one of these. Specifically, the claim is that they were united under the term originally used only for WOOD. It follows that the cases of 'tree/wood' polysemy are due to the metonymic extension of the type PART FOR WHOLE from WOOD to TREE. True, they also concede that

> [d]ocumentation of extensive 'wood'/'tree' polysemy alone is not, of course, certain indication that 'tree' usually develops through referential expansion from 'wood.' It is just as indicative of the reverse possibility, that 'wood' develops from 'tree.' Witkowski, Brown and Chase (1981:3)

However, they insist on the former rather than latter interpretation, in spite of some of their evidence that may be suspicious. We think that, as will become clear in the paragraphs that follow, both possibilities are plausible, but that imposing the former interpretation on Germanic and Slavic languages (and probably on many other Indo-European languages) does not square with actual data.

Before we take a closer look at Witkowski, Brown and Chase's evidence, let us explain why we think that a cognitively plausible account of this type of polysemy arises is based on a metonymic mapping of the type WHOLE FOR PART, whereby the whole plant, i.e. tree, is used to refer to one of its salient "parts", i.e. wood as the material gained from trees.We propose to consider specific types of trees. Moving to a more specific level from concept 3 above, we come to ways of referring to the wood of certain tree types, i.e. as material. A major division might be of course into fruit-trees and the rest. Let us first take a look at non-fruit-trees.

Spanish relies on metonymy, because one and the same label can be used to refer to the treee and its wood. Consider the following dictionary entries for *encina* 'oak' and *haya* 'beech':

(1) encina
1 Árbol de tronco grueso y corteza grisácea...
2 Madera de esta árbol: *En su habitación tiene un armario de encina.*

(2) haya
1 Árbol de gran altura, con y lisotronco grueso, ramas altas que forman una copa redonda y espesa, hojas alargadas, de punta aguda y borde dentado…
2 Madera de esta árbol: *Tengo una cofra de haya que no pesa nada y es muy resistente.*

English can either add *–wood* to the name of the tree to form a compound and thus reduce polysemy, or make use of metonymic polysemy reclassifying the plant to mass material noun:

(3) Our ranges include modern and traditional styles *made from oak* and *pine.*
(4) Wands were made from *oakwood* from trees which had been struck by lightning in the belief that lightning would not strike twice in the same place, ...

In German and Hungarian alike we witness a distributed use of metonymy or compounds with *–holz* and *–fa*, respectively.

We find a systematic pattern of suffixation instead of the metonymic use of the noun in Croatian with names of trees used as material. Whereas *oak* could in English be used to refer to the plant, i.e. oaktree, the wood of that tree, and objects made of its wood, the wood of that tree is usually referred to in Croatian using a suffixed noun, *hrastovina*, while *hrast* normally means just the tree, though it can be in appropriate contrastive contexts used to denote the wood as well. There is again a whole series of analogous suffixations:

(5) bukovina 'wood of beech', brezovina 'wood of birch', borovina 'wood of pine', orahovina 'wood of nut tree', cedrovina 'wood of cedar', vrbovina 'wood of willow'

Russian presents us with a mixed picture. Tree names can be used metonymically to denote the material, particularly in wood processing/carpentry context:

(6) Однако *паркет из бука* очень чувствителен к смене влажности, поэтому изделия из него надо покрывать лаком.
'However, beech parquetry (literally parquetry from beech) is very sensitive to changes in humidity, and therefore products made out it [beech] must be varnished.'

Otherwise, we also find adjectives derived from the tree names used attributively in front of the noun *древесина* 'wood':

(7) *Дубовая древесина* хорошо поддается окраске.
'Oak wood paints well.'

The noun *древесина* 'wood' can also be postmodified by the name of the tree in the genitive:

(8) *Древесина клена* легко режется и полируется.
'The wood of field maple can be easily cut and polished.'

Note that the noun is itself a suffixation from *дерево* 'wood' (cf. Table 1). The same suffix is found in some isolated formations like *сливина* 'plumwood.'

Slovene is again very close to Croatian. It has a whole series of suffixations in *-ovina* denoting the wood of various tree species:

(9) smreka 'pine' - smrekovina 'pinewood', hrast 'oak' – hrastovina 'oakwood', bukva 'beech' – bukovina 'beechwood', lipa 'linden tree' – lipovina 'linden tree wood'

Some of these are, however, polysemous. *Smrekovina*, in addition to wood as material, can also be used to refer to furniture made from that wood. *Hrastovina*, in addition to being used to refer to wood and furniture, can also be used to refer to the forest. *Bukovina* is used for wood and forest.

There are also metonymic conversions in appropriate contexts, following the preposition *iz* 'from':

(10) Tradicionalno je izdelana *iz hrasta.*
traditionally AUX made from oak
'It is traditionally made of oak'

In Polish, we find that in addition to some metonymic polysemy (e.g. *sosna* can mean 'pine', but also 'the wood of pine', as shown by the entry *drewno tego drzewa* (c.f. http://www.sjp.pl/sosna) and phrases of the type "wood + tree$_{GEN}$," e.g. *drzewo brzozowe; drewno brzozowe* 'the wood of birch', there are also some special suffixations in *–(n)ina/yna*:

(11) buk 'beech' – buczyna 'beech wood', lipa 'linden tree' – lipina 'linden tree wood', sosna 'pine' – sośnina 'pine wood', klon 'maple' – klonina 'maple wood'

Two important facts emerge from these examples. First, if we concentrate on specific tree types in Slavic languages, we see that in many cases the lexeme used to denote the wood from that tree type is morphologically complex, i.e. it is derived by some suffix from the lexeme used to denote tree as a plant. In other words, polysemy is resolved in such a way that it indicates that the WOOD concept is secondary to the TREE concept. Second, in Russian the same suffixation pattern is also found at the basic level term, i.e. with *древесина*, which is clearly derived from *дерево*, i.e. again the WOOD concept is secondary to the TREE concept.

We can also compare this with the situation with analogous cases of polysemy linking an object concept with a material concept. Conceptually very close to the case of TREE/WOOD polysemy, we find the ANIMAL/MEAT polysemy.

The term "animal grinding" has been used to refer to specific cases of the phenomenon variably called logical metonymy, logical polysemy, or regular polysemy, i.e. to those cases in which one and the same label can be used to refer holistically to the animal species or specimen as well as to the flesh of the animal in question, not necessarily conceived of as foodstuff, i.e. as meat of that animal, as illustrated in the following examples:

(12) a. "I would not eat *cat*," he murmurs.
b. Henry VIII was the first English King to enjoy *turkey* and Edward VII made turkey eating fashionable at Christmas.

In cognitive linguistics this is again treated as a subtype of WHOLE FOR PART metonymy, specifically OBJECT-FOR-MATERIAL-CONSTITUTING-THE-OBJECT within the *Constitution ICM* (Radden and Kövecses 1999: 32). The lexical item labelling the concept of the whole animal stands here only for a particular aspect of the whole animal, i.e. its bodily substance/flesh/meat as processed and used as foodstuff. Even literally, the substance that we use as food is only part of the whole animal's body, as animals are skinned, boned, etc., and usually it is not the whole carcass that is meant, but rather some smaller portion of it.

The noun is now reclassified in grammatical terms as a concrete, but non-count, or mass noun (this phenomenon is called partial conversion in Quirk et al. 1985). This contrasts with the majority of animal denoting nouns used in their primary sense, where they are countable, capable of overtly indicating plural at both token and type/species level (though there are some notorious exceptions, e.g. *fish*, which is capable of taking plural only at the type/species level). Their countable status is indicated by the fact that they may take the whole range of determiners, indefinite, definite and zero. Replacing the verb *eat* by a verb such as *kill* in (13) a. and b., would result in *cat* taking a definite or an indefinite article, and *turkey* probably taking the indefinite article:

(13) a. "I would not kill *a/the cat*," he murmurs.
b. We did not always kill *a turkey* for Christmas dinner.

Although this metonymy may appear fairly productive as far as English is concerned, which is also suggested by terms used in some unification frameworks, such logical metonymy/polysemy, or regular polysemy, an examination of corpus data reveals a slightly different picture. While this sort of conceptual conversion is certainly more or less always available in theory as an open pattern, the fact is that it is not so regularly made use of. Of course, one of the factors diminishing the productivity of this metonymy is the well-known historical incident in the course of which a number of lexical items were borrowed from Norman French that denoted the meat of certain domestic and wild animals, thus effectively blocking the polysemy from kicking in in the case of a series of native Anglo-Saxon items:

(14) cow – beef
calf – veal
pig – pork
sheep – mutton
deer – venison

Apparently, English uses lexical substitution as a way of getting around metonymyic polysemy, but metonymy can be found at a more specific level. Depending on its age, an instance of the sheep species may be classified as *lamb*, *hogget*, or as *sheep*. Under one year it is called *lamb*, and the same lexeme is used to refer to its meat. *Hogget* is a sheep of one year, and the lexeme can be used not only to cover the meat of this animal, but also to refer to its wool. The same item, *hogget*, is, according to the Oxford English Dictionary (OED), also used to denote a young boar of the second year, and a year-old colt.

It is also interesting to point out that the lexical item *mutton* is used in some parts of the English-speaking world, such as the Caribbean islands or some parts of Asia, and in particular in India, to cover both goat and lamb meat. *Chevon* is another lexical item sometimes used to refer to goat meat.

As for other animals whose meat is (or was) regularly eaten by humans, we note that although OED does not indicate that *horse* can be used metonymically to refer to meat, some examples can be found on the Google, though we almost consistently find only *horse meat*:

(15) Gordon Ramsay is to shatter the last taboo of English cuisine by urging the public to eat *horse meat*. (Telegraph, May 7, 2007)

(16) The celebrity chef we all love to hate, Gordon Ramsay, is causing some waves. The chef is saying that the British should eat more *horse*, this on the tail-end of the Kentucky Derby.
Ramsay, who recently served up *horse* on his new TV show, said it was "slightly gamey" & "packed with protein."
While it is not illegal to eat *horse meat* in the UK, it has not widely been eat[en] since the 1930s. (http://www.bfeedme.com/celebrity-chef-gordon-ramsay-let-them-eat-horse/)

To sum up, what we have seen in English in the case of the culturally most salient animals whose meat is used as food boils down to the following: polysemy is blocked due to language contact facts. In many other cases, we have a replacement by a N + N combination, with *meat* as the second noun.

As might have been expected, French exhibits regular metonymy in the case of the exceptional items listed for English in (15) above. *Mouton* is used, similarly to English *hogget*, to refer to sheep, its meat, and also to sheepskin.

Vache in addition to functioning as the counterpart of *cow* is also used to refer to 'cowhide', *boeuf*, on the other hand, not only means 'beef' but also 'ox/steer' (note also the use of both *viande de vache* 'meat of cow', and *viande de boeuf* 'meat of ox'). *Porc* denotes both 'pig' and 'pork' (next to *viande de porc*, 'meat of pig'). Finally, *veau* denotes both 'calf' and 'veal' (next to *viande de veau* 'meat of calf').

The situation in Spanish is in one respect quite close to what we have established for French. Nouns used to denote animals can also be used metonymically to refer to their meat, but they can also be preceded by *carne de* 'meat of' (endocentric compounds being relatively rare in Spanish):

(16) (carne de) cerdo '(meat of) pig', cordero '(meat of) lamb'
(17) a. El medico me ha prohibido el *cerdo*.
'The doctor forbade me to eat pork.'
b. Me encanta el *cordero asado*.
'I love roast lamb.'

This pattern of the replacement of metonymy by means of N + *meat* combination, realized as a compound noun, is quite wide-spread in German (with *Fleisch* 'meat' as the second constituent, i.e. as the compound head):

(18) Schweinefleisch 'pig-meat', Rindfleisch 'cattle-meat', Schafsfleisch 'sheep-meat', Ziegenfleisch 'goat-meat'

However, names of some animals can also be used metonymically in German to refer to their processed meat. Certainly, these are even more acceptable if we have an ostensive context in which a dish made out of the meat of one of these animals is contrasted with dishes made of the meat of other animals.

Danish and Swedish employ the same N + N compound strategy with *kød* 'meat', and *kött* 'meat', respectively:

Danish:

(19) svin 'pig' - svinekød 'pig-meat', oks 'ox' – oksekød 'ox/cow-meat, beef', kalv 'calf' – kalvekød 'calf-meat/veal', kalkun 'turkey'– kalkunkød 'turkey-meat', kylling 'chicken' – kyllingkød 'chicken-meat'

Swedish:

(20) får 'sheep' – fårkött 'sheep-meat/mutton', gris 'pig' – griskött 'pig-meat/pork', kalv 'calf' – kalvkött 'calf-meat/veal', kalkon 'turkey' - kalkonkött 'turkey-meat'

Hungarian exhibits the same behaviour, i.e. it has compounds with *hús* 'meat' as the second constituent, i.e. as the compound head:

(21) sertéshús 'pig-meat', marhahús 'cattle-meat', birkahús 'sheep-meat', kecskehús 'goat-meat'

A language such as Croatian, which unlike Germanic languages or Hungarian, as good as lacks compounding as a word formation process, resorts to other syntactic or morphological means. First of all, for all cases of animals that are, culturally speaking, less usual or unusual as potential foodstuffs, there is the phrasal expression *meso (od)* X_{gen} 'meat of X'. However, the most important strategy is suffixation, with a cluster of related suffixes, *-ina*, -V*tina* and –V*vina*, used to derive names of meat of various animals:

(22) a. govedina 'beef'
b. svinjetina 'pork', prasetina 'piglet meat', teletina 'veal', ovčetina 'mutton', koz(l)etina 'goat meat', guščetina 'goose meat', piletina 'chicken meat', konjetina 'horse meat', jeletina 'deer meat', jaretina 'kid meat', janjentina 'lamb meat', zečetina 'rabbit meat', nojetina 'ostrich meat'
c. tunjevina 'tuna meat', veprovina 'boar meat', kitovina 'whale meat', jelenovina 'deer meat'

It is interesting to note that, if need be, names for the meat of almost any animal can be coined by using one of these suffixes, e.g. *tigrovina* 'meat of a tiger', *zmijetina* 'meat of a snake', *slonetina* "meat of an elephant', i.e. it is at least as productive as reclassification in English.

Metonymic conversion seems, however, to be more acceptable in Croatian in the case of (smaller) fish:

(23) Zatim obožavam *smuđa*. Mogla bih ga svaki dan jesti.
'Then I adore pikeperch. I could eat it every day.'

The metonymic conversion is more viable here because it is difficult to conceptualize the flesh, i.e. the meat of smaller fish types separately from the whole animal. In view of the size of chunks of their flesh that are available and which still bear some relatively obvious resemblance to the fairly well-articulated body of the whole fish, they simply fail to qualify as amorphous enough, which is a requirement for their reclassification as mass nouns. In other words, their meat is not considered to be different enough to be conceptually divorced from the animal name by means of suffixations in Croatian. Russian data below clearly show that the size of the fish is decisive here. However, it also transpires that the cut-off point for the choice between suffixation and metonymic conversion seems to be subject to cross-linguistic parametrization.

Slovene, Polish and Russian are very similar to Croatian: they use the same suffixation pattern to replace metonymic conversion. As can be seen from the examples below, the suffixes are apparently cognate with the Croatian ones: Slovene:

(24) svinja 'pig' – svinina 'pork', prase 'piglet' – prasetina 'piglet meat', jagnje 'lamb – jagnjetina 'lamb meat', tele 'calf' – teletina 'veal', ovca 'sheep' – ovčetina 'mutton'

Polish:

(25) owca/baran(ek) 'sheep' – baranina 'mutton', wieprz/świnia 'pig' – wieprzowina/świnina 'pork', cielę 'calf' – cielęcina 'veal', źrebię 'colt' - źrebięcina 'colt meat'

Russian:

(26) баранина (baranina) 'lamb meat', свинина (svinina) 'pork', говядина (govjadina) 'beef', телятина (teljatina) 'veal', курятина (kurjatina) 'chicken meat' палтусина (paltusina) 'halibut meat',

In addition to these, Russian exhibits a whole range of suffixations in *–(ов)ина* (-(ov)ina) denoting meat of various saltwater and freshwater fish types, in (27) a., and and for whale, as in (27) b.:

(27) a. сомовина (from *сом* (som) 'catfish'), лососина (from *лосос* (losos) 'salmon'), осетрина (from *осетр* (osyatr) 'European sea sturgeon/Baltic sturgeon (Acipenser sturio)), севрюжина (from севрюг (sevryug) 'starry sturgeon'), белужина (from *белуга* (beluga) 'European sturgeon, beluga (Huso huso)'), сиговина, (from *сиг* (sig) 'whitefish'), щучина (from *щука* (ščuka) '(northern) pike'), палтусина (from палтус (paltus) 'halibut')
b. китятина (from кит (kit) 'whale')

It should be noted that they are all fish that can reach notable size, e.g. beluga can grow up to 6m in size, while a catfish can be up to 3m and over 150 kilos. Halibut weighs on average between 11 and 14 kilos, but it can grow to weigh as much as 190 kilos.

As an interim conclusion, we can remark that Slavic languages again resolve the potential polysemy in a fairly similar way, by adding a suffix which is very similar to the suffix used in the case of trees and wood. It follows again that the object concept is primary, and the material concept (i.e. meat/flesh) secondary.

Returning to the realm of plants, we just point to a single English example pointing to a similar conclusion. *Flower* and *flour* are often used as stock examples of homonymy, i.e. as two nouns that have the same (phonological) form, but are not related in terms of their meaning. However, on closer inspection, it turns out that they are historically related in the similar way as tree and wood, i.e. by means of a metaphor and the PART FOR WHOLE metonymy. According to OED, flour is originally the "flower" or finest quality of meal; hence, the finer portion of meal (whether from wheat or other grain) which is separated by bolting. Also, in modern use, the ordinary name for the meal or farina of wheat as opposed to that obtained from other grain. "Flower", in turn, is "the best, choicest, most attractive or desirable part or product of anything, material or immaterial; the essence, quintessence; also 'the gist' (of a matter)."

Let us now consider what evidence Brown, Witkowski and Chase present in support of their account. We shall start from the diachronic (and etymological) evidence, then proceed towards considering the path of semantic change leading to polysemy and its resolution as they see it. The diachronic strand of their argumentation is mostly based on reconstructions of protolanguages from which several groups of genetically related modern languages developed. They quote as the best explored case Proto-Uto-Aztecan (of 4000-5000 years ago), ancestral to the Uto-Aztecan language family of central and northern Mexico and the western United States. Fowler (1972), a study of botanical terms in languages of the Numic branch of Uto-Aztecan also brings a large number of specific tree and plant names reconstructed for Proto-Numic. However, she found little evidence supporting reconstruction of botanical life-form terms. Specifically, she notes that the concept of tree was most likely not encoded at all.

Some contemporary Uto-Aztecan languages in central-northern Mexico that do not belong to the Numic branch do not apparently encode this concept either, e.g. Huichol (cf. Price 1967). Similar comparative investigations also suggest that Proto-Austronesian, approximatively spoken 5000 or more years ago, lacked a lexeme for the 'tree' life-form (cf. Blust 1974).

As far as English is concerned, Brown and Witkowski (1981) state that the contemporary English word *tree* designated both TREE and WOOD in Old English (*trēow*) and in Middle English (*tre*). Apparently, the modern word *tree* does not carry the latter meaning, which is now expressed by the lexeme *wood.* Although they implicitly count English among those languages in which the concept TREE is more recent than WOOD, the fact that they cite, i.e. that the two meanings were united at a point, does not constitute real evidence in support of their claim. In fact, if we should rely on the records in OED the TREE concept was encoded by the lexeme in question slightly before the WOOD concept. The first three entries for the former stem from 825, 890 and 897, while the first three records for the latter are from 890, 1122 and 1290. The same appears to be the case with *wood* (Old English *widu*, and later *wodu*).

What they claim to have happened is that the host lexeme, originally designating just WOOD, was at some point made more versatile in that it also came to express another, related concept, i.e. TREE. The original concept of the host lexeme was eventually dropped altogether, and the lexeme continued to denote only the grafted concept:

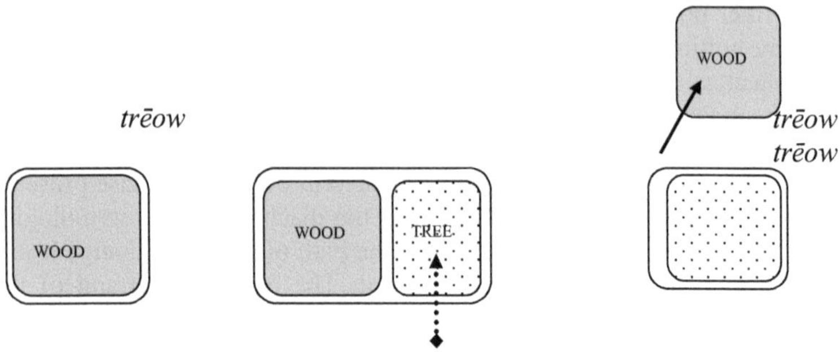

Figure 1. The path of the semantic change with *trēow*

The path of semantic change that Brown, Witkowski and Chase propose does not look very natural. It rather appears that we are dealing here with a case of semantic expansion from a specific tree to trees in general, and not from wood to specific tree, and then to trees in general. The third possibility, to have a specific simplex word for the wood of a specific tree, which would then be expanded to cover all trees, would be extremely odd as we might then expect that there should be, if not a specific word for the wood of every tree type, but then at least several such words for wood.

However, we may take the lead from Brown, Witkowski and Chase and proceed further into the past. They point out that:

> Incidentally, English *tree* derives from a Proto-Germanic form which also meant 'wood'/'tree.' This form in turn derives from a Proto-IndoEuropean antecedent which meant either 'wood'/'tree' or 'oak' (Friedrich 1970: 143-9).
> Brown, Witkowski and Chase (1981:8)

Indeed, etymological sources agree that the word can be traced back from PIE **deru-/*doru-* "oak." There is no doubt that oak had special status in the mythology of many ethnic groups, cf. Welsh *derwen* "oak," Albanian *drusk* "oak". Similarly, *dub*, an archaic Croatian word for tree in general, was originally used to designate a type of oak.

For the Tarahumara language, a member of the Uto-Aztecan language family, Brown, Witkowski and Chase cite several reports indicating that there is no widely shared term for 'tree,' but also some studies reporting variation in the occasional use of certain terms for that concept (Thord-Gray 1955). Town dwelling Tarahumara speakers sometimes employ the *termgusi,* derived from *gu* 'wood' and a suffixing element *-si,* in reference to 'tree.' However, the same study notes that in the countryside speakers sometimes extend the referential range of words for locally important types of tree, such as 'ocote pine,' or the word for a type of 'oak,' which are extended to designate 'tree in general.'

The same pattern has been observed for Indo-European languages above, and this cannot be pure co-incicdence. In fact this is one of the natural paths of semantic extension/expansion that can be expected within a usage-based model, whereby a prototypical instance undergoes schematization and comes to occupy the basic-level category status. Some specific trees are apparently more salient than the rest for reasons that are culture-dependent and body-based, as the basic-level is the level at which humans can optimally interact with their environment.

Whether the TREE or the WOOD concept will be more basic, and serve as the starting point for a possible expansion incorporating the other concept, therefore depends on the human environment and their interaction with it. In principle, as we hope to show in Part 3, there is no a priori reason why only one should be possible or more natural, as can be implicitly concluded from Brown, Witkowski and Chase. The point of our arguments, mostly based on evidence from Germanic and Slavic languages, is not to disprove their claims, but just to show that the opposite is also possible, what is more, that it is more plausible in these latter languages. In the remaining part of this paper we provide an account of the motivation for the paths of semantic expansion based on embodiment in the broadest sense of the term.

3. Motivating the path of semantic expansion: from the physical environment to societal complexity

To sum up Brown, Witowski and Chase's argument, "while 'tree' and other botanical life-forms appear to be relatively recent additions to certain lexicons, 'wood' has probably always been a salient category." It is certainly true that a word for the material, i.e. WOOD, is extremely useful for people using this material for certain purposes. But as soon as we start thinking about the variety of purposes to which wood can be put, we are bound to start doubting the soundness of Brown, Witkowski and Chase's argument.

People using wood for fuel are less likely to be concerned with the type of tree this material comes from than people also using this material as tools and for construction purposes. The latter people would certainly be quite interested in whether the wood in question is a hard-type or a soft one, and the distinction can only be sensibly made if one can also distinguish between various types of trees. The degree of their interaction with the environment also counts, people less actively interacting with their environment in search for wood as fuel (and also perhaps as the material for tools and for construction), i.e. people who just tend to collect some suitable pieces of wood would also be less likely to possess extensive knowledge of types of trees.

This is now also where the type of human environment in general comes into the picture. Specifically, people living in environments such as deserts, steppes, or jungles can be expected to have reasonably different sorts of concepts of WOOD and TREE, the most important factor being the diversity of trees. In areas both largely lacking trees (such as deserts), and in those abounding in them (e.g. jungles), one is more likely to be indifferent to specific types of trees than in an environment in which wooded areas are separated by pastures, fields, and cleared areas in general.

Another environmental aspect of the WOOD/TREE distinction, making trees far more salient, concerns the role of trees in orientation. They are extremely useful as reference points in space in certain types of environments, specifically in those exhibiting a variety of trees (in which types of trees are more likely to be distinguished). On the other hand, referring to individual trees (for which the TREE concept would be a precondition) in the process of orienting oneself or others in a jungle is of doubtful value.

We certainly do not intend to question Witkowski, Brown and Chase's language sample, but let us just remark in passing here that thinking just in terms of the absolute number of languages exhibiting a certain trait, as is often done in linguistic typology, without taking into account the number of speakers, may lead to a certain degree of misrepresentation. As our specific example seems to indicate, this danger may be particularly acute in lexical typology, where the type of environment significantly determines conceptual structures as far as realia are concerned. Certain types of natural environments, such as deserts or jungles, are an obvious barrier for the development of large-scale societies. In other words, we will find a higher number of small-scale communities, separated by natural barriers. A highly likely linguistic outcome of this is a higher number of different (though perhaps related) languages spoken by a smaller number of speakers.

This may be coupled or paralleled with the lifestyle of the people in question in the sense that we can distinguish between people that settle in a more or less fixed area and predominantly raise stock on the one hand, and hunter-gatherers, on the other. In terms of the certainty with which they can count on the supply of food in their immediate environment, the former are quite likely to stay for a longer period of time in one place, and because their activity is intensive, they require less space. This means that communities that engage in the same type of activity can co-exist quite close to each other. One of the consequences of this is also a larger linguistic community (though possibly with dialectal variation). Hunter-gatherers, on the other hand, due to the unpredictability of the outcome of their hunting activity, may be forced to move from one place to another, and require more space to sustain themselves. Consequently, communities that engage in this type of activity cannot co-exist very close to each other, i.e. they are more likely to be isolated from other groups for longer periods of time. A linguistic outcome is a higher degree of diversity, i.e. less mutual intelligibility, and in the long run, more languages.

As far as our pair of concepts is concerned, i.e. WOOD and TREE, it is now easy to see a possible danger, smaller languages spoken in areas in which it is not so useful (or even impossible) to distinguish various tree types may actually outnumber languages spoken by a large number of speakers, thus producing a tweaked typological (and anthropological) picture.[1]

From what we have said above, we may expect not only that there will be a correlation between the type of environment and the path of semantic expansion concerning the WOOD/TREE polysemy, but that a similar correlation will also hold with respect to some other environment-related concepts. Further, we may also expect that there will be a correlation between these paths of semantic expansion and the degree of societal complexity, as environment motivates here the path of semantic expansion.

As for the latter correlation, it is clearly shown in Brown, Witkowski and Chase (1981:8) that some such a correlation indeed exists.They use Marsh's (1967) scale of societal complexity in terms of size and integration of political units and degree of social stratification. In his system, a primary scale ranging from 0 (low) to 7 (high) applies to societal units other than contemporary national societies; the latter range from 8.6 to 109.4 (highest). When the sixty-six languages in Brown, Witkowski and Chase's sample are matched with the societal complexity scores from Marsh's index and divided into those languages in which the WOOD/TREE polysemy is present or absent, the following picture obtains:

Table 2. Association between societal complexity and the occurrence of 'wood'/'tree' polysemy (Brown, Witkowski and Chase 1981:8)

societal complexity	polysemy	
	Present	absent
high (above 8)	8	15
low (0-7)	36	7
gamma = .81	P<.001	N=66

We can further hypothesize that as far as those languages are concerned in which the polysemy is present, the semantic path SPECIFIC TREE → TREE → WOOD will be associated with a higher score for societal complexity, and the path WOOD → TREE with a lower score. The first part of the hypothesis is confirmed by Russian

1 Our intention here is certainly not to plead for a sort of linguistic neo-imperialism, but just to warn of potential dangers of utterly disregarding the number of speakers of languages exhibiting a certain typological feature. In this context, we point out to the weight of Slavic languages, although they do not match Germanic languages in importance, they are spoken by a sizeable portion of the total population of the world in spite of their relatively low number (less than 20 of them).

(41.4), Croatian (28.2), and Polish (45.8). On the other hand, some of the language in which the tree concept is expressed by adding an overt marker (often an affix-like element) to the lexeme associated with wood (just the opposite of the pattern established for Slavic languages), such as those in (28), generally exhibit a lower March societal complexity score (between 1 and 20.5):

(28)

	'wood'	'tree'	Marsh score
Hawaiian (Austronesian)	laa'au	*kumu*laa'au	5
Indonesian (Austronesian)	kaju	*pohon*kaju	20.5
Lao (Kam-Tai)	maj	*kok*-maj	9.6
Mongolian (Altaic)	мод	модон	5
Tagalog (Austronesian)	kahoy	*punung* kahoy	7
Thai (Kam-Tai)	h)mai	*f)ton*h)mai	13.7
Tibetan (Tibeto-Burman)	śi'n	śi'n-*sdő'n*	7
Vietnamese (Mon-Khmer)	gô	*cay*gô	16.8
Zuni (U.S. Southwest)	ta	ta-*tta*	1

We now consider whether there is a correlation with respect to some other environment-related concept. We have already discussed the example of animal grinding in Part 2, but we have focused on the items below the basic-level category, i.e. on specific animals and their meat. Let us now check whether there are cases of similar polysemy at the basic level (ANIMAL/MEAT).

Boyeldieu (2008) notes that the identity of the terms referring to both "meat" and "animal" was pointed out by Greenberg (1983) as a property of certain African languages. In other words, a number of African languages, and particularly those belonging to the Niger-Congo philum, have only one term referring both to "animal" and "meat". According to Greenberg (1983), the meaning "meat" is first.

Cf. some examples:

(29) a. Ngbandi *sa* "animal, game, meat"
b. Proto-Gbaya **sa'di* "meat, animal"
c. Ngbugu *gi* "animal, meat"
d. Laal *mèìrm* "animal, game, meat, flesh"
e. Proto-SBB (Central Sudanic) **yida* "meat, animal"

The same phenomenon is reported for Lahu and Jinghpaw (two Tibetan languages) in Matisoff (1978), and for Walpiri in Wierzbicka (1992). It turns out, however, that the cultures in which this phenomenon is attested are predominantly hunting cultures, just as expected under our hypothesis. Witness the following quotation from Matisoff (1978):

> For most modern Western languages it does make sense to have two separate items for *animal* and *meat/flesh*. In the hunting cultures of backwoods SEA [South-Eastern Asia], these two concepts are often expressed by exactly the same word...
>
> Matisoff (1978:132)

Further evidence comes from the fact that those Niger-Congo languages that have separate words for ANIMAL and MEAT, the latter is often associated with GAME (Boyeldieu 2008: 304), as in:

(30) a. Fula *barogel* or *kulle* "animal" vs. *teewu* "meat, flesh, game",
b. Wolof *mala* "animal" vs. *yapp* "meat",
c. Bambara *dantafen* or *bagan* "animal" vs. *sogo*, *sogobu* "meat, flesh" and *kungosogo* "game",
d. Kasim *vfrīm* "animal", *gāvarū* "wild animal" vs. *nnī* "meat, flesh, game",
e. Moore *dunga* "animal" vs. *neemdo* "meat, flesh, game"

It is understandable that the MEAT concept should be more basic in a hunting community, and that animals are conceptualized so to say as running pieces of meat. Of course, they generally have quite elaborate vocabulary as far as individual species of animals are concerned, but the two most important things about them are: i. whether they provide edible meat or not, ii. whether they are available. According to Wierzbicka (1992: 8), edible animals are distinguished from non-edible ones, and actually the word for edible animals is the same as the word for meat. As for their availability, there is an element of haphazardness, as there is normally no guarantee that will be continuously available, which is certainly a matter of concern for a hunting community. The choice of the type of meat is normally not an option in such a society (meat from different species may be available but not more than one at a time), and therefore all kinds of meat are quite likely to be referred to by a single word.

But even here, the MEAT concept does not appear completely out of the blue: it is often an expansion from a name for a specific animal. It often derives from the name of the most salient animal. For example, in Jinghpaw *shan* is used to refer not only to "flesh," but also to "deer," and "large game in general" and "animal, as used for food" (Matisoff 1978: 132). Boyeldieu reports that

> In southern Gbaya the reflex of **sa'di* disappeared and was substituted by the term *nam* "meat, animal", which stems from proto-Gbaya **nam* "buffalo" (the latter is related to proto-Bantu **nyama* "meat, animal": southern Gbaya would actually have come back to the original meaning). As we saw earlier, in Jinghpaw, *shan* has the different meanings "flesh; deer; large game animal, as used for food". One last example is provided by the Beja (Cushitic) term *ša* "meat" that may be linked to *š¿a* "heifer, cow".
>
> Boyeldieu (2008: 306)

Communities that focus on raising stock rather than hunting animals are in a slightly different situation: meat is more likely to be in more or less permanent stock, available in a more predictable manner. This lifestyle also makes it possible for farmers to focus on animals themselves, and also on different types of meat as meat from different species may be available simultaneously at least in certain quantities.

On a final note, let us just point out what may happen to a word for a specific type of meat when it is "transplanted" from a language used by a community that can be characterized as primarily stock-raising rather than as a hunting community to community that at roots belongs to the other cultural type. The case in point is *bif* in Krio (a West-African creole language, primarily spoken in Freetown, the capital of Sierra Leone). The lexeme is obviously a borrowing from English, i.e it is a replica of English *beef.*

Nordlander (2007) reports at least five metonymically related senses of *bif* in Krio. The first meaning listed in Fyle and Jones (1980) is "beef, meat, animal flesh of any kind". This means that the lexeme can be used in the literal sense of the English model to refer to a specific type of meat. But it can also be used metonymically to refer to other types of 'meat' and 'edible flesh', including 'beef', 'poultry', 'pork' and 'mutton'. In other words, an instance of the category MEAT has been metonymically generalized/schematized to a higher level category (while *mit*, from *meat*, is used only in idiomatic expressions). From this *bif* is further extended to cover all kinds of flesh, edible and non-edible ones.

(31) O-o, dohg noh it dohg *bif ehn?*
Oh-oh, dog no eat dog *beef,* yes?
'Oh-oh, dogs don't eat dog flesh, right?' (Nordlander 2007: 274)

As might have been expected, *bif* can also be used metonymically to refer to "forest animal hunted for food," by means of the OBJECT-FOR-MATERIAL-CONSTITUTING-THE-OBJECT metonymy (32), from which it cn be even further generalized into 'any large animal (33):'

(32) (Dehn Grik pipul de kohl am Dohkas we min, *bif we* dehn kohl dia.)
(PLUR Greek people CONT call OBJ Dorcas REL mean, *beef* REL them call deer)
'(The Greeks call her Dorcas, which means the *animal* they call deer.)'

(33) A de tehl una igen se, i tranga foh mek da big *bif* we nem kamehl pas insay nidul yay, ...
I DON'T tell you (pi) again COMP, he strong for make the *beef* REL name camel pass inside needle eye, ...
'And again I say unto you, it is easier for a camel to go through the eye of a needle, ...'

4. Concluding remarks

We would like to conclude our considerations with two types of remarks. First, we make a general, methodological point, and then proceed to more specific ones.

Metonymy is, unlike metaphor, in principle reversible, so both directions in the motivation of extension are possible in principle if polysemy happens to be induced by metonymy. This means that we face different types of problems with polysemy based on metaphor and polysemy based on metonymy. The former type of polysemy may be opaque, and the main problem may be to reconstruct the path (i.e. retrace the steps back to utimate the origin). If the link is very opaque we also have the problem of distinguishing between polysemy and homonymy as the limiting cases. With polysemy based on metonymy, however, the problem may be caused by the reversibility, specifically we may have to decide between two equally plausible paths of extension.

What we propose in cases when various types of evidence/data are in conflict is a step by step, or cyclic examination of both data and the theoretical claims on which the hypothesis rests. Specifically, we propose to consider a wider range of data in as meticulous a way as possible, and then if necessary, refine and upgrade the initial hypothesis so that it fits the data. The theoretical construct resulting from this process may of course be occasionally less elegant than the initial one, but this is of course unproblematic in a usage-based model of language. We are thus warned against jumping to conclusions and deciding a priori exclusively in favor of one direction of semantic extension in motivating cases of polysemy.

Moving to the more specific level of what and how of motivating extensions that lead to polysemy, our discussion has shown that embodiment as a general motivation should be understood in the widest possible sense, extending radially from the body to immediate the physical environment to the social and cultural environment. As pointed out by Rohrer (2007: 29),

> "[e]mbodiment" is also used to refer to the social and cultural context in which the body, cognition, and language are perpetually situated. For example, such context can include factors such as governmental language policy, cross-cultural contact/aversion, or the influence of historical scientific models and theories on individual language learners (Geeraerts and Grondelaers 1995). Similarly, the context can include the cultural artifacts that aid and manifest cognition—many of which are not only constrained by but are also extensions of the body (Hutchins 1995, 2005; Fauconnier and Turner 2002).

If embodiment is understood in this way, then we are also warned to be very careful when sampling languages in lexical/semantic typology. Taking into account the absolute number of languages exhibiting a certain trait, linguistic typology often fails to take into account the number of speakers and the environment in which a language is spoken, and this may lead to a certain degree of misrepresentation. As our discussion in Section 3 indicates, this danger may be particularly acute in lexical typology, where the type of environment significantly determines conceptual structures as far as realia are concerned in terms of the absolute number of languages exhibiting a certain trait. Certain types of natural environments, such as deserts or jungles, are an obvious barrier for the development of large-scale societies. In other words, we will find a higher number of small-scale communities, separated by natural barriers. A highly likely linguistic outcome of this is a higher number of different (though perhaps related) languages spoken by a smaller number of speakers.

Finally, assuming that in a usage-based model schemas arise from instances and categories grow around prototypes (central instances), our discussion seems to indicate that prototypes may over time become basic-level categories, and start functioning as category labels. In view of this, as well as of the reversibility of metonymy, it is not always warranted to put the basic-level category cart before the instance horse which drives the category, as Witkowski, Brown and Chase (1981) do it, i.e. exclude from considerations a possible history of a basic-level category, during which it may have functioned as an instance.

References

Boyeldieu, Pascal. 2008. From semantic change to polysemy. In Martine Vanhove (ed.), From polysemy to semantic change (Studies in Language Companion Series 106), 303-315. Amsterdam & Philadelphia: John Benjamins

Blust, Robert A. 1974. Proto-Austronesian syntax: The first step. *Oceanic Linguistics* 8. 1-15.

Cuyckens, Hubert & Britta Zawada (eds.). 2001. *Polysemy in cognitive linguistics. Selected papers from the Fifth International Cognitive Linguistics Conference, Amsterdam, 1997*. Amsterdam & Philadelphia: John Benjamins.

Fauconnier, Gilles, & Mark Turner. 2002. *The way we think: Conceptual blending and the mind's hidden complexities*. New York: Basic Books.

Fowler, Catherine Louise Sweeney. 1972. Comparative Numic ethnobiology. Pittsburgh: University of Pittsburgh dissertation.

Fyle, Clifford N., & Eldred D. Jones. 1980. A Krio-English dictionary. Oxford & New York: Oxford University Press.

Geeraerts, Dirk. 1993. Vagueness's puzzles, polysemy's vagaries. *Cognitive Linguistics* 4(3). 223-272.

Geeraerts, Dirk, & Stefan Grondelaers. 1995. Looking back at anger: Cultural traditions and metaphorical patterns. In John R. Taylor and Robert E. MacLaury (eds.), *Language and the cognitive construal of the world*, 153–179. Berlin: Mouton de Gruyter.

Greenberg, Joseph H. 1983. Some areal characteristics of African languages. In Ivan R. Dihoff (ed.), *Current approaches to African linguistics*, 3-21. Dordrecht: Foris Publications.

Hutchins, Edwin. 1995. *Cognition in the wild*. Cambridge, MA: MIT Press.

Hutchins, Edwin. 2005. Material anchors for conceptual blends. *Journal of Pragmatics* 37(10). 1555–77.

Ibarretxe Antuñano, Iraide. 1999. *Polysemy and metaphor in perception verbs: A cross-linguistic study*. Edinburgh: University of Edinburgh dissertation.

Koch, Peter. 2001. Lexical typology from a cognitive and linguistic point of view. In Martin Haspelmath, Ekkehard König, Wulf Oesterreicher, & Wolfgang Raible (eds.), *Language typology and language universals. An international handbook*. Vol. 2.2, 1142-1178. Berlin & New York: Mouton de Gruyter.

Langacker, Ronald W. 1987. *Foundations of Cognitive Grammar. Volume 1. Theoretical prerequisites*. Stanford: Stanford University Press.

Langacker, Ronald W. 1988. A usage-based model. In Brygida Rudzka-Ostyn (ed.), *Topics in cognitive linguistics*. (Amsterdam Studies in the Theory and History of Linguistic Science. Series IV. Current Issues in Linguistic Theory 50), 127-161. Amsterdam & Philadelphia: John Benjamins.

Marsh, Robert M. 1967. *Comparative sociology: A codification of cross-societal analysis*. New York: Harcourt, Brace & World.

Matisoff, James A. 1978. *Variational semantics in Tibeto-Burman*. Philadelphia: Institute for the Study of Human Issues.

Nerlich, Brigitte, Zazie Todd, Herman Vimala, David D. Clarke (eds.). 2003. *Polysemy: Flexible patterns of meaning in mind and language*. Berlin & New York: Mouton de Gruyter.

Nordlander, Johan. 2007. The metonymic element in Krio conceptualization: The cases of BIF and BUSH. In Krzysztof Kosecki (ed.), *Perspectives on metonymy. Proceedings of the International Conference 'Perspectives on Metonymy', held in Łódź, Poland, May 6-7, 2005*, 271-287. Frankfurt am Main: Peter Lang.

Price, P. David. 1967. Two types of taxonomy: A Huichol ethnobotanical example. *Anthropological Linguistics* 9(7). 1-28.

Quirk, Randolph, Sidney Greenbaum, Geoffrey Leech, & Jan Svartvik. 1985. *A comprehensive grammar of the English language*. London: Longman.

Radden, Günter, & Zoltán Kövecses. 1999. Towards a theory of metonymy. In Klaus-Uwe Panther, & Günter Radden (eds.), *Metonymy in language and thought*, 17-59. Amsterdam & Phildelphia: John Benjamins.

Ravin, Yael, & Claudia Leacock (eds). 2002. *Polysemy: Theoretical and computational approaches*. Oxford: Oxford University Press.

Rohrer, Tim. 2007. Embodiment and experientialism. In Dirk Geeraerts & Hubert Cuyckens (eds.), *The Oxford handbook of cognitive linguistics*, 25-47. Oxford & New York: Oxford University Press.

Thord–Gray, Ivor 1955. *Tarahumara-English, English-Tarahumara dictionary and an introduction to Tarahumara grammar*. Coral Gables, Fl.: University of Miami Press.

Tuggy, David. 1993. Ambiguity, polysemy, and vagueness. *Cognitive Linguistics* 4(3). 273-290.

Wierzbicka, Anna. 1992. *Semantics, culture and cognition. Universal human concepts in culture-specific configurations*. Oxford & New York: Oxford University Press.

Witkowski, Stanley R., Cecil H. Brown, & Paul K. Chase. 1981. Where do tree terms come from? *Man. New Series* (16)1. 1-14.

Metaphors and metonymies in Serbian proverbs containing names of animals[1]

Diana Prodanović-Stankić, University of Novi Sad

1. Introduction

This article is based on the findings of a corpus-based study of conceptual metaphors and metonymies in Serbian proverbs containing names of animals. The first part of the article is devoted to the analysis of metaphors and metonymies found in the corpus, whereas the second focuses on the issue of understanding of metaphorical and metonymic expressions. The basic idea behind this analysis was to determine the nature of cognitive mechanisms used in proverbs, the extent to which they are universal and the role played by social and cultural factors in the process of conceptualization.

The most frequent metaphors that can be found in Serbian proverbs containing names of animals are PEOPLE ARE ANIMALS, PROBLEMS ARE ANIMALS, THINGS ARE ANIMALS, and SUCCESS IS AN ANIMAL. When it comes to metonymies, it must be said that in most examples they serve as the basis for metaphorical mapping. They are less frequent in the overall corpus than metaphors and usually we have metonymies of the kind PART FOR WHOLE.

The second part of the article deals with the problem of recognizing and interpreting metaphors in proverbs. Although it is widely assumed that no particular mental effort is required for interpreting a certain linguistic expression as metaphorical, research conducted on students indicates that they had problems with understanding conceptual metaphors in Serbian proverbs. The findings of this particular research show that automatic and unconscious understanding of metaphorical language cannot always be assumed. Presuming that people are less able to access conceptual metaphors when interpreting some new instances of metaphorical expressions, adequate theoretical preparation certainly makes this task easier.

In the past three decades we have witnessed a growing interest in the study of conceptual metaphor and metonymy worldwide, however, not always from the same perspective (e.g. Lakoff and Johnson 1980; Lakoff and Turner 1989; Gibbs 1994; Sperber and Wilson 1995; Kövecses and Radden 1998; Fauconnier and Turner 2002; Barcelona 2003; Turner and Fauconnier 2003). Although it has been proved that many conceptual mappings are universal because they are based on bodily experience common to all human beings, there are many

---

1 I wish to express gratitude to Prof. Tvrtko Prćić for his many invaluable comments and insightful suggestions during the preparation of this paper.

metaphorical expressions across different languages that show specific paths of conceptualization. Thus, the dimension along which languages and cultures shape metaphors represents a fertile ground for linguistic research.

Following the cognitive approach, the main objective of this paper is to determine conceptual metaphors and metonymies and the nature of conceptual mappings that can be found in Serbian proverbs containing names of animals. In this article, metaphor is regarded as "understanding one domain of experience in terms of a very different domain of experience, or, a mapping from a source domain to a target domain" (Lakoff 1993: 206–207), and metonymy as a "cognitive process in which one conceptual entity, the vehicle, provides mental access to another conceptual entity, the target, within the same domain" (Kövecses 2002: 145). Proverbs, on the other hand, are defined as "a specific type of idioms, in the form of a sentence, used to comment on a specific situation. A proverb is a short and witty traditional saying which expresses a moral, general truth or a popular belief" (Prćić 1997: 128). Known to be a result of creative and spontaneous use of language, they are no exception when it comes to cognitive mechanisms – metaphors and metonymies are an inseparable part of them. These highly economical linguistic expressions are packed with traditional symbolism and elements of a particular culture, and it seems that the inherited traditional formulae accompanied by conceptual mappings ensured the survival of proverbs. As a part of each nation's cultural heritage, proverbs were passed down from generation to generation; hence they had to be based on mnemonic techniques and conceptual metaphors. However, metaphors did not influence their form significantly, since slight transformations in the form of a proverb are allowed as long as the main structure of the conceptual mapping is undisturbed.

Assuming that metaphors and metonymies in proverbs are mainly universal, and that the traditional and culture-specific elements that account for specific variations should be known to native speakers of a language, it seems that a native speaker of a language will have no difficulties in recognizing and comprehending metaphorical expressions when he/she encounters them in proverbs. Extensive research in the field of psycholinguistics (Gildea and Glucksberg 1983; Glucksberg and Keysar 1990; Gibbs 1993, 1994) has proved that people do not need to "translate" metaphorical expressions into literal ones in order to understand them and that given sufficient and appropriate context people need the same amount of time to grasp both metaphorical and literal expression. However, judging by a small-scale study which we conducted on university students, native speakers are not always able to understand the metaphorical meaning of a proverb in their own language if they are not taught to look for metaphors, or if they are not fully familiar with the proverb.

This analysis of metaphors and metonymies is based on a corpus of 380 Serbian proverbs containing names of animals. The most significant collection of Serbian proverbs is the one compiled by Vuk Stefanović Karadžić (Karadžić 1964

[1836], [1849]), whereas other dictionaries of proverbs (Kekez 1973; Đorđević 1996; Jovanović 2006) usually represent revised editions of Karadžić's seminal work. The corpus includes only traditional sayings (epigrams, aphorisms, rhymes and similes were not taken into account) that contain conceptual metaphors with an animal as a source domain – it is remarkable that in the total of 380 proverbs, only 1 % of them are not metaphorical. Those "literal proverbs" in most cases express a kind of direct instruction in the form of an imperative sentence. In the process of identifying and excerpting metaphors and metonymies from the corpus we followed the methods suggested by Barcelona (1997), Steen (2002) and Stefanowitsch (2006).

2. Conceptual metaphors from the corpus

Viewing proverbs as "a simplest form of poetry", Lakoff and Turner (1989: 160) demonstrate that two metaphors are essential for understanding them: the GENERIC IS SPECIFIC and the GREAT CHAIN metaphor (Lakoff and Turner 1989: 162–170). Whereas the GENERIC IS SPECIFIC provides mapping from one specific-level schema onto the generic-level schemas, preserving at the same time the generic-level structure of the target domain and allowing us to interpret a particular proverb in many different specific situations, the GREAT CHAIN metaphor offers access to the hierarchical order of the universe. Namely, this metaphor links attributes and properties of an entity from one level to attributes and properties of another entity belonging to a different level in this strictly arranged hierarchy, and at the same time it picks the most relevant aspects of the source domain schema. (The number and directions of possible mappings within the five levels of the Great Chain are discussed in detail in Krzeszowski 1997: 161). Furthermore, each level entails the so-called metaphorical propositions within schemas (Lakoff and Turner 1989: 194). Accordingly, it is assumed that dogs are loyal, dependable and dependent, foxes are clever, cats are fickle and independent, etc. (ibid.).

In other words, it can be said as Kövecses has it "that each conceptual metaphor has its main meaning focus" (Kövecses 2003: 82). As a result, in accordance with the prototypical folk understanding of the conceptual domain of animals, most people all over the world will share the same knowledge about a particular animal. Nevertheless, cross-cultural variations are inevitable, given the fact that we all live in different geographical regions and that we share different cultural models. Still, proverbs containing names of animals are very frequent across various languages and cultures, and some studies show that most of these proverbs are similar, especially in Indo-European languages, or that people frequently tend to use animal names metaphorically (Yusuf 1997; Martsa 1999; Baider and Gesuato 2003; Kiełtyka and Kleparski 2005; Moreno 2005).

The starting point of this analysis was the source domain – an animal. In order to determine the range of possible target domains to which this source domain applies, or the scope of metaphor in Kövecses's terms (Kövecses 2003: 80), the metaphorical mappings found in the proverbs were grouped according to the type of target domain. The most frequent metaphor found in Serbian proverbs is the complex metaphor PEOPLE ARE ANIMALS. The main meaning focus that leads to this particular mapping is definitely based on the shared knowledge people have about animals. First of all, both human beings and animals belong to the same category of living creatures, generally speaking, they have a similar life cycle, and there are many aspects along which human beings and animals can be linked. Furthermore, all these specific aspects can be highlighted in various simple submetaphors, and they can be expressed in some other verbal or non-verbal ways. To that extent, the so-called animal metaphors are often reflected in different parts of speech (in English: *chicken out of something*, *a foxy lady*; in Serbian *namagarčiti nekoga* 'make a donkey of somebody [to cheat someone]', *ona je dobra riba* 'she is a good fish [she is an attractive girl]') or this complex conceptual metaphor may be depicted in drawings, cartoons and similar (as in the case of Mickey Mouse, Garfield and many other cartoon characters).

However, the activation of some particular aspects of the target domain PEOPLE in the conceptual mapping results in simple metaphors that highlight each of these particular aspects. For example, if we focus on someone's appearance, behaviour, or character, different animals will be used as a source domain. For each of these aspects of the target domain there are specific animals that are prototypically used as a source domain in the metaphorical mappings, although some variations are possible both in the choice of animal and in the very form of the proverb. To illustrate this point, we can take some examples from the corpus. In most Serbian proverbs, when the distinction between two sexes and the roles they play in the household is highlighted, animals that serve as a source domain are those that are prototypically kept in pairs on a farm, such as bulls and cows, or cocks and hens. Thus, in proverbs such as *Bolje je biti pjevac jedan dan nego kokoš mjesec* 'It is better to be a cock one day than a hen for a month' and *Bolje je za godinu volom nego sto godina kravom* 'It is better to be a bull for one year than a cow for a hundred of years' the same idea is expressed by the means of different animals (cocks / hens, bulls / cows) because the focus is on the role they have in the farm / household. However, when both men and women are regarded as individuals on their own, they are usually metaphorically represented as horses / stallions, dogs, mares, or bitches respectively (*Sedlan konj ne može biti kao neosedlan* 'A saddled horse is not the same as an unsaddled one', *Koja se vrca ona i vrše* 'A mare that is wriggly is hardworking too', *Dok kuja repom ne vine, neće pas za njom potrčati* 'As long as the bitch does not wag its tail, the dog will not follow her'). In a similar way, several forms of one proverb may exist, yet the conceptual metaphor on which they are based is the same (*Čuvaj se pseta koje mučki kolje* 'Beware of dogs which gnaw cruelly', *Podmuklo pseto najpre će ujesti* 'A mean dog is first to bite', *Od podmukla pseta čuvaj se* 'Beware of mean dogs').

As can be seen in Table 1 below, various animals are used in Serbian proverbs when different aspects of the conceptual domain HUMAN BEING are metaphorically described. Moreover, in most cases the choice of a particular animal is in accordance with the metaphorical propositions that are universal across many languages and cultures, though the most favored ones are dogs, horses and donkeys. These animals are always used when people are regarded as representatives of their species, or as members of a society. The choice of these animals is definitely based on the prevailing assumptions that the great majority of people are deprived and underprivileged, and that they have to work very hard in order to survive in this cruel world.

Table 1. PEOPLE ARE ANIMALS metaphors used in Serbian proverbs

PEOPLE ARE ANIMALS	
CHILDREN ARE KITTENS / PUPPIES	*Nema ni kučeta ni mačeta* 'He / she does not have any puppies nor kittens' *Štogod maca omaci, ono sve miše lovi* 'Whatever comes of a cat catches mice'
A MAN IS A COCK / BULL / DOG / HORSE A WOMAN IS A HEN / COW / BITCH / MARE	*Gdje god krava vodi, doma se teli* 'If the cow manages the household, nothing is going smoothly' *Bolje je biti pjevac / vo jedan dan nego kokoš / krava mjesec* 'It is better to be a cock / bull one day than a hen / cow for a whole month'
AN OLD PERSON IS A DOG / HORSE / GOAT	*Kad star pas laje, valja vidjeti šta je* 'If an old dog is barking, one should give heed to it' *Ti si koza stara, ti si vuka znala* 'You are an old goat, you should have known the wolf'
A LAZY PERSON IS A SHEEP / HEN	*Kojoj ovci svoje runo smeta, onđe ni runa ni ovce* 'If a sheep minds its own fleece, it is neither a true sheep, nor the fleece is proper' *Ne treba kvocati nego jaja nositi* 'Don't cluck, lay eggs'
A STUPID AND NAIVE PERSON IS A DONKEY / GOAT	*Prva pamet magarcu pod rep* 'A donkey can put its whole brain under its tail' *Čim se koza dičila, tim se ovca sramila* 'What made the goat proud, made the sheep ashamed'
A SHREWD AND CUNNING PERSON IS A CAT / FOX	*Kad lisica predikuje, pazi dobro na guske* 'When the fox is preaching keep watch over the geese' *Komu se mačka umiljava, onoga i ogrebe* 'If the cat shows affection for you, it may scratch you as well'

AN EVIL PERSON IS A DOG / WOLF	*Čuvaj se pseta koje mučki kolje* 'Beware of dogs that gnaw cruelly' *Vuk dlaku menja, al ćud nikada.* 'A wolf may change its fur but never its temper'
VAIN AND CONCEITED PEOPLE ARE DONKEYS / GOATS	*Koza lipsuje, a rep ne spušta* 'Even when it is dying, a goat won't lower its tail' *Digni magaretu klašnje, njemu lašnje* 'Pull the donkey's socks up, and it will feel better'
GREEDY PEOPLE ARE DOGS / PIGS / HENS	*Kad je najviše strvine, onda najviše psi bjesne* 'Dogs are most angry when there is a lot of carrion around' *Lakoma kvočka ne može pilića izvesti* 'A greedy hen cannot raise chicken'
POOR AND UNDERPRIVILEGED PEOPLE ARE MICE / DOGS / DONKEYS	*Mački je do igre, a mišu do plača* 'The cat feels like playing and the mouse like crying' *Što pas na zvijezde laje to Bog ne sluša* 'A dog may bark at stars in the sky, yet God is not listening to it' *Jako je magare, al dva tovare* 'A donkey may be strong, but two men are loading sacks on its back'
RICH AND POWERFUL PEOPLE ARE HORSES / CATS	*Gde nema mačke, tu miševi kolo vode* 'When the cat's away mice will play' *Gde se konji igraju, onde magarci mrtvi padaju* 'When horses are playing, donkeys drop dead'
HARD-WORKING AND NOBLE PEOPLE ARE HORSES	*Za dobrim se konjem prašina diže* 'A good horse leaves clouds of dust behind' *Dobar se konj i u štali nađe* 'A good horse can be found even in a stable'

It is also very interesting that proverbs that comment on some positive things or that express approval are very rare. In these examples people are usually regarded as horses. As in some other languages (Yusuf 1997; Baider and Gesuato 2003), women are metaphorically represented in Serbian proverbs as small and/or edible animals. In addition to that, women are mainly metaphorically described as cows, and their role as a prototypical mother and housewife is highlighted.

In addition to the metaphor PEOPLE ARE ANIMALS, some other metaphors may be included in a particular proverb. For example, if the focus is on someone's position in the society, other metaphors are used as well, for example, MORE IS UP, IMPORTANT IS BIG, CONTROL IS UP, etc. Hence, in the following proverbs, the size of the animal and popular belief about the behavior of some animal species stands for the importance and power people have in the society and the relation they have toward other people:

Gde se konji igraju, magarci mrtvi padaju
'When horses are playing, donkeys drop dead'

Teško je pseto s mačkom i kurjaka s ovcom pomiriti
'It is hard to make peace between a dog and a cat or a wolf and a sheep'
Da je na pasju volju, niđe konja ne bi bilo
'Dogs would like to see the day when all horses become extinct'.

Other metaphors found in the corpus are THINGS ARE ANIMALS, PROBLEMS / DIFFICULTIES ARE ANIMALS, and SUCCESS IS AN ANIMAL. Though less frequent, since they occur in 18 per cent of the total number of proverbs, these metaphors account for all cases where the target domain is either a thing or an abstract concept, such as a problem, an obstacle, success, luck or an opportunity. When talking about things and possessions, people usually use some small animals or their offspring as a source domain in the metaphorical mapping. Those animals have a certain financial value for people or they are used in the diet. On the other hand, problems or difficult situations are mostly viewed as cats, dogs or donkeys, since these animals can not always be dealt with easily and their behaviour may often be unpredictable. As in the case with proverbs that express some positive traits of someone's personality, proverbs that mention good luck, happiness, or positive outlook are extremely rare, probably because of the fact that in Serbian culture talking about such things is believed to bring bad luck. Perhaps this attitude can be regarded as a kind of confirmation of the Great Chain of Being – the folk theory that helped people understand their own position in the universe, as well as the position and role of God, or any other supreme form of being. Being positioned in this hierarchy below the Creator, people are not allowed to discuss divine laws and will.

Table 2. Other types of conceptual metaphors used in Serbian proverbs

THINGS ARE ANIMALS	*Dok je ovaca biće i novaca* 'As long as you have enough sheep, you'll have enough money as well' *Mačka se u vreći ne kupuje* 'Don't buy a cat in a bag'
A PROBLEM OR A DIFFICULTY IS AN ANIMAL	*Čuvaj se da ne povučeš garova za rep* 'Be careful not to pull a black dog's tail' *Pošao mu magarac niz brdo* 'His donkey went downhill' *Isterao mu mačku na odžak* 'His cat was driven out on the chimney'
SUCCESS / AN OPPORTUNITY IS AN ANIMAL	*Blago tomu komu konj uz vodu ide* 'You are lucky if your horse is walking upstream' *I ćorava koka nađe zrno* 'Even a blind hen can peck at a crumb'

Even in the examples which imply that someone has succeeded in doing something, the role of personal responsibility and the sense of accomplishment is diminished, as in the proverb *I ćorava koka nađe zrno* 'Even a blind hen can peck at a crumb', where blindness of the hen implies, in accordance with the metaphor KNOWING IS SEEING, that those who know nothing may by pure chance get something good in their lives.

3. Conceptual metonymies from the corpus

Conceptual metonymies, which are quite frequent in the analyzed corpus, are mostly used as a basis for a metaphorical mapping, drawing our attention to a particular aspect of the concept. In this corpus they account for 38 per cent of examples, and in most cases these metonymies are 'source-in-target metonymies' as described by Ruiz de Mendoza Ibáñez (2003: 121–122). For example, the proverb *Šut se s rogatim ne može bosti* 'The hornless cattle cannot fight against the horned' means that it is impossible for people who are not equipped enough to fight against those who are more powerful and strong. The source in this metaphorical mapping is represented by two groups of animals, those which are horned (*rogati*) and those which are not (*šuti*). The PART FOR WHOLE metonymy plays a crucial role here and in Serbian it is expressed even on the level of the lexeme, referring to the whole class. As Ruiz de Mendoza Ibáñez has it, the main function of this metonymy is "to develop the source of the metaphor for full interpretation to be possible." (Ruiz de Mendoza Ibáñez 2003: 122). Hence, in this particular example, a hypothetical situation is assumed in which the 'horned' would fight against the 'hornless' and this source metaphorically maps on the target that denotes a kind of general situation in which powerful people always stand a better chance than the inferior.

Due to the nature of the very concept of animals, metonymies in this corpus are mostly of the kind PART (OF THE BODY) FOR WHOLE. In other words, it is usually the tail, skin, teeth, or in some examples the voice of a particular animal, which stands for the whole animal. In that way, metonymies structure the source domain and enable further metaphorical mapping. In the case of proverbs, metonymies appear to be even more mentally economical than metaphors, making it possible for proverbs to be expressed in a pithy and poetic manner:

Gde je više repova, tu je manje glava
'If there are more tails around, there are few heads available'
Čuvaj se da ne povučeš garova za rep
'Be careful not to pull a black dog's tail'
Više je jagnjećih koža na pazaru nego ovčijih
'There is more lamb than sheep hide at the market'
Ko na zvijezde laje ispašće mu zubi
'If you bark at the stars, you will lose your teeth'
Magareći glas na nebo ne ide
'A donkey's braying voice cannot reach the sky'

What is interesting, though, is that proverbs where conceptual metonymies are sources for metaphorical mappings usually describe the underprivileged, deprived people, or individuals who are not allowed to ask for more. It seems that, metaphorically speaking, by reducing an individual or a group to its part, one can emphasize their position and role in society.

4. Understanding of proverbs

Within the cognitive framework, it is generally assumed that the process of understanding a particular metaphorical expression should not require a considerable mental effort, bearing in mind the fact that conceptual metaphors are behind all metaphorical expressions found in a language. Some psycholinguists (Glucksberg and Keysar 1990; Cacciari 1998) claim that in order to perceive a particular metaphorical expression as metaphorical, one has to recognize that metaphor was intended in the first place. However, when it comes to proverbs, it seems that they have lost their metaphoricity in the sense that people tend to regard them as some "frozen semantic units." (Gibbs 1994: 21). According to Gibbs, "people do not need to activate their preexisting metaphorical knowledge each and every time they read or hear the expression." (Gibbs 1994: 19). Besides, some authors (Gildea and Glucksberg 1983) show that the amount of contextual information surrounding a metaphorical expression does not have to be different from the context surrounding a literal expression, in order to understand the metaphorical expression appropriately. Nevertheless, applying these findings to class work with students has thrown new light on this issue. While teaching translation courses (from English into Serbian and vice versa) at the English Department at the University of Novi Sad, we noticed that students had difficulties in recognizing metaphorical expressions in the first place, no matter whether they had to deal with texts written in English or in their mother tongue, i.e. Serbian.

In light of testing students' ability to recognize and understand metaphors and metonymies, a survey was carried out on 40 second-year and 40 fourth-year students at the Department for English Language and Literature. The second-year students did not have any theoretical or practical instruction in CTMM (The Cognitive Theory of Metaphor and Metonymy). On the other hand, fourth-year students were introduced to the basics of CTMM in their Lexicology course. Prior to taking the test they were told that all proverbs in the test contained some metaphors and metonymies which they were supposed to recognize. With both groups of students, the notion of conceptual metaphor and metonymy was briefly introduced. The test was designed in accordance with several objectives. First of all, their task was to specify the source and target domain, yet, since not all of them were familiar with these specific terms, they had to determine what the animals mentioned in the proverbs designate. Secondly, the proverbs given in the test contained several metaphors PEOPLE ARE ANIMALS, THINGS ARE ANIMALS and PROBLEMS ARE ANIMALS. Some examples contained metonymy-based metaphors, for example, *Ako se ne uhvati za glavu, za rep nikad* 'If you don't catch hold of his head, you'll miss the tail as well'. In some cases, there were two animals mentioned in the proverb, like in *Gde nema mačke tu miševi kolo vode* 'When the cat's away the mice will play'. There were ten proverbs given without the context and five proverbs taken from newspaper articles, embedded in a short text. Students' answers were grouped into three categories: the first contained completely correct answers, the second partially correct, and finally the third consisted of incorrect answers. In the case of proverbs given without an appropriate context, students were expected to mention at least one specific target domain, and in the case of proverbs given within the context, they had to determine the target that was indirectly mentioned in the text. To illustrate this: in the proverb *Prvi se mačići u vodu bacaju* 'The first kittens are thrown in the water' students were supposed to determine that *mačići* 'kittens' refers to any activity or job that was done unsuccessfully for the first time, whereas in a short text about a famous, yet widely criticized director, students had to notice that in the proverb *Za dobrim se konjem prašina diže* 'A good horse leaves clouds of dust behind', *konj* 'horse' referred to the director himself.

The results show that almost 20 per cent of students who were more familiar with CTMM did better in the test when compared to the other group of students. Not only did they have more correct answers, but they also had less partially correct answers. Still, for both groups around 30 per cent of proverbs were problematic. The easiest examples were proverbs that contained only one source domain and the PEOPLE ARE ANIMALS metaphor, especially in the case when the proverb contained only a simple metaphor that referred to someone's behavior or character, rather than to the domain of gender, or to people in general. Moreover, students had difficulties in determining target domains which were abstract or which denoted something inanimate. For example, in most cases they were unable to recognize and understand that an animal could stand for a concept such as 'a problem.' Metonymies represented another problematic issue, since around half of the students in both groups failed to notice them. Furthermore, only fourth-year students took advantage of the context in the examples where the proverb was embedded in it. Namely, context provided them with relevant information and helped them have more correct answers and less partially correct answers than students from the other group.

To sum up, it appears that students still have problems with recognizing and understanding metaphorical expressions in proverbs. One of the reasons for this might be their insufficient exposure to proverbs in everyday use of language. There are no relevant studies that would show how frequent and prevailing proverbs are in the modern Serbian language, though some studies in English (Black 1999; Pasamanick 2000) have proved that proverbs are still frequent in modern use. Proverbs containing names of animals are probably even less familiar to younger generations due to the process of urbanization. Being less in touch with animals in general, it is difficult for younger generations to find correspondences between human beings or some abstract concepts and the animal world. However, apart from the fact that the level of familiarity influences the process of understanding, it seems that to some extent, students can be taught to recognize and understand metaphors and metonymies better if they are theoretically and practically prepared to do so. Automatic understanding cannot always be assumed, no matter how universal or culturally specific some metaphors may be.

5. Conclusions

Analyzing proverbs within the theoretical framework of cognitive linguistics offers a considerable insight into the way people conceptualize abstract domains. Besides, it is remarkable to see to what extent folk understanding of the world and popular beliefs are reflected in proverbs. The most frequent conceptual metaphor in Serbian proverbs is PEOPLE ARE ANIMALS, a complex metaphor that encompasses several simple metaphors. Those simple metaphors highlight different aspects of the target domain 'human being' and hence, we get metaphors such as AN EVIL PERSON IS A DOG / WOLF, VAIN AND CONCEITED PEOPLE ARE DONKEYS / GOATS, GREEDY PEOPLE ARE DOGS / PIGS / HENS, etc. Other frequent conceptual metaphors found in proverbs are THINGS ARE ANIMALS, PROBLEMS ARE ANIMALS or SUCCESS IS AN ANIMAL. Apart from these metaphors where the source is always a particular animal, various orientational, ontological or structural metaphors can play an important role in proverbs, depending on the given scenario. Moreover, most of the metonymies found in the corpus had the function of developing the source domain for the metaphorical mapping. Since in this case the source was a particular animal, all the examples had the PART FOR THE WHOLE metonymy, which acted as a source-in-target metonymy within the metaphor. The animals that usually serve as a source domain in Serbian proverbs are dogs, horses, donkeys, cats, sheep, pigs, goats, hens and cows. The metaphorical propositions that we link to these animals are partly universal, yet to some extent they are culture specific.

Generally speaking, proverbs are based on the universal metaphors GENERIC IS SPECIFIC and the GREAT CHAIN metaphor, which enable us to understand particular metaphors we find in proverbs. Combined with culture-specific elements that native speakers of a language share, these universal cognitive mechanisms should ensure full understanding of all metaphorical expressions we encounter. Yet, although it is not hard for us to understand that animals in proverbs actually refer to people, i.e. their behaviour, appearance or character traits, sometimes we have difficulties in understanding metaphorical expressions in these proverbs. It appears that understanding of metaphorical expressions can be enhanced by adequate theoretical and practical training; however, this hypothesis should be verified by further research in this field.

References

Baider, Fabienne H. and Sara Gesuato. 2003. Masculinist metaphors, feminist research. *Metaphorik.de* 05/ 2003. http://metaphorik.de/05/baidergesuato.pdf. (accessed 10 September 2004).

Barcelona, Antonio. 1997. Clarifying and applying the notions of metaphor and metonymy within cognitive linguistics. *Atlantis* 19(1). 21–48.

Barcelona, Antonio (ed.). 2003. *Metaphor and metonymy at the crossroads: A cognitive perspective*. Berlin and New York: Mouton de Gruyter.

Black, Johnatan Charteris. 1999. The survival of English proverbs: A corpus-based account. *De Proverbio* 5/2. http://www.deproverbio.com/DPjournal/DP,5,2,99/BLACK/SURVIVAL.html. (accessed 3 March 2000).

Cacciari, Cristina. 1998. Why do we speak metaphorically? Reflections on the functions of metaphor in discourse and reasoning. In Albert Katz, Cristina Cacciari, Raymond W. Gibbs, Jr., and Mark Turner (eds.), *Figurative language and thought,* 119–157. Oxford and New York: Oxford University Press.

Đorđević, Slobodan (ed). 1996. Vukove narodne poslovice s registrom ključnih reči [Vuk's traditional proverbs with a lexical index]. Beograd: Nolit.

Fauconnier, Gilles and Mark Turner. 2002. *The way we think: Conceptual blending and the mind's hidden complexities*. New York: Basic Books.

Gibbs, Raymond W. 1993. Process and products in making sense of tropes. In: Andrew Ortony (ed.) *Metaphor and thought*. 252–277. Cambridge: Cambridge University Press.

Gibbs, Raymond W. 1994. *The poetics of mind: Figurative thought, language, and understanding.* Cambridge: Cambridge University Press.

Gildea, Patricia and Sam Glucksberg. 1983. On understanding metaphor: The role of context. *Journal of Verbal Learning and Verbal Behaviour* 22. 577–590.

Glucksberg, Sam and Boaz Keysar. 1990. Understanding metaphorical comparisons: Beyond similarity. *Psychological Review* 97(1). 3–18.

Jovanović, Jelena. 2006. Knjiga srpskih narodnih poslovica [The book of Serbian traditional sayings]. Belgrade: Naučno društvo za negovanje i proučavanje srpskog jezika.

Lakoff, George and Mark Johnson. 1980. *Metaphors we live by*. Chicago: The University of Chicago Press.

Lakoff, George and Mark Turner. 1989. *More than cool reason*. Chicago: Chicago.

Lakoff, George. 1993. The contemporary theory of metaphor. In: Andrew Ortony (ed.) *Metaphor and thought*. 202–252. Cambridge: Cambridge University Press.

Kekez, Jovan. 1973. Bibliografija poslovica i zagonetaka [Bibliography of proverbs and riddles]. *Zbornik za slavistiku* 5. Novi Sad: Matica srpska. 182–220.

Kiełtyka, Robert and Grzegorz A. Kleparski. 2005. The ups and downs of the great chain of being: The case of canine zoosemy in the history of English. *Skase Journal of Theoretical Linguistics 2/ 1.* http://www.skase.sk/Volumes/JTL02/03.pdf. (accessed 3 June 2004).

Kövecses, Zoltán and Günter Radden.1998. Metonymy: Developing a cognitive linguistic view. *Cognitive Linguistics* 9. 37–77.

Kövecses, Zoltán. 2003. The scope of metaphor. In Antonio Barcelona (ed.) *Metaphor and metonymy at the crossroads: A cognitive perspective*, 79–92. Berlin and New York: Mouton de Gruyter.

Kövecses, Zoltán. 2002. *Metaphor: A practical introduction.* Oxford and New York: Oxford University Press.

Krzeszowski, Tomasz P. 1997. *Angels and devils in hell: Elements of axiology in semantics.* Warszava: Wyndawnictwo Energeia.

Martsa, Sándor. 1999. On exploring the conceptual structure of folk knowledge: The case of animal terms. *Linguistica e Filologia* 9. 73–87.

Moreno, Ana Ibáñez. 2005. An analysis of the cognitive dimension of proverbs in English and Spanish: The conceptual power of language reflecting popular beliefs.

Skase Journal of Theoretical Linguistics 2/ 1. http://www.skase.sk/Volumes/JTL02/04.pdf. (accessed 17 May 2004)

Pantić, Miroslav. 1964. Vuk Stefanović Karadžić i naše narodne poslovice: Predgovor u knjizi Srpske narodne poslovice [Vuk Stefanović Karadžić and our traditional proverbs: Preface in Serbian traditional proverbs] *Srpske narodne poslovice, Sabrana dela Vuka Karadžić IX* [Serbian traditional proverbs, The complete works of Vuk Karadžić IX]. Beograd: Prosveta.

Pasamanick, Judy. 2000. Watched pots do boil: Proverb interpretation through contextual illustration. *DeProverbio* 11/12 http://www.deproverbio.com/DPjournlal/DP,6,1,00/PASAMANICK/WATCHEDPOT.html. (accessed 23 September 2003).

Prćić, Tvrtko.1997. *Semantika i pragmatika reči* [Semantics and pragmatics of the word]. Sremski Karlovci and Novi Sad: Izdavačka knjižarnica Zorana Stojanović

Ruiz de Mendoza Ibanez, Francesco. 2003. The role of mappings and domains in understanding metonymy, 109–132. In: Antonio Barcelona (ed.) *Metaphor and metonymy at the crossroads: A cognitive perspective*. Berlin and New York: Mouton de Gruyter.

Sperber, Dan and Deidre Wilson. 1995. Relevance: Communication and cognition. Second Edition. Oxford: Blackwell.

Steen, Gerard J. 2002. Identifying metaphor in language: A cognitive approach. *Style* 36. 368–407.

Stefanowitsch, Anatol. 2006. Corpus-based approaches to metaphor and metonymy. In Anatol Stefanowitsch and Stefan Th. Gries (eds.) *Corpus-based approaches to metaphor and metonymy,* 1–17. Berlin and New York: Mouton de Gruyter.

Turner, Mark and Gilles Fauconnier. 2003. Metaphor, metonymy, and binding, 132–145. In: Antonio Barcelona (ed.), *Metaphor and metonymy at the crossroads: A cognitive perspective*. Berlin and New York: Mouton de Gruyter.

Yusuf, Yisa Kehide. 1997. The sexist correlation of women with the non-human in English and Yoruba proverbs. *De Proverbio* 5/ 1997. http://www.deproverbio.com/DPjournal/DP,3,1,97/YORUBA_PROVERBS.html. (accessed 4 October 2004).

Emotions as causes of human behavior in Polish and Slovene[1]

Agnieszka Będkowska-Kopczyk, University of Bielsko Biała

1. Introduction

According to Dirven (1997:55), one way of understanding concepts related to the category of emotions is to examine how emotions are conceptualized in the context of emotional causality. With reference to English prepositional expressions, Dirven (1997) and Radden (1998) have pointed out that it is the extensions of spatial concepts such as CONTAINER, STARTING POINT, COMPANION, PATH that serve to express the idea of emotions as abstract causes of human responses. In this paper I examine selected prepositional and case constructions with emotion nouns in Polish and Slovene[2] in order to show that the structure of EMOTIONS, understood as CAUSES OF HUMAN CONTROLLABLE AND UNCONTROLLABLE BEHAVIOR (i.e. PHYSIOLOGICAL RESPONSES AND ACTIONS), is shaped in Slavic by two source domains, CAUSE and FORCE, and that the following ontological correspondences play an important role in the understanding of emotions as causes: 1) emotions are causes of human reactions, 2) causes are forces, and 3) emotions are forces.

Following the assumption adopted by cognitivists that concepts arise from both physical and cultural experience, I shall suggest that it is human experience pertaining to the action of natural and supernatural forces that influences the structure of emotional causes. This is reflected in the primary metaphoric schema EMOTIONS ARE FORCES, perspicuously elaborated by Kövecses (2000:61–86), which may be more precisely defined as EMOTIONS ARE INTERNAL or EXTERNAL FORCES WHICH ARE AGENTS OF THE PERSON'S PHYSIOLOGICAL RESPONSES AND ACTIONS. Studies of the conceptual structure of the concept CAUSE, conducted from the cognitive perspective by Talmy, Lakoff and Johnson, have in turn shown that this concept is understood with reference to the category of FORCE, expressed in the conceptual metaphor CAUSES ARE

---

1 I would like to thank Carl Humpries and Marcin Grygiel who did the final proof reading of the manuscript and Ljiljana Šarić for all her valuable comments on the quoted examples.

2 The examples presented in the paper come from the Polish phraseological dictionary (SFJP), computer dictionary of Polish language (KSJP) and Slovene dictionary of literary language (SSKJ). Other data are based on the IPI PAN corpus of Polish and Nova beseda corpus of Slovene (abbreviations IPI PAN and NB, respectively). Since my analysis is inspired by Dirven's (1995, 1997) and Radden's (1998) works on English prepositional expression, the Slavic constructions under question will be also disscused together with relevant phenomena in English.

FORCES (Talmy 1988, 2000, Lakoff 1987, Lakoff and Johnson 1980, 1999). This also applies to abstract causes, which include emotions.

Kempf (1978, 2007), analyzing the Indo-European languages, has shown that phenomena of the spatial world influence our understanding of categories, and this can be studied within the framework of a localist understanding of cases. In the Proto-Indo-European language three types of constructions were predominant in articulating the relationship of causality: the instrumental of cause (labelled as *Instrumentalis causae* by Kempf) and, in terms of directional understanding, two cases indicating movement from something: the ablative and elative (Kempf 2007: 69). The instrumental of cause currently occurs only in Polish, whereas the ablative case is expressed in Polish and Slovene by means of the preposition *od* 'from' + genitive, and the elative is expressed in Polish by means of the preposition *z(ze)* 'from', 'out of' + genitive and in Slovene by the preposition *iz* 'out of' + genitive. In this paper I shall confine the analysis to examples that are realizations of these three grammatical structures.[3] As Johnson (1987) has pointed out, the concept of FORCE is multi-dimensional and has the structure of a complex gestalt, which reflects the different characteristics and action patterns of force in the spatial dimension. I assume that each of the examined constructions expresses a different structure of FORCE. Instrumental constructions express EXTERNAL FORCE structured as an INSTRUMENT which causes physiological responses, ablative constructions express EXTERNAL FORCE structured as SOURCE and INITIAL POINT of physiological responses, elative constructions express EXTERNAL or INTERNAL FORCE structured as CONTAINER-SHAPED SOURCE and INITIAL POINT of both physiological responses and actions. Following Radden and Dirven, I shall assume that the choice of an appropriate prepositional phrase also in Slavic depends on whether the link between the effect and its emotional cause is controlled by the person experiencing emotion or not. If EMOTION is conceptualized as AN EXTERNAL FORCE acting independently of his/her will, the link between the emotion and the response (which in most cases is a physiological one) is uncontrollable. If EMOTION is conceptualized as AN INTERNAL FORCE, and thus as resulting from his/her will, the link between the emotion and the action is controllable.

Proponents of the cognitive theory of language (Tyler and Evans 2003, Dirven 1995, 1997, Radden 1998, Przybylska 2002, Šarić 2008) have demonstrated that the meanings of prepositions are based on conceptual schemas and they are created on the basis of metaphorization and metonymization. Bacz (1997) points out that in Slavic languages, unlike in English, meanings of prepositions in a prepositional phrase should be analyzed along with the basic

3 Beside the construction types mentioned above emotional causality can be expressed by constructions with other prepositions, e. g. English *in, for, through, by* as shown in Radden (1998), Dirven (1997:58–68), Croatian *u* 'in' + locative (Šarić 2008:25), Polish *w* 'in' and Slovene *v* 'in' + locative (Będkowsk–Kopczyk (2009:38–39), and Old-Polish *przed* 'before' + instrumental (Kempf 2007:43).

meaning of the case which this preposition combines with and whose meanings also reflect spatial relations, transferred via metaphorical extension onto abstract domains (cf. Janda 1993, 2004, Rudzka-Ostyn 2000, Przybylska 2002, Šarić 2008). The meanings of prepositions and cases form radial structures, with the prototypical spatial meaning located at the center of the semantic category of particular prepositions. Tyler and Evans (2003:52) use the concept of a *proto scene*, meaning "an idealized mental representation across the recurring spatial scenes associated with a particular spatial particle (…)" in the semantic analysis of prepositions. In considering the examples analyzed here I will refer to ideas related to experiences with natural forces which may have influenced the development of metaphorical extensions of the FORCE concept to the domain of CAUSES and EMOTIONS. Such images may not have the same explicatory force as *proto-scenes*; however, they do enrich the account, not only with elements of real-world physical experience but also by highlighting culture-dependent aspects of our experience which, in my opinion, have been an important factor in shaping the folk-understanding of emotions as causes. Likewise, I refer to the conclusions of Janda (2008), who suggests that grammatical structures reflect culturally specific interpretations of physical phenomena in the surrounding world.

Langacker (1987) states that an expression's meaning is not just the conceptual content it evokes but the way the content is construed by a symbolic structure. Out of the many possibilities which linguistic conventions have provided us with, we choose those with which we can build our utterance as aptly as possible or in the manner that best suits our communicative needs. Some of the constructions analyzed here – especially those relating to emotions as uncontrollable causes of human actions – have been conventionalized, and have the structure of phraseological units. The choice of a prepositional phrase is, however, not arbitrary, and depends on the speaker's intention, which, depending on what aspect of an emotion he/she wants to elicit, involves choosing between competing constructions (e.g. with the preposition *od* 'from' or *iz* 'out of' in Slovene). In other, non-conventionalised expressions, the same criteria determine the appropriate choice of preposition. Following Tabakowska (2001:7–8), I assume that the analysis of prepositional and case structures should be considered holistically, i.e. in terms of the Langackerian categories of "scene" or "event", where this then enables us to arrive at a complete characteristisation of the mechanisms involved: those that condition the speaker's selection of a particular manner of constructing the scene. In the analysis of the material presented here the semantics of the verb plays the major role – it is the verb that expresses the kind of reaction evoked in people under the influence of an emotional cause. The kind of reaction (i.e. physiological response or action), in turn, "cooperates" with the conceptualization of an emotion, and consequently with the selection of a specific preposition and case.

2. Emotions as causes in Slovene and Polish

2.1. *The Polish instrumental case: uncontrollable causes of states*

In Polish there is a group of phraseological units in the form of case constructions with the instrumental which, as with English *with*-phrases (Dirven 1998, Radden 1997), express intense emotions of the sort that elicit uncontrollable human responses such as the following: a physiological state (1a), a response that is visible across the body (1b), one that is visible across one's face or eyes (1c), or behavior that can be considered unintentional (1d). These are illustrated below:

(1) a) Pol. *kipieć radością*
boil over joy-INST
'to boil over with joy'

b) Pol. *promieniować szczęściem*
beam happiness-INST
'to glow with happiness'

c) Pol. *twarz wykrzywiona złością*
face twisted anger -INST
'a face twisted with anger'

d) Pol. *ziać nienawiścią*
breathe hatred-INST
'to breathe with hatred'

In Polish these constructions are the remnants of instrumental of cause (*Instrumentalis causae*), which, according to Kempf (1978, 2007), was present in Indo-European languages (cf. Luraghi 2003), and is still in use in some Slavic languages (e.g. Russian and Czech).[4] Before we discuss the role of this construction in the expression of emotions as causes, let us present its role in the expression of causality in general.

Kempf (2007:68) explains the essence of the meaning of these structures in terms of approximating the category of CAUSE to INSTRUMENT (the category experienced by people as part of their daily life), which is an example of linguistic anthropomorphization of the category CAUSE. The concept of instrument is associated with the concept of "agent" – the person who operates

4 In Polish *Instrumentalis causae* began to fall out of use in the Middle Ages, and it has now been replaced with numerous prepositional constructions, inter alia, with prepositions *od* 'from' and *z* 'out of' + genitive (cf. later in the text) (Kempf 2007:71–72).

the instrument – and, in virtue of that (as Kempf argues), we encounter examples such as the following:

(2) a) Pol. *człowiek zginął mieczem*
man was killed sword-INST
'a man was killed by a sword'

and, equally:

(3) a) Pol. *on uderzył mieczem*
he hit sword-INST
'he hit with a sword'

The meaning of the instrumental, which expresses the concept of "originator of activity" and which may refer both to people and objects, is based on the following metaphor: AGENTS ARE INSTRUMENTS (Tabakowska 2002:379).[5] According to our folk knowledge, each event should have its agent or perpetrator. On the basis of examples quoted by the author, such as:

(4) a) Pol. *Nero (...), kiedy swą umierał ręką* (Kempf 2007:71)
Nero (...) when his own died hand-INST
'Nero (...) when he died by his own hand'

b) Pol. *tylko muzułmańską krwią te ognie gasną* (Kempf 1978:98)
only Muslim blood-INST these flames die out
'these flames are extinguished only by Muslim blood'

it can be concluded that in these examples "agent" can be expressed by a possessive pronoun (*swą* 'his own') or a possessive adjective (*muzułmańską* 'Muslim'), which describes the noun referring to an instrument. It then follows that "the agent" who causes a particular event as a result of which the state of the entity in the nominative case is changed, is conceptualized as OWNER OF THE INSTRUMENT. In these examples we observe the following metonymy: AN INSTRUMENT STANDS FOR THE ACTIVE AGENT TO WHOM THE INSTRUMENT BELONGS (cf. Luraghi 2003:32 who also argues that the extension from Instrument to Agent is based on metonymy).

In many other Old-Polish examples with *Instrumentalis causae* (Kempf 1978:97–98, 2007:71), such as:

(5) a) Pol. *umrzeć chorobą*
die illness-INST
'die because of illness'

---

5 This case is also referred to as the *instrumental of agent* (Luraghi 2003:32).

b) Pol. *pożarem zgorzała* (Kempf 1978:97)
fire-INST burnt-PERF she
'she was burnt by fire'

c) Pol. *pogubieni byli walkami od wielu narodów* (Kempf 1978:97)
were lost-PAST they wars-INST from many nations-GEN
'they were lost because of wars with many nations'

change of the state of the entity coded by the nominative is caused by an event. The concept of INSTRUMENT is extended to the abstract domains of EPIDEMIC ILLNESS, FIRE or WAR, which belong to the category of non-controlled states of affairs (Luraghi 2003). People tend to believe that there are some forces – be they natural or supernatural – that stand behind the instruments that exert control over them. The forces are defined by one's knowledge of the world and outlook, or by the religious system predominant in a particular culture. This type of event-cause is referred to – even in Polish legalese – as so-called force majeure.[6] In the expressions above there is no human causer (instigator). We assume that EVENT-INSTRUMENT is conceptualized as NATURAL FORCE, which takes control over people and changes their state. People are helpless when confronted with this force.

Finally, the concept of INSTRUMENT is extended to the abstract domain of EMOTIONS, which are understood as CAUSES of physiological processes and changes of the state of the person experiencing emotions. This is illustrated by an Old-Polish example: *drżeli strachem* [they trembled fear-INST] 'they trembled with fear' (Kempf 1978:97–98, Kempf 2007:71) as well as contemporary examples, mentioned in (1). These expressions mirror involuntary physiological changes sensed by people and observable across their body or changes of state such as disease or death. Lakoff and Johnson note that each force must apply to something producing some effect as it is applied (Lakoff and Johnson 1999:207). Its effect might be the entity's movement as well as the entity's change of appearance: "In the most basic kind of causation, a physical force is applied to move something or change its appearance" (Lakoff and Johnson 1999:213). The examples under consideration do not refer to the image of movement, but rather to the change of the epxperiencer's appearance. Each change, like a movement (cf. Mandler 1992 as shown later in the text), may result from the will of the one who changes (himself/herself) (*self-change*) or can be independent of the will (*caused-change*). In the case of physiological changes we are only dealing here with a caused change (one cannot control his/her internal physiological processes), and what evokes changes is an indefinite external force, which transmits energy to the person experiencing

6 With regard to the Old Testament it is God's will that is the cause of misfortunes such as hunger, plague, crop failure or natural distasters.

emotion and makes changes in his/her organism. This FORCE is structured as an INSTRUMENT, which – similarly to the examples with EVENT-INSTRTUMENT – is identified with an indefinite CAUSATIVE FORCE, it takes control over the person and evokes responses in him/her. In other words, the person is the object of the action of EMOTIONS-INSTRUMENT, and his/her involuntary responses are the effect of this action.[7] Thus, it might be assumed that the aforementioned Old-Polish *drżeli strachem* and the examples in (1) arose as a result of the integration of several domains: EMOTION, CAUSE, FORCE, INSTRUMENT and on the basis of the following extensions: EMOTION IS CAUSE, EMOTION IS INSTRUMENT, INSTRUMENT IS FORCE, EMOTION IS FORCE, CAUSE IS FORCE. This type of construction (and others, such as *Instrumentalis activus*, where a person is conceptualized as an instrument) reflect human ideas dating from the times of the naturalistic interpretation of the world, where man was the instrument of natural forces rather than a free agent (Kempf 2007:45).

Interestingly, the examples in (1) can be contrasted with the Polish and Slovene expressions, which only seemingly express a similar conceptualization of emotions. For example:

(6) a) Pol. *pokonać zło miłością*
conquer evil-ACC love-INST
'to conquer evil with love.'

b) Slo. *zlo premagati z ljubeznijo*
evil-ACC conquer with love-INST
'to conquer evil with love'

The verbs in (6) express the effects of one's intentional activity. EMOTION is conceptualized as INSTRUMENT (MEANS), which people use in order to achieve a particular goal. The examples in (6) imply an understanding of emotions as the causes of intentional activities, and refer to emotions that, in terms of their categories, border on propositional attitudes. In these examples we are dealing with the integration of the domain of EMOTION and INSTRUMENT. I therefore believe that the examples in (1) and (6) may be regarded as instances of homonymy.[8]

---

7 The entity in the nominative case does not refer to the active causer (like in the sentence. *Jan kroi chleb nożem* John-NOM is cutting bread-ACC knife-INST 'John is cutting bread with a knife.'), but is ascribed the role of an archetypical experiencer. The experiencer undergoes the action of the agens, expressed via instrumental.

8 A similar meaning is expressed in Polish by means of the construction with the preposition *przez* 'through' + accusative, where *przez* expresses an INSTRUMENT, which is used to achieve a goal.

Polish, Slovene and English also have constructions with the preposition *z* 'with' + instrumental, based on the COMPANION schema (ASSOCIATED ENTITIES ARE COMPANIONS, cf. Radden 1987:279). However, these constructions demonstrate a different aspect of emotions than the examples in (1) and (6). Verbs in this type of construction do not express people's physiological responses but their more or less controllable and intentional action:

(7) a) Pol. [...] *z miłością opiekuje się chorym ojcem.* (IPI PAN).
with love-INST take care REFL sick father-INST.
'She takes care of the sick father with love.'

b) Slo. *Pogledam ga z gnusom v očeh.* (NB)
is.looking-1sg he-ACC with aversion-INST in eyes-LOC
'I am looking at him with aversion in my eyes.'

Examples in (7) describe the way in which an action is carried out. For instance, taking care of someone with love means being attentive, patient, understanding, whereas looking at someone with aversion is to look at them in a way that an observer would interpret as revealing loathing, etc. These expressions describe a person's behavior from the perspective of the observer; behavior that, owing to the metonymy AN ACTION STANDS FOR THE EMOTION, is conventionally linked to an emotion expressed by a noun. The coexistence of actions and emotions explains why in these examples there is the preposition *z* 'with' + instrumental, whose spatial meaning is based on the COMPANION schema, and more precisely on the image of an accompanying person or thing (cf. Eng. *to sing with joy*). This meaning articulates the relation expressed by the so-called *sociativus* or *comitativus* (cf. Maciejewski 1996:117). Examples in (7) do not express emotional causality, because the way in which an action is performed by the person expressed by the nominative does not have to be dictated by his/her emotional commitment – one can take care of someone with love (that is, carefully, patiently), without actually loving him/her at all. Hence, the question that more aptly corresponds to the sentences above is "in what way" one behaves rather than "why". The metaphor ASSOCIATED ENTITIES ARE COMPANIONS is extended to the abstract domain of MANNER.

In the light of the above statements it can be assumed that English phrases with the preposition *with*, such as *to tremble with fear*, *to boil with disgust*, *to laugh with agony*, as analyzed by Dirven (1998) and Radden (1997), and expressing emotions as causes of uncontrollable physiological responses, have also arisen on the basis of the FORCE schema, structured as an instrument which influences one's physiological condition, thus causing his/her involuntary responses. They would therefore be a continuation of the same Indo-European construction as *Instrumentalis causae*, described by Kempf. Obviously, Dirven

(1995, 1997) and Radden (1998) associate the preposition *with* with the causative meaning, but they interpret it with reference to the COMPANION schema, which reflects a different spatial relationship: coexistence of a physiological response and an emotion (the physiological state is constructed as the one that accompanies an emotion) rather than the action of a force structured as an instrument. In their opinion, it is the metonymic relationship between physiological responses and emotions that expresses the causative relationship between an emotion and a state: "a physiological or psychological state is construed as an accompanying an emotional state which, due to the metonymy AN EMOTIONAL STATE STANDS FOR THE CAUSE ORIGINATING FROM THAT STATE, is also understood as bringing the reaction about" (Radden 1998:282).

According to Dirven, the preposition *with* may also express the part-whole relationship of two companions and it is this meaning that serves as the source domain for the metaphorical extension to an inseparable link between an emotional cause and the physiological state. Undoubtedly, the basis of *with*-phrases is constituted by the master COMPANION metaphor, but this does not negate the fact that in examples such as *to tremble with fear* an emotion is simultaneously conceptualized as an instrument, as with the examples in (1). The COMPANION schema is based on the idea that an instrument can be conceived as acting together with an agent, in order to bring about a state of affairs (Lakoff and Johnson 1987:135).[9]

Hence examples such as *to tremble with fear* can be interpreted as the realization of the metaphors EMOTION IS COMPANION and EMOTION IS INSTRUMENT (FORCE), as opposed to e.g. *to sing with joy* or *made with love,* which simultaneously express COMPANION (the so-called comitative) and MANNER (there is the same difference between the Polish examples in 1 and 3). The conceptual difference between the constructions under consideration in relation to their spatial motivation may be testified by a 19th century meaning explication of the preposition *with* with reference to Latin by Grant (1823:154): "*with* is found before the *cause, manner* and *instrument*, and is a sign of ablative: as, he killed him with his own hand, *manu sua occidit. With*, denoting *in company with*, or *together with*, is made by *cum*: as, he entered with a sword, *cum gladio ingressus est*." Elsewhere the author then explains the use of the Latin ablative in the example of *Palleo metu*, which is translated into English by means of the preposition *with* as: 'I am pale with fear', that is, "not only with fear, but for fear, fear being not only an accompanying circumstance, but the cause of paleness" (Grant 1823:8). I shall argue that in this case, too, we can talk about homonymy, i.e. the homonymy of the instrumental (*to tremble with*

9 As Luraghi argues (2003), the analysis of relations between the comitative and instrumental has ensured that comitative markers are seen as a possible source for instrumental markers in the Indo-European languages: the Indo-European instrumental case derived its instrumental meaning from an earlier comitative meaning.

fear) and sociative (*to sing with joy*), as noted by Maciejewski (1996:117). This applies, inter alia, to English, but also to Slovene, where the instrumental, in addition to occurring with a morphological marker, is always accompanied by the preposition *z* 'with'. In Polish it is the instrumental itself that performs the function of an instrument, whereas the function of accompanying (the so-called sociativus or comitativus) has been taken over by the preposition *z* 'with' + instrumental.

2.2. *Slovene and Polish* od *'from' + genitive: uncontrollable causes of states and actions*

In English *from* denotes the cause both of a physiological response and human action, which is neither conscious nor controllable (Dirven 1997:64). The Slovene ablative construction with the preposition *od* 'from' + genitive primarily conveys the idea of emotions as uncontrollable causes of physiological responses. As with Polish instrumental constructions, these expressions involve conventionalized phraseological units:

(8) a) Slo. *tresti se od jeze*
tremble from anger-GEN
'to tremble with anger'

b) Slo. *postati rdeč od jeze*
get red from anger-GEN
'to get red with anger'

c) Slo. *umreti od strahu*
die from fear-GEN
'to die from fear'

d) Slo. *izbruhniti od jeze*
explode from anger-GEN
'to explode with anger'

e) Slo. *sopihati od jeze*
breath hard from anger-GEN
'to breath hard with anger'

The types of responses evoked by an emotion are similar here to the responses expressed in (1) by Polish instrumental expressions, and to the English examples involving *with*-phrases mentioned earlier.

In Polish we may also encounter examples where the combination of *od* 'from' + genitive conveys the idea of emotions as causes of physiological responses. For example:

(9) a) Pol. [...] *gadała prędko, ledwie już dysząc od złości.* (IPI PAN)
is talking-3rd PAST fast hardly breathing from fear-GEN
'She was talking fast hardly breathing from fear.'

The Polish preposition *od* 'from' + genitive in conjunction with names of emotions has now been superseded by the preposition *z* (*ze*) 'out of' + genitive (as shown later in the text), even though dictionaries published at the beginning of the 19th century (Linde 1814 [1860]) and the 20th century (Karłowicz, Kryński, Niedźwiedzki 1927 [1953]) record both constructions on an equal basis: "*pienić się, zaczerwienić od złości* a. *ze złości*" (literal translation: 'foam, blush from anger or out of anger') (Karłowicz, Kryński, Niedźwiedzki 1927 [1953]:531).

Slovene and Polish phrases with the preposition *od* 'from' + genitive are prototypically related to the idea of the spatial human or non-human SOURCE or INITIAL POINT OF MOTION, whereas the shape of this source is not precise, e.g.

(10) a) Pol. *Wychodzę od Piotra.*
leave-1SG from Peter-GEN
'I am leaving Peter.'

b) Slo. *Prišel je od nekod iz Nemčije.* (NB)
came-3SG PAST from somewhere-GEN out of Germany-GEN
'He came from some place in Germany.'

As Šarić states, the meaning of the phrase *od* 'from' + genitive bears the meaning of the source in many spatial uses, the meaning being triggered by the verb indicating withdrawal or removal from the object designated by the genitive noun. In that spatial relation, a figure is moving away from the location[10] or is being removed from it (Šarić 2008: 24).

Mandler (Mandler 1992:593–595, after Peña 2003:144–145) distinguishes between a *self-instigated motion*, where a figure starts to move on its own as if it were animate, and a *caused motion*, where one entity sets the other in motion.[11] The self-motion results from the moving entities's will, whereas the caused motion is instigated by an external causative factor, which is some

10 To describe spatial configurations cognitive linguists, including Šarić (2008) and Peña (2003), use the Langackerian terms *trajector* and *landmark* (Langacker 1987). The trajector refers to a located object (a foreground figure), while the landmark refers to the location of the figure (it is a reference point entity).

11 Cf. the Langackerian billiard-ball model (Langacker 1991).

causative force or other. Peña (2003: 146–148) shows that the concept of COMPULSION, one of the types of FORCE image schema defined by Johnson (1987: 145), is essential to an understanding of the category of caused motion. She notes that "If any external force (either in the form of an emotion or of any other subject entity) is seen as if endowed with will-power, it will be able to cause any passive subject to move and to exert control over this subject" (Peña 2003: 146). However, the examples in (8) – (9) do not refer to an image of motion but, as with the examples in (1), to a change in the appearance or state of the person experiencing emotions. We assume that the action of natural forces such as the sun, rain, wind or lightning has played an important role in the formation of such expressions. Such "visible" forces would affect (e.g. hit) objects, thus changing their appearance: e.g., objects starting to catch fire. Besides this, there are "invisible" yet noticeable forces, such as frost or high temperature, described by Radden (1998:280) as "external circumstances", which also engender both physiological responses (e.g. sweating, trembling, getting numb) and changes of a person's appearance. Causes of this type are expressed in modern Polish and Slovene by the preposition *od* 'from' + genitive, as in the following examples:

(11) a) Pol. *skostnieć od mrozu*
grow numb from frost-GEN
'to grow numb from the frost'

b) Slo. *biti trd od mraza*
grow numb from frost-GEN
'to grow numb from the frost'

The same is true for English *from*-phrases (e.g., *to shiver from the cold*), as Dirven (1995: 109) has shown.

The metaphorical extension of the SOURCE concept to the metaphorical meaning of EMOTION as CAUSE of uncontrollable responses is again related to the idea of EXTERNAL FORCE, which influences people independently of their will.[12] This FORCE is SOURCE and simultaneously the INITIAL POINT of one's uncontrolled behavior and state. The examples in (8) – (9) were formed again as a result of the integration of several domains: EMOTION, CAUSE, FORCE, SOURCE and INITIAL POINT OF CHANGING PROCESS and on the basis of metaphorical and metonymic extensions: EMOTION IS CAUSE, SOURCE and INITIAL POINT REPRESENTS EMOTION, and EMOTION IS FORCE (COMPULSION) (CAUSE IS FORCE).

---

12 People have no control over physical forces, therefore, the link between an emotion and a response evoked by the emotion is conceptualized as uncontrollable.

In Slovene the preposition *od* 'from' + genitive may also express emotions that are causes of actions. (However, these constructions are extremely rare in this language.) For example:

(12) a) Slo. *Od jeze je porinil mater.* (NB)
from anger-GEN pushed-3SGPAST mother-ACC
'He pushed his mother from anger.'

The example in (12) in contrast to the examples in (8) is not a phraseological unit, and the preposition *od* 'from' + genitive is chosen because the speaker interprets the destructive action of a person, who is under the influence of a strong and intense emotion, as uncontrollable behavior resulting from "an outburst of emotions". In this example, the verb expresses the purposeful action of a person experiencing emotions. Normally, it is the preposition *iz* 'out of' + genitive (as explained later in the text) that combines with such verbs in the domain of emotional causality. The use of the preposition *od* 'from' + genitive, which, in conjunction with emotion nouns, denotes the causative action of EXTERNAL force, suggests that the example in (12) was formed on the basis of metaphorical and metonymic extension, as with the examples in (8) – (9).

2.3. *The Slovene* iz *'out of' + genitive and Polish* z(ze) *'out of' + genitive: controllable causes of actions*

The English "source" preposition *out of* denotes the rational motivation for an action. Emotion in the *out of*-phrases is seen both as a controllable source of action and a rational explanation for it (Dirven 1987:61). The same meaning is represented by Polish and Slovene equivalents of the preposition *out of* – the Slovene *iz* and the Polish *z(ze)* – both combined with the genitive. That applies both to intense short-lasting emotions (anger, fear) and long-lasting ones (love, hatred). For example:

(13) a) Pol. *Bóg stworzył człowieka z miłości.* (IPI PAN)
god created a man-ACC out of love-GEN
'God created a man out of love.'

b) Pol. (…) *wy drwicie ze złosci* (…) (IPI PAN)
you mock out of anger-GEN
'(…) you are mocking out of anger (…)'

c) Slo. *Firmo lahko vodiš tudi iz ljubezni.* (NB)
a company- ACC you can run also out of love-GEN
'You can run the company also out of love'

d) Slo. *Gole sem zabijal iz jeze.* (NB)
goals scored-1SGPAST out of anger-GEN
'I scored goals out of anger.'

The verbs in (13) express actions (creating, scoring goals) which people perform consciously and according to their will. It can therefore be argued that in these examples the emotion is perceived as the cause (or even motivation) of one's intentional activity. It is worth noting that among these links there are no phraseological units, and this proves that emotions can be a motivation for any conscious action. EMOTION is conceptualized here as a spatial SOURCE (and simultaneously) INITIAL POINT. However, in this case it has the shape of a CONTAINER. The prepositions *iz* or *z* (*ze*) 'out of' indicate that the figure departs from or is removed from the container-shaped location (Šarić 2008:24). As with the *od* 'from' + genitive phrases discussed earlier, the image of a figure's motion to outside of the container is associated with *self-instigated* or *caused-motion*. Since verbs express intentional actions, it can be argued that the examples in (13) were formed on the basis of the metaphorical extension of the figure's spontaneous motion to outside of the container. This image may stem from our experience of going out from container-shaped objects, where this happens in accordance with our will and is driven by an internal force (cf. Radden 1998:287–288).[13] The examples in question are the result of the integration of the following domains: EMOTION, CAUSE, FORCE, CONTAINER-SHAPED SOURCE and INITIAL POINT OF MOVEMENT. Moreover, they are based on the following metaphorical and metonymic extensions: CAUSE IS CONTAINER-SHAPED SOURCE or INITIAL POINT OF MOVEMENT, and SOURCE REPRESENTS EMOTION and EMOTION IS INTERNAL FORCE (CAUSE IS INTERNAL FORCE).

2.4. *The Polish* z/ze *'from, out of' + genitive: uncontrollable causes of states*

Interestingly, in contrast to Slovene and English the preposition *z*(*ze*) 'out of' + genitive is used – mainly in phraseological units – to also express emotions as causes of uncontrollable states and processes. The types of response evoked by emotions here are the same as in English *with*-phrases and Slovene *od* 'from' + genitive phrases, and they are also similar to those in Polish instrumental expressions. For example:

(14) a) Pol. *trząść się ze złości*
tremble out of anger-GEN
'to tremble with anger'

13 People usually control the actions they undertake in accordance with their own will, so the link between the emotion and the behavior caused by the emotion is conceptualized as controllable.

b) Pol. *posiwieć ze strachu*
become grey out of fear-GEN
'to become grey out of fear'

c) Pol. *oszaleć z miłości*
go mad out of love-GEN
'to go mad from love'

d) Pol. *kipieć ze złości*
boil out of anger-GEN
'to boil with anger'

e) Pol. *dyszeć z gniewu*
breath hard out of anger-GEN
'to breath hard with anger'

However, this raises the question of why the same pattern is involved in Polish in the conceptualization of two different types of events – intentional actions and uncontrollable physiological responses. We assume that, as with (1), (8) and (9), the basis for the examples in (14) is to be found in the image of causal changes to people (i.e. caused changes of appearance), but in this case the source of these changes is a FORCE hidden in a CONTAINER. This image may result from observing the causes of natural phenomena, such as water flowing out of a spring (water flows out but the force that causes this action is unidentified because it is hidden in the container). The idea of mysteriousness conveyed by the interior of a container, especially a closed one, may correspond to what Kempf has, with reference to the preposition *za* 'behind' + instrumental, labelled a "hidden pre-source", understood to give rise to the phenomenon in question (Kempf 2007: 73). To sum up, the examples in (14) have arisen on the basis of metaphoric and metonymic extensions analogous to (8) – (9), the difference being the CONTAINER-shaped SOURCE and INITIAL POINT of the person's changes, while the FORCE which brings about people's response is undefined (invisible).

3. Conclusions

EMOTIONS as CAUSES of physiological changes or human behavior are understood with reference to the concept of FORCE, which may be structured as follows:

1) INSTRUMENT, which influences people, takes control over them, and regardless of their will, triggers physiological processes in them (caused changes) which, for an external observer can be seen across the body and in

the voice of the person experiencing emotions (Polish Instrumental case – *Instrumentalis causae*).

2) EXTERNAL FORCE (COMPULSION), which influences people from a source of unspecified shape, and whose action is the starting point for changes (caused changes): physiological processes, visible across their body and in their voice (Slovene and Polish *od* 'from' + genitive).
3) EXTERNAL FORCE (COMPULSION), which influences people from CONTAINER-SHAPED SOURCE and whose action is INITIAL POINT OF CHANGES (caused changes): physiological processes, visible across theri body and in their voice (Polish *z*(*ze*) 'out of' + genitive).
4) INTERNAL FORCE, which influences people from CONTAINER-SHAPED SOURCE and whose action is INITIAL POINT OF CONSCIOUS AND INTENTIONAL ACTION (self-movement) (Slovene *iz* 'out of' + genitive, Polish *z*(*ze*) 'out of'+ genitive).

Referring again to Kempf (1978:8), who argues that formal instrumental, ablative and elative constructions were used in the Proto-Indo-European language to articulate the relationship of causality, one might be tempted to presume that there should be similarities in the expression of emotional causes between the Slavic languages analyzed here, and English. Today, in these languages, elative constructions function in the most approximate manner, and serve to articulate the idea of emotions as causes (and motivations) for people's conscious and intentional activity. In this respect, however, Polish is less precise because elative expressions with *z*(*ze*) 'out of' are also used to articulate the idea of emotions as causes of physiological changes, independent of a person's will. Ablative constructions are used to express emotions conceptualized as causes of uncontrollable physiological changes in Slovene and Polish (rarely today), and in English. In Slavic, ablative constructions with the preposition *od* 'from' serve to articulate the idea of emotions as causes of actions, but this refers to actions performed under strong emotion or, as Dirven (1997:63) puts it with regard to English *from*, to an action whose motives the person experiencing the emotions is not properly aware of.[14] Analyzing it from a historical perspective lends support to the conjecture that Polish instrumental structures may, like English *with*-phrases (such as *to tremble with fear*), be interpreted as a continuation of Indo-European's *instrumentalis causae*. Similarities and differences in the ways of expressing emotions-causes in Polish and Slovene in comparison with English are listed in the Table 1.

---

14 Dirven (1997:63) interprets the example *Had I refused out of fear as well as from resentment?* in the following way: the speaker is not sure whether his actions are inspired by the emotion or by other less conscious motives.

Table 1. Similarities and differences in the ways of expressing emotions-causes in Polish and Slovene in comparison with English

		Instrumentalis causae	*Instrumentalis sociativus**	*Ablativus*	*Elativus*
Type of FORCE and reaction		EXTERNAL FORCE	INTERNAL FORCE	EXTERNAL FORCE	INTERNAL FORCE
Image schema		INSTRUMENT	COMPANION	(SOURCE, INITIAL POINT OF STATE)	(CONTAINTER-SHAPED SOURCE, INITIAL POINT OF ACTION)
Construction type and its meaning	Polish	*Instrumentalis causae*	*z* 'with' + instrumental	*od* 'from' + genitive	*z(ze)* 'out of' + genitive
	type of action	caused changes	intentional action or manner	caused changes	intentional action or caused changes
	Slovene	-	*z* 'with' + instrumental	*od* 'from' + genitive	*iz* 'out of' + genitive
	type of action			caused changes	intentional Action
	English	*with*-phrase		*from*-phrase	*out of*-phrase
	type of action	caused changes	manner	intentional action or caused changes	intentional action

* Non-casual constructions.

It ought to be emphasized that all the constructions considered here, expressing caused physiological changes which one experiences under the influence of emotional causes (i.e. Slovene and Polish *od* 'from' + genitive phrases, and Polish *z(ze)* 'out of' + genitive phrases, as well as Polish instrumental constructions), are overwhelmingly conventionalized in the languages in question, and function as phraseological units. In all of them we are dealing with the metonymy AN EMOTIONAL STATE STANDS FOR THE CAUSE ORIGINATING FROM THAT STATE (Radden 1998).

With regard to the scientific model, emotions are not causes of physiological responses but their consequences (Le Doux 1996). The occurrence of physiological responses in the body is determined by external stimuli (events). Stimuli, which we record by means of the senses, at the same time stimulate the so-called amygdale and the cerebral cortex. In practice, that means that the body's response to a stimulus commences before the neocortex has begun to process information (i.e. before we realize that we are experiencing a particular emotion). From the anatomical point of view, in the case of situations such as those involving danger, the emotional system acts largely independently from the neocortex and the higher cognitive processes stimulated by it. Therefore, the aforementioned phraseological units have been formed on the basis of a (folk) image, according to which it is emotions rather than particular stimuli that trigger physiological changes in the functioning of people's bodies. Human beings have a tendency to metonymically associate the symptoms of these processes with emotions/feelings: the reddening of one's face is conventionally associated with shame or anger, a trembling body with fear or anger, etc. It is the situational context that enables us to "match" observable changes in people's appearances with the emotions accompanying them. Without this, a trembling body could be interpreted as a symptom of both anger and fear.

As Malinowski (2000:39) puts it, the basic grammatical categories, universal for each and every language, can only be understood through their relation to the pragmatic outlook of primitive man, so that by using language these – as the author calls them – "barbaric" primitive categories must have had a profound impact on the subsequent development of mankind's philosophical orientation. I believe that this analysis complements our knowledge concerning the ideas underlying not just the conceptualization of EMOTIONS in Slavic languages but also the three basic ways of structuring the category of FORCE (INSTRUMENT, SOURCE AND CONTAINER-SHAPED SOURCE) involved in the conceptualization and expression of causality in the languages discussed here.

References

Bacz, Barbara. 1997. On the meaning of prepositional cases: The locative and the accusative in Polish expressions of time. *Canadian Slavonic Papers*. 39(3–4), 417–438.

Będkowska–Kopczyk, Agnieszka. 2009. W „pojemniku uczuć". Analiza wybranych konstrukcji przyimkowych w języku polskim i słoweńskim. [In the container of emotions. Analysis of selected prepositional constructions in Polish and Slovene] In Ilona Generowicz, Elżbieta Kaczmarska & Ignacy M. Doliński (eds.), *Świat ukryty w słowach, czyli o znaczeniu gramatycznym, leksykalnym i etymologicznym* [World hidden in the words, or about grammatical, lexical and etymological meaning], 27–44. Warszawa: Wydawnictwo Uniwersytetu Warszawskiego.

Dirven, René. 1995. The construal of cause: The case of cause prepositions. In John R. Taylor & Robert E. MacLaury (eds.), *Language and the Cognitive Construal of the World*, 95–118. Berlin & New York: Mouton de Gruyter.

Dirven, René. 1997 Emotions as cause and the cause of emotions. Suzanne Niemeier & René Dirven (eds.), *The Language of Emotions: Conceptualization, Expression, and Theoretical Foundation*, 55–84. Amsterdam/Philadelphia: John Benjamins Publishing Company.

Grant, John. 1808 [1823] *Institutes of Latin Grammar.* 2nd edn. London.

IPI PAN. 2004. *Korpus Języka Polskiego IPI PAN* [The IPI PAN Korpus of Polish]. Warsaw: IPI PAN.

Janda, Laura Alexis. 1993. *A Geography of Case Semantics: The Czech Dative and the Russian Instrumental*. Berlin: Mouton de Gruyter.

Janda, Laura Alexis. 2004. Concepts of Case and Time in Slavic *Glossos* 3 at http://seelrc.org/glossos/issues/3/. (accessed 2 May 2009)

Janda, Laura Alexis. 2008. From Cognitive Linguistics to Cultural Linguistics. *Slovo a smysl/Word and Sense* 8(2008). 48–68.

Johnson, Mark. 1987. *The Body in the Mind: The Bodily Basis of Meaning, Imagination, and Reason.* Chicago: Chicago University Press.

Karłowicz, Jan, Kryński Adam & Niedźwiedzki Władysław. 1927 [1953]. *Słownik języka polskiego* [The Dictionary of Polish Language] (vol. VIII Z –Ż). Warszawa: Państwowy Instytut Wydawniczy.

Kempf, Zdzisław. 1978. *Próba teorii przypadków. Część I.* [An attempt at a theory of cases. Part I.] Opole: Towarzystwo Przyjaciół Nauk.

Kempf, Zdzisław. 2007. *Próba teorii przypadków. Część II.* [An attempt at a theory of cases. Part II.] Opole: Wydawnictwo Uniwersytetu Opolskiego.

KSJP. 1995 *Komputerowy słownik języka polskiego*. [Computer Dictionary of Polish Language]. Warszawa: Wydawnictwo Naukowe PWN.

Kövecses, Zoltan. 2000. *Metaphor and Emotion: Language, Culture, and Body in Human Feeling*. Cambridge: Cambridge University Press.

Lakoff, George. 1987. *Women, Fire, and Dangerous Things. What Categories Reveal about the Mind.* Chicago: University of Chicago Press.

Lakoff, George & Johnson, Mark. 1980. *Metaphors we live by*. Chicago: University of Chicago Press.

Lakoff, George & Johnson, Mark. 1999. *Philosophy in the Flesh. The Embodied Mind and its Challenge to Western Thought.* New York: Basic Books.

Langacker, Ronald W. 1987. *Foundations of Cognitive Grammar*, vol. I. *Theoretical Prerequisites*. Stanford: Stanford University Press. Langacker, Ronald W.

1991. *Concept, Image, and Symbol: The Cognitive Basis of Grammar*. Berlin. New York: Mouton de Gruyter.
Le Doux, Joseph. 1996. *The Emotional Brain. The Mysterious Underpinnings of Emotional Life*. New York: Simon and Schuster.
Linde, Samuel, Bogumił. 1995. Reprint. *Słownik języka polskiego* [The Dictionary of Polish Language] (vol. VIII). 2nd edn. Lwów: Zakład Narodowy im. Ossolińskich. Warszawa: Wydawnictwo "Gutemberg Print". Original edition: 1814.
Luraghi, Silvia. 2003. *On the Meaning of Prepositions and Cases. A Study of the Expression of Semantic Roles in Ancient Greek*. Amsterdam/Philadelphia: John Benjamins Publishing Company.
Maciejewski, Witold. 1996. *O przestrzeni w języku. Studium typologiczne z językiem polskim w centrum*. [About space in the language. Typological sudy with the Polish language at the center.] Poznań: Wydawnictwo Naukowe UAM.
Malinowski, Bronisław. 2000. Problem znaczenia w jezykach pierwotnych. [The problem of meaning in primitive languages] In Krystyna Pisarkowa (ed.) *Językoznawstwo Bronisława Malinowskiego* Linguistics by Bronislaw Malinowski], 5–48. Kraków: Universitas.
Mandler, Jean. 1992. How to Build a Baby: II. Conceptual Primitives. *Psychological Review*, 99 (4). 587–604.
NB. 2000–2009 *Nova beseda* [New word corpus of Slovene]. Ljubljana: Fran Ramovš Institute of Slovenian Language.
Peña Cervel, Sandra M. 2003. *Topology and Cognition: What Image-Schemas Reveal about Metaphorical Language of Emotions*. München: Lincoln Europa.
Przybylska, Renata. 2002. *Polisemia przyimków polskich w świetle semantyki kognitywnej*. [Polisemy of Polish prepositions in the light of cognitive semantics.] Kraków: Universitas.
Radden, Günter. 1998. The conceptualization of emotional causality by means of prepositional phrases. In Angeliki Athanasiadou & Elżbieta Tabakowska (eds.), *Speaking of Emotions: Conceptualization and Expression*, 273–294. Berlin & New York: Mouton de Gruyter.
Rudzka–Ostyn, Brygita. 2000. *Z rozważań nad kategorią przypadka*. [Reflections on the category of case.] (Trans. Elżbieta Tabakowska). Kraków: Universitas.
SFJP. 2002. *Słownik frazeologiczny języka polskiego* [Polish phraseological dictionary] (vols. 1–2). Stanisław Skorupka (eds.) Warszawa: Wiedza Powszechna.
SSKJ. 1970–1991 *Slovar slovenskega knjižnega jezika* [Slovene Dictionary of Literary Language]. Ljubljana: SAZU in DZS.
Šarić, Ljiljana. 2008. *Spatial Concepts in Slavic. A Cognitive Linguistic Study of Prepositions and Cases*. Wiesbaden: Harrasowitz Verlag.
Tabakowska, Elżbieta. 2001. Kognitywizm: Obrazki z polskiej sceny. [Cognitivism: Pictures from the Polish scene] *Glossos* 1 at http://seelrc.org/glossos/issues/1/. (accessed 6 May 2009).
Tabakowska, Elżbieta. 2002 What can the Polish Instrumental be instrumental in? In Barbara, Lewandowska–Tomaszczuk & Kamila, Turewicz (eds.), 375–396. *Cognitive Linguistics today*. Lódź Studies in Language. Frankfurt am Main: Peter Lang.
Talmy, Leonard. 1988. The relation of grammar to cognition. In Brygita Rudzka Ostnyn (ed.), 165–206. *Topics in Cognitive Grammar*. Amsterdam: John Benjamins.
Talmy, Leonard. 2000. *Toward a Cognitive Semantics* (2 vols.). Cambridge, MA: MIT Press.
Tyler, Andrea & Evans, Vyvyan. 2003. *The Semantics of English Prepositions. Spatial Scenes, Embodied Meaning and Cognition*. Cambridge: Cambridge University Press.

III.
Cognitive discourse analysis and applied linguistics

Cognitive discourse analysis: local discourse structure[1]

Andrej A. Kibrik, Institute of Linguistics of the Russian Academy of Sciences

1. Introduction

In this paper I propose a cognitively oriented analysis of Russian spoken discourse. The analysis suggested here builds upon the approach to discourse structure and grammar, developed in the collective monograph Kibrik and Podlesskaya eds. 2009, recently published in Russian. The empirical basis of the monograph is the first systematically transcribed corpus of Russian spoken discourse. The kind of discourse is personal stories – narratives told by children and adolescents about their night dreams; hence the name of the corpus – "Night Dream Stories". One of the main goals of that project was to design such a discourse transcription for the bulk of spoken language that would represent cognitive processes in the speaker. We reconstruct how thought unfolds in the course of online language production, and represent by means of a transcript how it gets converted into speech. Of course, in this short paper I can only address very few of the problems discussed in the mentioned study, so an interested reader is referred to the full text of the monograph.

I set three separate, even though closely related, goals for this paper. First, it is necessary to introduce cognitive discourse analysis as a linguistic discipline, since it is not as entrenched a field as, for example, cognitive semantics. In accordance with this goal, sections 2–4 are of a rather general nature. In section 2 I outline the main concerns of discourse analysis as a linguistic discipline. Section 3 lays out some specifics of linguistic study of spoken language. In section 4 basic assumptions of the cognitive approach to discourse are formulated. The second goal, central to this paper, is to formulate the main issues in local discourse structure, and this is done in section 5, beginning with the concept of elementary discourse unit. The third goal is to discuss more specific aspects of local discourse structure. Sections 6 and 7 treat two such aspects: subclausal discourse units and sentences. Main findings and further directions are discussed in the concluding section 8.

---

1 This study has been supported by the Russian Foundation for the Humanities grant #08-04-00165a and by the Russian Foundation for Basic Research grant #09-06-00390a. The approach developed in Kibrik and Podlesskaya 2009 and employed in this paper was a joint venture with a group of colleagues, and I owe a lot to them – in the first place Vera Podlesskaya, as well as Nikolay Korotaev, Alla Litvinenko, Olga Savel'eva-Trofimova and others.

2. Linguistic discourse analysis

Discourse analysis is a huge interdisciplinary field (see Van Dijk ed. 1997, Schiffrin et al. eds. 2001, Renkema 2004, Biber et al. 2007, inter alia). A variety of disciplines, including linguistics, psychology, sociology, anthropology, and many others, contribute to discourse analysis. Here we are interested in *linguistic discourse analysis* – a part of linguistics belonging to the paradigm of constituent-oriented fields, such as phonetics/phonology, morphology, and syntax. Linguistic discourse analysis deals with linguistic constituents of the maximal, unlimited size, that is, whole discourses.

Linguistic discourse analysis has three main issues of concern (Kibrik 2003). The first question, usual in the study of any natural phenomenon, is the question of classification, or *taxonomy*: what types of discourse occur? The second question concerns the internal organization of discourse: what is its *structure*? The third question is: how is discourse related to more local, smaller-scale linguistic phenomena, or, more specifically, how *discourse factors influence such smaller-scale phenomena* (grammatical, phonetic, etc.)? It is useful to briefly overview these three parts of linguistic discourse analysis.

2.1. *Taxonomy of discourses*

There are several ways in which particular discourses can differ from each other. The first major taxonomy is based on the *mode* of discourse, that is, the opposition between spoken and written discourse. (Sometimes one also speaks about the internal mode of discourse (Vygotsky 1934/1994) and about the electronic mode (Herring 1996, Baron 2000), although the latter is actually a submode within the written mode.) A systematic comparison of spoken and written discourse began only in the last several decades, see Zemskaja et al. 1981, Chafe 1982, Miller and Weinert 1998. Written discourse is secondary to the oral use of language in all respects, and thus must be viewed as an adaptation of the basic features of language to the graphic/visual mode of presentation. In discourse studies it is useful to always control for whether data comes from spoken or written language use, as discourse processes may be sensitive to mode. Specifics of spoken discourse analysis are discussed in section 3 below.

The second central taxonomy of discourse is a classification into *genres* (Bakhtin 1953/1986, Swales 1990). Discourse genres are classes of discourses that correspond to certain standard communicative goals, typical of particular discourse communities. Discourse genres crosscut the modes: for examples, the genre of story can appear in the spoken mode and in the written mode, still being the same genre. Genres can be defined in terms of underlying genre schemata – templates that generalize the order of meaningful components, or "moves", in a token of the given genre. For example, according to Chafe (1994:128), stories told by conversationalists follow the following schema: (a) orientation; (b) complication; (c) climax; (d) denouement; (e) coda.

It has been suggested sometimes that genres can be identified on the basis of their lexico-grammatical peculiarities. As the study by Biber (1989) has revealed, this is unlikely: from the point of view of lexico-grammatical characteristics, discourse of one genre can be very heterogeneous. Much more homogeneous are the so-called *passages* (see e.g. Longacre 1983, Smith 2003) – sections of discourses, such as narrative, descriptive, expository, argumentative, and instructive.

Other differences between kinds of discourse, cross-cutting the ones just discussed, relate to the so-called *functional style* and *degree of formality*. The notion of functional style, developed in Russian linguistics (see e.g. Solganik 2003, Dolinin 2004), is defined in terms of typical social domains, such as lay, official, commercial, political, learned, etc. Degree of formality depends on the kind of social relationship between discourse participants, including their relative status, gender, age, etc., and is closely related to the phenomena often subsumed under the notion of politeness (Brown and Levinson 1987, Scollon and Scollon 2001, Bergel'son 2007). Both functional styles and formality are reflected in many lexical, grammatical, and phonetic choices made by the speaker.

The diversity of discourse types is not exhausted by the distinctive features briefly discussed above. The variation of discourse is vast, and its comprehensive discussion is far beyond the goals of this paper. An overarching term sometimes used to capture all possible discourse types (identified on the basis of any or all of the distinctive features) is *register* – see for example Biber 2006. For a more detailed account of discourse taxonomy see Kibrik 2009.

2.2. *Discourse structure*

The second major issue in linguistic discourse analysis is discourse structure. It is useful to distinguish between the global and the local structure of discourse (van Dijk and Kintsch 1983). Global discourse structure is the segmentation of discourse into its immediate constituents or large chunks, such as *paragraphs* in an article (see e.g. Brown and Yule 1983) or *groups of adjacent and interrelated turns* in a conversation (Schegloff 1973). Local discourse structure is the structure consisting of minimal units that belong to the level of discourse; such *elementary discourse units* are discussed in detail in subsection 5.1 below.

There is of course no firm boundary between global and local discourse structures, these are simply two poles in a continuum. Thinking top down, global discourse structure is gradually broken into smaller and smaller units that eventually lead to the local structure. There are some frameworks that provide a unified account for discourse structure, without making a strong distinction between global and local structure. One remarkable framework of this kind is *Rhetorical Structure Theory* developed by Mann and Thompson (1988), also see Taboada and Mann 2006, that represents discourse as a hierarchical network of nodes ranging from one clause to a large discourse chunk; nodes are connected by one and the same kind of discourse-semantic ("rhetorical") relations, irrespective of the size of the node.

2.3. *Influence of discourse factors upon smaller linguistic constituents*

Finally, the third central issue in discourse analysis is the influence of discourse factors upon more local, small-scale linguistic constituents: grammatical, lexical, and phonetic. There is a great variety of such discourse-driven, relatively local phenomena, including word order, choice between finite and non-finite verb forms, use of articles, use of connectors and other cohesion devices, discourse markers, location and direction of discoursal pitch accents, location and length of pauses, etc., etc. In addition, there are also non-vocal linguistic devices, in particularly illustrative gestures, that are also largely motivated by discourse context (McNeill 1992).

Influence of discourse factors upon small-scale linguistic phenomena can be described with the help of the theoretical notion of *choice* (Kibrik 2006). When a speaker is using language, grammar cues him/her to particular choices: which word order to use, where to place a discourse marker and which one, etc. Thus grammar actually is a system that guides speaker's choices. Some choices are relatively rule-based, whereas other choices are rather probabilistic. Discourse-related choices mostly belong to the latter kind: a certain option is not strictly required or strictly ruled out, and more than one option is to a certain extent permissible. One of the clearest examples of discourse-driven choice is *referential choice* – the choice of a lexically full vs. a reduced referential device when mentioning a referent in question; see Kibrik in press.

3. Study of spoken discourse

Linguistics is generally biased toward the study of written language (Linell 1982). This bias is due to several factors, including the ready availability of written language to analysis, as well as the cultural prestige of written word. It is obvious, however, that spoken discourse is the primary, fundamental form of language use. Spoken language is original both in philogeny and in ontogeny, many languages remain unwritten, and even in the most literate cultures people speak more that they write. Therefore, the reasonable design of the science of language must be such that spoken language is studied in the first place, and only thereafter, on the basis of such study, writing is understood as a secondary adaption to the graphic medium. Linguists' knowledge of natural language is doomed to be incomplete and distorted if spoken language is not taken into account.

Scientific research in spoken discourse as a distinct kind of language use began in the 1970s and 1980s (e.g. Ochs 1979, Chafe ed. 1980, Chafe 1982, Biber 1988). There is a fair record of spoken language studies in Russian linguistics, see e.g. Zemskaja ed. 1973, 1983, Zemskaja et al. 1981, Lapteva 1976, Sirotinina 1974, Sirotinina ed. 2003, Miller and Weinert 1998, inter alia. Kibrik and Podlesskaya 2006, Kibrik and Podlesskaya eds. 2009 are cognitively oriented studies of spoken discourse, as it is produced by speakers in real time.

A technical problem immediately arising for anyone interested in spoken language is that, before any analytical study, it must be converted into a certain graphic form, or *transcribed*. An objective graphic rendering of spoken discourse is usually called *discourse transcription* (Du Bois et al. 1992, Edwards 1995, 2001, Makarov 2001 inter alia). There is no single correct kind of discourse transcription; different transcriptional systems vary greatly in terms of the number of phenomena they represent, and this depends primarily on transcribers' research goals. However, apart from the purely segmental, verbal component of talk, the elements typically addressed in discourse transcription include: pausing, prosody, disfluencies, division of the flow of speech into units and other aspects of local discourse structure. The most basic transcription conventions used below are summarized at the end of this paper; for a full presentation of discourse transcription employed here see Kibrik and Podlesskaya eds. 2009. For prior approaches to transcribing Russian speech see Zemskaja and Kapanadze 1978, Svetozarova et al. 1988, Kodzasov 1996b, Odé 2008.

In the remaining part of this paper I only discuss phenomena of spoken language, without mentioning this every time. Of course, this does not imply that written language is less prone to cognitive analysis. Also, of particular interest is a contrastive study of cognitive processes employed in speaking and in writing (see e.g. Chafe 1982, 1994).

4. Cognitive approach in discourse analysis

Cognitive linguistics is the study of how language relates to the human mind. Definitional for this line of research is the so-called *cognitive commitment*, formulated by Lakoff (1990:40). This is the commitment to coordinate linguistic research with what is known about mind and brain from the neighboring sciences also exploring cognition, in particular psychology and neuroscience.

Other important (and early) formulations of similar ideas in modern linguistics belong, inter alia, to Chafe 1974, Kacnel'son 1972, Zvegincev 1996 (written in the 1970s), A. E. Kibrik 1983, van Dijk and Kintsch 1983. Actual work done under the official heading of Cognitive Linguistics does not always live up to the standards of the cognitive commitment, but it seems that having this criterion in mind is extremely important. After all, boundaries between sciences are often accidental, arbitrary, of a historical nature, while the object of study – the mind – is one and undivided.

Language has two major functions and two corresponding modes of existence that can be called, using the computer metaphor, *on-line* and *off-line*. The on-line mode of language is communicative transfer of various kinds of information between individuals. The central phenomenon belonging to this mode is natural discourse, as it unfolds dynamically in real time. The off-line mode of language is information storage. One of the central phenomena characteristic of this mode is the relatively stable system of lexical semantics. Grammar is also often viewed in an off-line way, as a system of mappings between forms and functions.

Cognitive Linguistics, as an established trend of thought in modern science has mostly addressed off-line phenomena. This is true of the well-known work of Lakoff (1987) and Langacker (1987/1991) that is considered foundational for Cognitive Linguistics. Most of the time, the practice of Cognitive Linguistics has ignored natural discourse data and has not been interested in discourse phenomena. Two recent introductions to Cognitive Linguistics (Croft and Cruse 2004, Evans and Green 2006), 355 and 830 pages long respectively, do not mention "discourse" in their subject indexes. (Some exceptions to this tendency have been collected in Goldberg ed. 1996, van Hoek et al. eds. 1999; Németh ed. 2001.)

In turn, those practicing discourse studies quite rarely use explicit cognitive explanations of observed phenomena. For example, in the 851-page-long handbook Schiffrin et al. eds. 2001, terms containing the epithet "cognitive" appear on only 13 pages (according to the handbook's index, see p. 820); among these, cognitive linguistics and cognitive psychology are mentioned once each.

This mutual neglect of cognitive linguists and discourse analysts may suggest that the communicative on-line use of language somehow is "less cognitive" than information storage and conceptualization. But this is obviously false. Discourse is produced in and by a speaker's cognitive system, and a prerequisite for producing it is the speaker's normal assumption that the addressee will properly process it in his/her cognitive system. On-line linguistic phenomena are as cognitive as are off-line phenomena. Excluding discourse processes from the agenda of cognitive linguistics is not justified at all. The adherence of mainstream Cognitive Linguistics, as we know it, to off-line phenomena, is a mere historical accident, due to the genetic connection of its founders to generative linguistics. Despite such limitation in scope, the crucial role of Cognitive Linguistics has been calling attention to cognitive explanation of linguistic phenomena as the central type of explanation.

In fact, the only way to adequately understand discourse processes is through understanding the underlying elements of the cognitive system, such as memory, attention, consciousness, knowledge representation, categorization, etc. Some remarkable examples of cognitively oriented explorations of discourse are provided by such studies as Chafe 1994 (a consistent account of discourse structure and a number of lexico-grammatical and prosodic phenomena as a reflection of information flow in the speaker's consciousness or working memory) and Tomlin 1995 (an experimental study building an important connection between the choice of grammatical subject and the cognitive process of attention focusing). The cognitive pespective is also characteristic of the work by Teun van Dijk, one of those responsible for the formation of discourse analysis as a discipline, see e.g. van Dijk and Kintsch 1983. Also see Nuyts 2007, Sanders and Spooren 2007 for recent discussions of cognitively-minded analyses of discourse.

In my earlier works (Kibrik 2001, Kibrik 2003) I proposed the research field that can be called *cognitive discourse analysis*. In those studies, as well as in this paper, I aim at demonstrating that cognitive discourse analysis is:

- as legitimate as the cognitive approach in semantics
- useful for both the cognitive linguistic agenda and the theory of discourse
- building new links with related disciplines, including cognitive psychology and cognitive neuroscience
- shedding light on fundamental cognitive phenomena such as memory, knowledge representation, on-line discourse planning, etc.

An interesting recent development, building links between cognitive linguistics, discourse analysis and psycholinguistics is a series of experimental studies by Olga Fedorova and her associates, see for example Fedorova et al. 2010. Studies by Fedorova are based on Russian discourse material and address such issues as referential choice, collaborative creation of understanding, and local discourse structure.

5. Local discourse structure: main issues

Like any natural phenomenon, discourse has a certain structure. Discourse structure is hierarchical and ranges from the level of micro-, or local, structure all the way to macro-, or global structure. In this paper I only concentrate on local discourse structure. The discussion of local discourse structure starts off with the notion of elementary discourse unit – a building block of spoken discourse.

5.1. *Elementary discourse units*

Local discourse structure, our central concern in this paper, consists of those units that can be considered minimal, or elementary, with respect to the level of discourse. Spoken discourse is produced not as a smooth flow, but as a series of *quanta* – or, in alternative wordings, segments, steps, portions, pulses, or spurts. Linearly discourse is organized as a sequence of such quanta. Segmentation into quanta is both a theoretical and a practical issue. Theoretically, it is of interest to understand what is the size of quanta and why it is the way it is, how boundaries between quanta are established, etc. From a practical point of view, criteria are needed allowing a transcriber to represent the quantized local discourse structure with a sufficient level of confidence.

In phonetically-oriented studies, quanta are variously called syntagms, intonational phrases, intonational groups, rhythmic groups, intonation units, prosodic phrases/units, etc. Shcherba 1955; Cruttenden 1986; Svetozarova et al. 1988; Chafe 1994: 57; Chafe 2001; Xitina 2004; Krivnova 2007; Stelma and Cameron 2007. In this paper, the quanta of discourse are termed *elementary discourse units (EDUs)*, after Kibrik 2000, 2001, Litvinenko 2000, Kibrik and Podlesskaya eds.2009; cf. also Carlson et al. 2003. I prefer this term because it emphasizes the constructional role of these unit with respect to discourse and because it provides a general framework for analyzing spoken and written language.

EDUs are identified primarily on prosodic grounds, including the following features (see Bryzgunova 1977, Nikolaeva 1977, Svetozarova et al. 1988: 146-147, Krivnova 1989, Levelt 1989: 308, Chafe 1994: 58-59):

- holistic tonal contour: most often in narrative discourse, beginning at the base fequency level characteristic of a given speaker's voice; rise; fall to the bottom of the speaker's frequency range
- presence of an accentual center: typically, a rhematic accent
- typical loudness pattern: lowering towards the end
- typical tempo pattern: acceleration at the beginning, deceleration towards the end
- typical pausing pattern: breathing/planning pauses at the boundaries, no pauses inside EDUs.

For more details on prosodic identification of EDUs see Kibrik and Podlesskaya eds. 2009: 57ff. Here I will only remark, by way of illustration, on one important feature from the above list: tempo pattern. Consider the following example from story 055z of the Night Dream Stories corpus[2].

---

2 EDUs are represented in the corpus and in examples below as individual lines. On the left from the text of a EDU two numbers are indicated: time, in seconds, from the beginning of the story to the beginning of the given EDU; number of the EDU within the given story. On other transcription conventions see the list at the end of this paper.

(1) 055z
5.8 6. ..(0.2) /Вышла с этой \ка-ареты',
7.1 7. ..(0.2) захожу в \ё-олку,

As is shown in Table 1, there is a significant deceleration towards the end of each EDU of this example. The final syllables of EDUs are up to two times longer compared to initial syllables. This effect has long been known to phoneticians and students of speech, see e.g. Bondarko 1977: 165, Zlatoustova 1981, Levelt 1989: 308, Krivnova 2007a inter alia, and it is among the major perceptual cues helping transcribers to identify EDU boundaries.

Table 1. Difference in the length of initial and final syllables in the two EDUs of example (1), in seconds

	First phonetic word	Average syllable length	Last phonetic word	Average syllable length
First EDU	вышла	0.08	кареты	0.17
Second EDU	захожу	0.14	в ёлку	0.25

There exists a remarkable congruence of physiological, cognitive, semantic, grammatical and prosodic aspects of EDU unity. Physiologically, an EDU is typically produced in one exhalation. Cognitively, an EDU is verbalization of one "focus of consciousness", to use the term of Chafe 1994. Semantically, an EDU typically reports about one event or state, and grammatically, typically coincides with a clause (see the next subsection). EDUs most often contain two, three, or four words – these kinds of EDUs comprise 53% of all EDUs in the corpus; EDUs of up to five words constitute 81% of all EDUs.

It appears that the quantized structure of discourse and its specific prosodic aspects have deep neurocognitive, behavioral and evolutionary roots. In interaction with a group of Moscow neuroscientists (led by Konstantin V. Anokhin), we have discovered close similarities between the patterns of speech and the patterns of much more basic, but nevertheless goal-oriented behavior, namely the exploratory movement of rodents in a new environment. This movement is also organized in quanta (spurts), and many features of such quanta are highly resemblant to what is known about speech: they are identified through initial acceleration and final deceleration (cf. the tempo pattern discussed above), they are targeted at an informationally rich goal (analog of rheme in discourse segments), they are separated by periods of freezing (analog of pauses in discourse), they may involve internal haltings, etc. (see e.g. Kafkafi et al. 2001, Mukhina et al. 2003). These facts suggest that this basic neurocognitive structure is among the factors shaping EDUs as fundamental quanta of linguistic behavior. It is likely that language has evolved around the requirements imposed by the mammals' neurophysiology and basic cognition.

EDUs are the minimal steps in which discourse progresses. They prototypically coincide with clauses (Ford, Holmes 1978, Levelt 1989, Chafe 1994, Pawley and Syder 2000, Thompson and Couper-Kuhlen 2005, Kibrik and Podlesskaya 2006). In this respect discourse can properly be understood as a network of clauses. This is true of both spoken and written discourse. Provided that EDUs have a deep neurocognitive basis, a semantic clause must be about how much the human cognitive system can process in one step. Since external reality can be segmented into units of different size (Zacks and Tversky 2001), it is likely that the semantic extent of clauses adapted to the cognitively and neurophysiologically feasible capacity in the course of language evolution.

5.2. *Canonical and non-canonical EDUs*

A classification of EDUs into more and less canonical has been proposed in Kibrik and Podlesskaya eds. 2009: 367ff. In the following excerpt all eight EDUs are clausal:

(2) 056z
2.9 4. на /бал меня пригласили,
4.1 5. ...(0.5) а у меня нет –\пла-атья,
5.8 6. ...(0.8) значит ...(0.6) \думаю:
7.9 7. «/**Ба-ал** ну всё-таки!
8.6 8. \Как я \/пойду?»
9.4 9. ..(0.4) /\Ну-у /одела там /юбку короткую,
11.2 10. /пришла ,
11.5 11. а они там все в таких вот /юбках там-м,

The fact of correlation between EDUs and clauses has been identified for a variety of languages, including English (Chafe 1994: 65-66; Croft 1995: 845), Japanese (Iwasaki 1993), Mandarin (Tao 1996), Thai (Iwasaki 1996), Sasak (Indonesia; Wouk 2008), among others. In the Night Dream Stories corpus clausal EDUs constitute over two thirds of all standard[3] EDUs.

Among the clausal EDUs, the most numerous is the subset of *canonical* EDUs – those that contain a lexically full finite verbal predicate. They constitute 48.9% of all standard EDUs in the corpus. Clausal, but *non-canonical* EDUs include those with a non-verbal predicate (cf. lines 5, 7, 11 in example (2)), those with a verb of being, as well as non-finite clauses. Frequencies of these three subclasses in the corpus are 12.1%, 6.0%, and 0.7%, respectively.

---

3 "Standard" EDUs are those that were produced by the main story teller (that is, not by the interviewer) and did not have any technical defects of recording. There are 3684 standard EDUs in the Night Dream Stories corpus.

Further away from canonical EDUs are those that are not clausal. These can be divided into two big classes: short and long EDUs. *Short EDUs* are a sizeable (26.0%) and internally heterogenous group comprising the following subclasses:

- subclausal EDUs (13.5%)
- regulatory EDUs (6.4%)
- truncated EDUs (5.1%)
- special illocutions, such as forms of address, sound imitations, etc. (1.0%).

Subclausal EDUs are discussed in greater detail in section 6 below. Regulatory are those EDUs that consist exclusively of discourse markers, the most frequent of which is the particle *вот*, see Kibrik and Podlesskaya eds. 2009: 146ff. Truncated EDUs are primarily false starts, resulting from the cognitive process of speaker's self-monitoring: when a speaker discovers that the EDU s/he is producing turns out infelicitous, s/he drops it without concluding and, typically, rephrases it anew (Podlesskaya and Kibrik 2006). What all short EDUs have in common is that the propositional content they convey is less than what is conveyed by an individual clause.

An opposite deviation from canonical EDUs is found in *long EDUs* – those that convey more propositional content than a typical clause. Long EDUs constitute 6.3% of all standard EDUs in the corpus. Reasons leading to the formation of long EDUs are diverse and include verb serialization, verb reduplication, infinitival constructions, epistemic and quotative constructions, etc.; see chapter 7 of Kibrik and Podlesskaya eds. 2009.

5.3. *Prosody*

There are many aspects of prosody that are immediately relevant to local discourse structure. The first phenomenon that can tentatively be included into the realm of prosody is pausing. *Pauses* are periods of time within speech when no verbalization occurs. There are two major classifications of pauses. According to the first one, there exist boundary and internal pauses. Boundary pauses occur at the beginnings of EDUs; cognitively, these are periods when the given EDU is being planned; see e.g. Goldman-Eisler 1972; Levelt 1989: 257ff. Internal pauses are much less frequent (1222 occurrences in the corpus, as opposed to 2304 boundary pauses) and are due to certain difficulties a speaker experiences in the course of EDU production, usually interpretable as hesitation. The second classification distinguishes absolute pauses from filled pauses. Filled pauses consist of an oral (shwa-type) or nasal (*m*-type) sound, or glottal creak. Filled pauses are generally due to hesitation in speech production. Example (3) demonstrates a canonical structure: absolute boundary pauses, not internal pauses. Example (4) illustrates a less canonical situation: a filled boundary pause and several internal pauses, both absolute and filled.

(3) 001z
1.9 3. ..(0.4) /Зайчик вот,
3.0 4. ..(0.4) он был в \лесу.
4.6 5. ....(1.2) А /я была у \р-речки.

(4) 101n
44.9 20 ээ(0.3) 'и когда я \хочу ...(0.7) мм(0.2) '''(0.6) уже /\нажать
. на кнопку /лифта,

Some words in discourse are pronounced more prominently than others – this phenomenon is known as discourse *accents*; see Shcherba 1955, Ceplitis 1974, Nikolaeva 1982, Kovtunova 1976, Kodzasov 1996c, Janko 2001, 2008, inter alia. For example, in (5) out of six words only two bear accents:

(5) 115n
16.9 13. у нашего /подъезда пожарная \машина стоит.

The use of accents is attributed to the cognitive function of new information in studies such as Chafe 1994 and Kodzasov 1996c; other authors, for example Janko 2008, prefer to connect accents primarily to communicative dynamism (theme vs. rheme) and syntactic factors.

Accents are typically realized in conjunction with pitch, that is, directed tone movement on the accented syllable. For example, in (5) the first accented word bears a rising tone and the second a falling tone. Direction of tone is closely connected to the discourse-semantic category of phase, discussed in the next subsection.

Among the accents present in a given EDU one accent typically has a privileged status. It is located on the word or constituent conveying the most important piece of information. This kind of accent can be called primary. In (5), for example, there are two accents, and the second of them is the primary one, as the communicative goal of this EDU is to introduce that firefighters' vehicle. In the first approximation, primary accent is associated with what is known as rheme.

Further kinds of prosodic phenomena, related to local discourse structure, include tonal register, accelerated tempo of speech, phoneme lengthening, phonetic reduction and others; see chapter 8 of Kibrik and Podlesskaya eds. 2009.

5.4. *Phase*

Kodzasov (1996a, 2002) introduced the important notion of phase. *Phase* is a discourse semantic category describing a discourse unit's relationship to its environment. Most generally, the meaning of phase can be formulated as 'non-final' vs. 'final'. A given discourse constituent can be either final or non-final, that is, creating an expectation of continuation. Similar notions have been used by other authors, including "transitional continuity" in Du Bois et al. 1992, textual (in)completeness in Bryzgunova 2003: 895, Janko 2006; the status "open" in Cruttenden 1986.

The rather abstract semantic category of phase can be observed at three hierarchical levels. At the highest level of communicative exchange one can think of illocutionary phase. For example, question is a non-final illocutionary act, calling for a continuation, while response is a final illocutionary act. (Cf. the notion of adjacency pair in Conversation Analysis, see e.g. Levinson 1983: 303ff., and the notion of illocutionary coercion in Baranov and Krejdlin 1992.)

The second-level phase distinction is observed within one illocutionary act – it is a relationship between individual EDUs. An EDU can be either non-final or final within an illocutionary act or sentence (see section 7 below).

Finally, elements of one EDU, such as theme and rheme, are also in the phasal non-final vs. final relationship.

The unity of phase as an abstract semantic category at all of these different levels can be seen due to an identical prosodic manifestation: 'final' is typically encoded by means of a falling pitch accent, and 'non-final' by means of a rising pitch accent. The meaning 'non-final' is associated with the rising pitch accent not inherently, but due to the principle of mirror-image anticipatory adaptation: given that a falling accent is anticipated on the forthcoming final element, the non-final element receives a rising accent by way of automatic adaptation.

Particularly important for local discourse structure is the intermediate-level phase, characterizing individual EDUs as final or non-final. Prototypically, rising pitch accent is combined with the EDU's primary accent, even though such combination is not always the case; see Kibrik and Podlesskaya eds. 2009: 161ff.; Janko 2008: 128-170.

Example (6) below contains a typical example of EDU-level phase: the clausal EDU in line 4 is illocution- (or sentence-) final and, accordingly, contains a falling primary accent (the so-called "period intonation"), while clausal EDUs in lines 2 and 3 are non-final and thus contain rising primary accents (prototypical "comma intonation").

(6) 051z
0.5 2. ....(1.0) мм(0.2) ..(0.4) /сплю я,
2.6 3. 'и /вижу,
3.3 4. ..(0.2) как ...(0.8) мм(0.1) ...(0.9) две-е /женщины какие-то
..(0.1) в /складе в каком-то ложатся \спать.

The importance of phase and its relationship to the notion of sentence is further discussed in Section 7.

6. Retrospective subclausal units

As has been mentioned in subsection 5.2, the most common type of short EDUs are *subclausal EDUs*. These are EDUs characterized by the following features:

- they convey propositional information
- they do not constitute a whole clause
- they are semantically and linearly attached to a certain clausal EDU, here called *base clause*.

Subclausal EDUs are highly frequent and discourse, but relatively little attention has been paid to them in the literature. For this reason it make sense to discuss them here in some detail.

There are three subtypes of subclausal EDUs. The first of these result from the phenomenon of *split* — division of a clause into two EDUs due to a foreign EDU wedging between these two parts. For example, in (7) a relative clause wedges between the subject and the predicate of the clause undergoing split:

(7) 108n
80.0 42. ...(0.5) {ЧМОКАНЬЕ 0.1} ..(0.3) *А /потом вот эти /девочки*
—
82.0 43. которые \стояли' около \двери,
83.6 44. — ...(0.6) ээ(0.1) /вошли ,

There are 44 instances of split in the corpus, and in each instance two subclausal EDUs emerge.

The second subtype of subclausal EDUs are *prospective* subclausal EDUs, or *topics*. An example can be seen in (3) above: the base clause is found in line 4, and the prospective topic in line 3. Topics are infrequent in the corpus: only 20 occurrences are found in the corpus, constituting 0.5% of all standard EDUs; see Kibrik and Podlesskaya eds.2009:108–114.

Much more common (433 instances, or 11.8% of all standard EDUs) is the third subtype of subclausal EDUs: *retrospective* subclausal EDUs. Contrary to topics, retrospective subclausal EDUs follow the base clausal EDU, for example:

(8) 001z
62.8 34. ..(0.3) и они \спали там.
64.0 35. ...(0.7) На \коврике.

Three kinds of retrospective EDUs can be differentiated: echo, increment, and parcellation. I will consider each of these kinds in turn.

6.1. Echo

Echo is a retrospective *elaboration on a realized constituent* of the base clause, for example:

(9) 007z
13.9 7. И /я <\враз> ..(0.2) /подо-ошёл к нему,
15.9 8. ..(0.4) *ну к этому /дереву,*
17.4 9. у которого /сверкало чего-то,

The base clause in line 7 contains an anaphoric pronoun. After the clause has been produced, the speaker judges the pronoun infelicitous: apparently, the referent is not activated enough to be recoverable for the addressee. For this reason the speaker volunteers a retrospective subclausal EDU in line 8 (with a further elaboration by means of a relative clause) that specifies reference. The phenomenon of echo chiefly results from the cognitive process of self-monitoring that is constantly performed by the speaker (Levelt 1989). However, sometimes a pre-planned echo construction is used, for example in the case of lists following a head word, or in the case of contrast, as in (10):

(10) 119n
113.9 96. «\Поведу,
114.3 97. только сегодня не в /тренажёрный зал,
115.4 98. а в \бассейн.»

A classification of the types of elaboration, provided by echos, was proposed in Savel'eva-Trofimova 2008 and Kibrik and Podlesskaya eds. 2009: 116-119. These types include: specification (providing additional information about the concept in question); anti-specification (opting for a more general description than in the base clause); modification (using a synonym or another analogous concept); contrast (replacing the concept in question by an alternative one, as in (10)); antitopic (providing a lexically full description instead of a reduced referential device, as in (9)).

6.2. Increment

Unlike echo, increment is *addition of a constituent* that was not present in the base clause. In the increment construction, a speaker completes a clausal EDU and subsequently realizes that some necessary semantic element is missing from that clause. S/he then "pretends" that the clause was not actually completed and supplies such an additional element in the form of a subclausal EDU. One example of increment was already shown in (8) above. Cognitively, increment is possible if two requirements are satisfied. First, the speaker's production program calls for verbalizing a certain semantic element, closely connected to that implemented in the already produced clausal EDU. Second, such additional information can possibly be built in the given clausal structure. The speaker takes advantage of such compatibility and produces an annex to the previously produced clause.

Most often, the added element is an adjunct, as in (8). The second most common kind of increment is an attribute, as in (11):

(11) 054z
16.9 11. ...(0.8) /Потом ..(0.2) нам встретился какой-то /мостик,
19.5 12. ...(0.6) очень ..(0.2) ' /\узенький,

More rarely, an argument is added to a clause – in such cases the predicate's valency structure is reinterpreted. There are also instances of adding a verbal predicate (nominal predicate is then reinterpreted as a subject) or a particle; see Kibrik and Podlesskaya eds. 2009: 121-122.

6.3. Parcellation

Very similar to increment in formal structure, but distinct from it cognitively is the phenomenon of parcellation. Parcellation happens when a speaker has planned a clausal structure that is too complex, especially in terms of the number of new concepts. Chafe (1994: Ch. 9) has formulated the *one new idea constraint*, according to which a normal EDU should not contain more than one element of new (previously non-activated) information. When a clausal structure with two or more new elements has been planned, a speaker can relieve his/her own as well as the addressee's cognitive tasks, spacing the new elements into separate EDUs. This is what happens in (12), where an adjunct is verbalized as a separate parcellated EDU:

(12) 038z
53.1 24. и я \проснул↑ся,
54.0 25. на /\самом там-м ..(0.2) интересном –месте.

Increment and parcellation, even though resembling each other in formal structure, result from two different cognitive processes: post-hoc addition of a component vs. planned division of a clause, respectively. There exist several criteria helping to tell increment and parcellation apart. First, prosodic properties of the base EDU often demonstrates that no continuation was originally intended, especially if that EDU was pronounced with the "period intonation". Second, the length of boundary pause is important: a null pause strongly suggests that the division was a planned one, while a lengthy pause (0.5 second or longer) is a point in favor of increment. Third, a filled hesitation pause at the beginning of the added component witnesses against the planned parcellation interpretation. Fourth, using an extra EDU between the base clause and a retrospective subclausal EDU suggests that we face an instance of increment. These and other criteria, in conjunction helping to distinguish increment from parcellation with a high degree of certainty, are discussed in greater detail in Savel'eva-Trofimova 2008 and Kibrik and Podlesskaya eds.2009:123-125.

The quantitative data on retrospective subclausal EDUs are provided in Table 2.

Table 2. Types of retrospective subclausal EDUs and their frequencies in the Night Dream Stories corpus (after Kibrik and Podlesskaya eds. 2009: 139)

	Number	Percentage of all standard EDUs
Echo	178	4.8
Increment	149	4.0
Parcellation	97	2.6
Unclear instances	9	0.2
TOTAL	433	11.8

Generally, retrospective subclausal EDUs constitute a highly salient feature of Russian spoken discourse. Apparently, cognitive conditions leading to the formation of such EDUs occur regularly in the course of discourse production. Retrospective subclausal EDUs deserve much higher attention than has been given to them both in discourse studies and in Russian grammar.

7. Is the notion of sentence valid in spoken discourse?

The notion of sentence is among the most central notions in linguistics. Typically, it is explicitly or tacitly assumed to be a basic linguistic unit. This is characteristic of many integral theories of language, and this view is usually borrowed by non-linguists speculating about language. However, if looked at critically, the notion of sentence is difficult to identify in spoken language – the primary form of language use. Spoken language does not contain obvious marks of sentence boundaries, such as periods or question marks. So a series of questions emerge: Does sentence exist in spoken language at all? If yes, how it can be identified? If not, how is it possible that sentence has become such a salient phenomenon both in written language and in linguistic theory?

I address these questions in this section; see Kibrik 2008a for more detail. Note that the validity of sentence is discussed here not in general, but only with respect to one type of discourse: personal stories. Stories are homogeneous in terms of illocutionary function – they are declarative. An intriguing question is whether speakers divide stories into sentence-type units, each of which belongs to the declarative illocutionary function.

As has been pointed out in subsection 5.3, there is a crucial difference between two prosodic patterns: rising and falling primary accents in EDUs. Example (6) provided there illustrates this difference clearly. If stories consisted of sequences such as (6) alone, the notion of sentence in this discourse type would be uncontroversial. It would be natural to propose that every EDU with a falling primary accent concludes a sentence, while EDUs with a rising primary accent are non-final. However, the actual picture is more complex: there are instances of falling primary accent that cannot be judged final.

7.1. *Non-final falling*

As the evidence of the Night Dream Stories corpus demonstrates, speakers systematically contrast two kinds of falling primary accents. Apart from the final falling, as in line 4 of example (6), there are clearly distinct occurrences of non-final falling. Like the rising primary accent, *non-final falling* conveys the discourse-semantic function of non-final phase. Consider an example:

(13) 009z
0.0 1. >>гда-а ··(0.1) ’’(0.1) ··(0.4) ’’(0.1) ’А-аня ··(0.2) забирала меня из детского /с-садика,
4.5 2. /мы с= || ехали на \автобусеw.
6.6 3. ···(0.6) /Я /первая села в \автобус.
9.8 4. ···(0.6) А /тогда уже д= || ··(0.1) закрывались \двери,
12.8 5. ··(0.1) и /’Аня не –успела \сесть.
14.8 6. ···(0.7) Иw мм(0.4) /\когда-а ··(0.2) ’’(0.3) ··(0.4) {ЧМОКАНЬЕ 0.2} ··(0.4) когда я приехала на нашу /остановку’,

In this excerpt, lines 1 and 6 are instantiations of the canonical comma intonation – the rising primary accent, while lines 2, 3, and 5 contain the standard period intonation – final falling. Line 4, however, demonstrates an instance of non-final falling; note the comma mark at the end of that line. A speaker of Russian “feels” that the kind of falling in line 4 of (13) is qualitatively different from final falling, it is somehow “less final”.

Non-final falling is frequent in spoken discourse – its frequency is commensurable with that of the canonical comma and period intonations. In the Night Dream Stories corpus, there are 960 instances of the rising primary accent, 748 instances of final falling, and 557 of non-final falling. These facts call the validity of sentence into question: if one cannot establish sentence boundaries on the basis of the direction of pitch, perhaps this notion becomes ephemeral.

However, in the remaining part of this section I argue that the notion of sentence can still be rescued, at least as a prototype. I first discuss the discourse contexts in which non-final falling occurs and then the prosodic criteria that allow one to distinguish the two kinds of falling.

7.2. *Discourse contexts of non-final falling*

The first, and the most specialized, context of non-final falling can be called *anticipatory adaptive falling*. It is a falling that adapts, in a mirror-image fashion, to the subsequent rising intonation. One example is found in the following:

(14) 064n
7.5 4. ....(1.8) Когда я \ус-слышала,
10.9 5. ...(0.6) что-оо /бомба гремит,
13.0 6. ....(1.8) то \я /побежала к командиру,
16.7 7. и \сказала это.

In this example, line 5, as well as line 6, contains a rising accent, adapting to the final falling in line 7. And the non-final falling in line 4 is a mirror-image adaptation to the rising accent in line 5, as these two lines form a syntactic group – an adverbial clause. Mirror-image adaptive fallings of this kind are a subcase of a more general behavior associated with the notion of phase, see subsection 5.4.

Much more common in the corpus are those instances of non-final falling that are due to the discourse construction that can be called *inset*. Inset is a temporary deviation from the main line of exposition pursued by a speaker. Suppose there is a sequence of EDUs X+Y+Z, such that X and Z a part of the main line, while Y is a specification of X. In such discourse contexts Y typically plays the role of inset. In terms of Rhetorical Structure Theory (Mann and Thompson 1988), X and Z in a story are usually connected by the symmetric relation such as sequence, while Y is attached to Z by an asymmetric relation such as elaboration. Insets vary along a number of parameters, including the size of Y (it can be subclausal, clausal, and multiclausal); see Kibrik, Podlesskaya In press; Kibrik and Podlesskaya eds.2009:140ff. For our current purposes it suffices to say that, most often, insets are prosodically marked as non-final fallings. In example (15) inset in line 2 is a subclausal specification to line 1 that is connected to line 3, belonging, together with it, to the main line of narration:

(15) 032z
0.0 1. /Входит это ...(0.5) /\ма-аль↑чик,
2.1 2. ' ..(0.1) ' ..(0.1) /\ну к \другому,
3.5 3. ..(0.1) *и \говорит*:

The following discourse context, in which non-final falling is instantiated, is *stepwise falling*. This is a situation in which a speaker targets a final falling as the goal of the structure s/he is producing, but reaches this goal in two or more steps. If a given EDU is a part of such a sequence but not its last element, non-final falling is observed. Two such instances are found in the following excerpt:

(16) 059z
0.8 2. ....(1.6) /\Озеро ...(0.5) \какое-то,
4.2 3. ..(0.3) (*Или /\речка,*
5.0 4 *или /\озеро,*
5.4 5. *но по-моему \озеро,*
6.2 6. потому что-о ..(0.2) как-тоw ...(0.6) **маленькое** такое,
8.8 7. \небольшое.)
9.7 8. ....(1.0) /\и-иh ...(0.7) через /него ..(0.3) как-то **бревно** какое-то,
13.8 9. типа \моста.

In (16), final fallings are found in lines 7 and 9. In the preceding parts, lines 5–6 and line 8 respectively, the speaker is preparing to this end, using the non-final falling prosody.

Still another type of non-final falling is found in the contexts that can be dubbed *underconfident falling*. This happens when a speaker, while producing an EDU, is planning a final falling, but changes his/her mind at the last moment and opts for a non-final prosody. Several examples of this behavior, particularly evident in lines 6 and 11, can be seen in (17):

(17) 041z
11.1 5. ...(0.9) /Мы встали на большой /плот,
13.3 6. ..(0.4) /и –переехали на другую \сто↑рону,
15.5 7. ....(1.1) /потом ....(1.6) /мы /вышли из \две↑ри,
19.8 8. ...(0.6) (**Там была /дверь,*
21.1 9. ...(0.8) *такая \/жёлтая,*
22.5 10. ...(0.5) мы \откры↑ли её,
24.0 11. ....(1.3) и \выш↑ли,
26.0 12. ....(2.5) и /мы ...(0.5) оказались на-а ..(0.1) /\ярмарке.

In the instances of inset, stepwise falling and underconfident falling, it is impossible to determine whether the sentence is completed relying on semantic or morphosyntactic criteria. So it is necessary to resort to prosodic diagnostics.

7.3. *Prosodic criteria*

There are several prosodic criteria helping to tell final and non-final falling apart. The first of these is *target frequency level*. In final falling, a speaker tends to eventually bring the EDU's tonal curve to the bottom of his/her voice's comfortable tone range. In contrast, in non-final falling the target frequency level is significantly higher – generally 2 to 4 semitones above the bottom level. Good examples of this criterion's application are found in (16) cited above. Figure 1 shows the oscillogram and the tonogram corresponding to lines 5–9 of that excerpt.

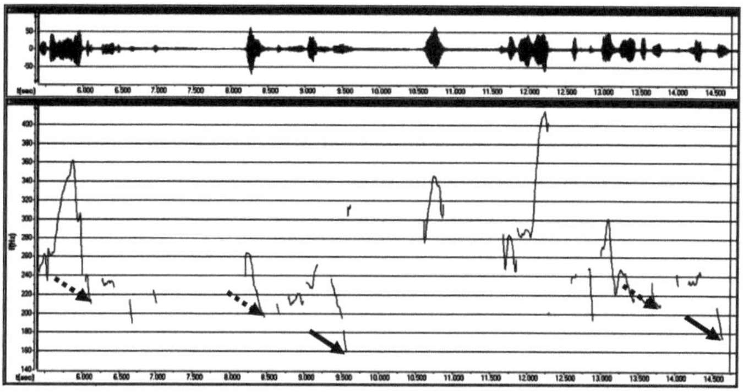

Figure 1. Target frequency levels of fallings in lines 5–9 of example (16)

As can be seen in Figure 1, final fallings in lines 7 and 9 are targeted at the level of 200 to 210 Hz, while non-final fallings in lines 5, 6, and 8, to the level between 160 and 170 Hz. In terms of semitones, the distance between these two levels, or bands, is between 2.5 and 4 semitones. In most speakers, target frequency levels corresponding to the two kinds of falling are clearly contrasted and can be identified with a high degree of confidence.

The second most important criterion is that of *post-accent behavior*. Many speakers, when producing a non-final falling, occasionally use the prosodic pattern that can be called V-curve: after the falling on the accented syllable the tonal curve rises and goes several semitones up; several examples of this behavior were found in (17) above. This is contrasted to final falling, in which lowering of the tonal curve steadily continues on post-accent syllables. An example of such contrast can be seen in (18):

(18) 025z
20.4 8. ....(5.7) /домикh ...(0.6) был /около \реч↑ки,
28.7 9. ....(3.3) /рядом были \-родник-ки,
33.6 10. ..(0.4) и \-ле-ес.

Acoustic graphs corresponding to (18) are found in Fig. 2.

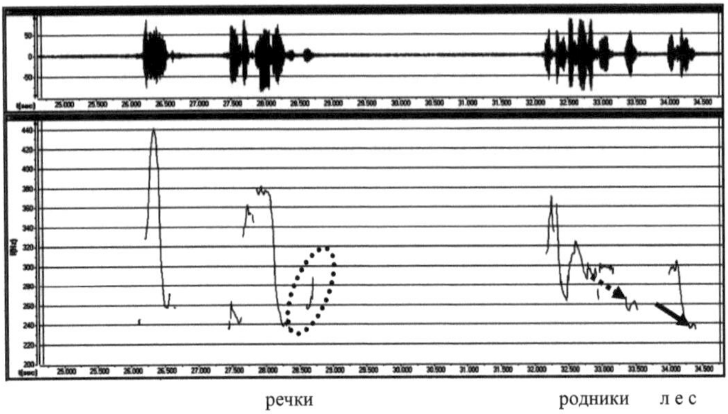

Figure 2. Three fallings in (18)

In Figure 2, a final falling to the level of 230 Hz is observed, and two instances of non-final falling: the falling in line 9 is of an already familiar kind – it differs from the final falling in being targeted at a level about 20 Hz higher. The falling in line 8, in contrast, is targeted at the level indistinguishable from that of the final falling, but there is a curve going 50 to 60 Hz up on the post-accent syllable.

These two major criteria help to tell final and non-final falling apart in the vast majority of instances. Sometimes additional criteria must be drafted, including the pausing pattern, tonal reset vs. latching at the beginning of the following EDU, steepness of falling, interval of falling, and the character of pitch accent; see Kibrik 2008a; Kibrik and Podlesskaya eds.2009: 159–160.

Thus a decision on the kind of falling observed in a particular instance must be made on systematic prosodic grounds. A discourse analyst should explore a given speaker's prosodic system (see Kibrik 2008b) and identify, in particular, his/her target frequency level in final fallings.

7.4. *The status of sentence*

Identification of non-final falling as a standard prosodic and discourse-semantic pattern allows one to maintain the validity of the notion of sentence in spoken language, at least in monologic narrative discourse. Speakers "know" when they are completing a sentence and when they are not, even though in both of these situations they can employ the falling intonation.

We can thus conclude that sentence is a viable notion with respect to the spoken form of language use. However, identification of sentence boundaries is possible only on the basis of a complex analytical procedure, involving construction of a given speaker's prosodic "portrait". In addition, between the prototypes of final and non-final fallings there exist intermediate instances, which makes sentencehood a matter of degree. Therefore, sentence, albeit relevant in spoken language, is a rather elusive, intermediate and non-basic linguistic unit.

This conclusion is further corroborated by high variability of sentences in terms of their content (Kibrik and Podlesskaya eds. 2009: 100). The most frequent kind of sentences consist of one EDU, but these constitute only 39% of all sentences in the corpus (442 out of 1128). Frequency of longer sentences monotonously decreases, all the way to the longest sentence consisting of 30 EDUs. Apparently speakers have significant freedom in the length of sentences they produce – much more freedom than the length of EDUs in words. As has been shown by Korotaev and Kibrik (2008), there is also a great variation between speakers in what length of sentences they prefer: some speakers prefer short, one-EDU-long, sentences, while others follow the clause-chaining strategy (Longacre 1983), in which a whole story is produced as a single sentence.

What is the function of sentence, why do speakers need them at all, especially in the kind of discourse that is homogeneous in terms of illocutionary function? Most likely, the main cause is the intention to ease both a speakers' and an addressee's cognitive processing. Sentences are useful as an intermediate node, potentially larger than a single EDU but shorter than a story's episode. This explanation is probably compatible with the hypothesis by Chafe (1994: 148), according to which the cognitive basis of sentence is the "superfocus of consciousness" – a volume of information exceeding the cognitive limits of simultaneous activation but allowing processing in the semiactive state.

8. Conclusions

In the scientific enterprise known as "Cognitive Linguistics" studies of discourse are not particularly salient. This applies to Slavic cognitive linguistics as well. In this paper I have attempted to outline a research agenda for cognitive discourse analysis.

Discourse analysis is a linguistic discipline addressing three major issues: taxonomy of discourses, discourse structure, and discourse factors of more local linguistic phenomena. This paper has dealt with the second of these issues, namely discourse structure. More specifically, local discourse structure has been discussed. Empirical evidence used here is spoken language; this paper is based on the materials of the systematically transcribed corpus of spoken Russian known as Night Dream Stories (Kibrik and Podlesskaya eds.2009). Analyses proposed here are cognitive in nature: observable linguistic phenomena are explained in terms of cognitive processes, such as on-line production, local planning, self-monitoring, etc. I argue that the cognitive approach is just as important and necessary in discourse studies as it is in the research on off-line semantic phenomena.

The most basic notion around which local discourse structure revolves is elementary discourse unit (EDU). EDUs are minimal steps by which speakers push discourse forward. They can be defined at a variety of levels, including physiological, cognitive, semantic, grammatical, and prosodic, and the remarkable convergence of these aspects suggests that EDUs are truly foundational in the organization of language. Moreover, EDUs have evolutionary precursors in animals' non-linguistic goal-directed behavior. It is very likely that the semantic and grammatical phenomenon of clause emerged as a linguistic adaptation to the quantized character of behavior.

Canonical EDUs are clauses with a finite verbal predicate, and they constitute about one half of all EDUs in discourse. Non-canonical EDUs are necessarily heterogeneous. In particular, there are short EDUs (smaller than a clause) and long EDUs (larger than a clause). Special attention has been given to one subclass of short EDUs – the so-called retrospective subclausal EDUs. This kind of EDU is highly frequent in discourse, but not in linguistic analyses. A cognitively motivated classification of retrospective subclausal EDUs has been proposed, including echo, increment, and parcellation.

Increment and parcellation both encode elements that semantically fit within the preceding clausal EDU but are cognitively distinct. Parcellations result from low-level discourse planning and occur when a planned clausal EDU turns out informationally overloaded; this relates to Chafe's (1994) "one new idea constraint". In contrast, increments appear not due to planning but rather to speaker's monitoring of the produced speech: when the speaker assesses the already constructed clause as informationally insufficient, s/he adds, by means of a post-hoc amendment, an additional element of propositional information.

Apart from the division into EDUs, other crucial aspects of local discourse structure include discoursal prosody (pausing, accents, direction of pitch in accent, etc.) and the discourse semantic category of phase. Particularly important is the phasal characterization of EDUs as final or non-final in the discourse configuration currently produced. This relates to the question of the validity of sentence with respect to spoken discourse. Although canonically there is a prosodic contrast between final (falling primary accent) and non-final (rising primary accent) EDUs, the actual situation is significantly more complex. Instances of non-final falling occur with a high frequency in discourse, and this calls the notion of sentence into question.

Non-final falling occurs in several typical discourse contexts, such as the use of insets or stepwise movement to final falling. In order to identify sentence boundaries in such instances, one need to employ prosodic criteria. It turns out that these criteria, in the first place target frequency level and post-accent behavior, help to discriminate between final and non-final fallings in the great majority of instances. This suggests that the notion of sentence remains valid in spoken language, at least as a prototype, even though prosodic appearance of non-final EDUs may differ strongly.

Still the notion of sentence must be recongnized as a complex and secondary unit. Speakers have a nearly unlimited freedom in the length of sentences they produce, and various speakers employ different strategies in this respect. The underlying function of sentence is to ease cognitive processing in discourse production and comprehension by forming an intermediate hiearchical level between EDUs and more global discourse units.

The most general claim of this paper is that the cognitive approach is a necessary prerequisite for a proper understanding of discourse, as it is produced by a speaker in real time.

Conventions of discourse transcription[4]

Notation	Discourse phenomenon
..(0.1), ...(1.0)	Absolute pauses (length in seconds)
мм(0.3), эээ(0.6), ''(0.1)	Filled pauses (length in seconds)
/	Rising accent
\	Falling accent
–	Level tone accent
/\, \/, /– etc.	Accent with a complex tone
а	Primary accent of an elementary discourse unit (EDU)
↑	Rising of tone curve after the primary accent of a non-final EDU
.	Declarative, final phase
,	Non-final phase
—	Boundary of a split
в \деревне у меня \было так	Lowered tonal register
вокруг	Accelerated tempo
говорит	Reduced pronunciation
{ВЗДОХ}, {ЧМОКАНЬЕ} и т. п.	Non-speech sounds, produced by a speaker

References

Bakhtin, Mixail M. 1953/1986. Problema rechevyx zhanrov [The problem of speech genres]. In Mixail M. Bakhtin (ed.), *Literaturno-kriticheskie stat'i.* Moscow: Xudozhestvennaja literatura, 428-472.

4 This is a partial list of conventions – only the most frequent ones are included. For the full list refer to Kibrik and Podlesskaya (eds.) 2009.

Baranov, Anatolij N. & Grigorij E. Krejdlin. 1992. Illokutivnoe vynuzhdenie v strukture dialoga [Illocutionary coercion in the structure of dialogue]. *Voprosy jazykoznanija* 1992(2). 84-99.

Baron, Naomi. 2000. *Alphabet to email: How written English evolved and where it's heading.* London: Routledge.

Bergel'son, Mira B. 2007. *Pragmaticheskaja i sociokul'turnaja motivirovannost' jazykovoj formy [Pragmatic and sociocultural motivation of linguistic form].* Moscow: Universitetskaja Kniga.

Biber, Douglas. 1988. *Variation across speech and writing*. Cambridge: Cambridge University Press.

Biber, Douglas. 1989. A typology of English texts. *Linguistics* 27. 3-43.

Biber, Douglas. 2006. *Dimensions of register variation: A cross-linguistic comparison.* Cambridge: Cambridge University Press.

Biber, Douglas, Ulla Connor, & Thomas A. Upton. 2007. *Discourse on the move: Using corpus analysis to describe discourse structure.* Amsterdam: Benjamins.

Bondarko, L.V. 1977. *Zvukovoj stroj sovremmennogo russkogo jazyka [The sound system of modern Russian]*. Moscow: Prosveshchenie.

Brown, Gillian & George Yule. 1983. *Discourse analysis.* Cambridge: Cambridge University Press.

Brown, Penelope, & Stephen C. Levinson. 1987. *Politeness: Some universals in language usage.* Cambridge: Cambridge University Press.

Bryzgunova, Elena A. 1977. *Zvuki i intonacija russkoj rechi [Sounds and Russian intonation]*. Moscow: Russkij jazyk.

Bryzgunova, Elena A. 2003. Intonacija i sintaksis [Intonation and syntax]. In: V.A.Beloshapkova (ed.) *Sovremennyj russkij jazyk*. Moscow: Azbukovnik, 869-903.

Carlson, Lynn, Daniel Marcu & Mary Ellen Okurowski. 2003. Building a Discourse-Tagged Corpus in the Framework of Rhetorical Structure Theory. In Jan van Kuppevelt & Ronnie Smith (eds.), *Current Directions in Discourse and Dialogue.* Dordrecht: Kluwer, 85-112.

Ceplitis, L. K. 1974. *Analiz rechevoj intonacii [The analysis of speech intonation]*. Riga: Zinatne.

Chafe, Wallace. 1974. Language and consciousness. *Language* 50. 111-33.

Chafe, Wallace (ed.). 1980. *The pear stories: Cognitive, cultural, and linguistic aspects of narrative production*. Norwood, NJ: Ablex.

Chafe, Wallace. 1982. Integration and involvement in speaking, writing, and oral literature. In D. Tannen (ed.), *Spoken and written language: Exploring orality and literacy.* Norwood: Ablex, 35-54.

Chafe, Wallace. 1994. *Discourse, consciousness, and time. The flow and displacement of conscious experience in speaking and writing.* Chicago: University of Chicago Press.

Chafe, Wallace. 2001. The analysis of discourse flow. In Deborah Schiffrin, Deborah Tannen & Heidi E. Hamilton (eds.), *The handbook of discourse analysis.* Malden, MA: Blackwell, 673-87.

Croft, William. 1995. Intonation units and grammatical structure. *Linguistics* 33. 839-882.

Croft, William, & D. Alan Cruse. 2004. *Cognitive linguistics.* Cambridge: Cambridge University Press.

Cruttenden, Alan. 1986. *Intonation.* Cambridge: Cambridge University Press.

Dolinin, K. A. 2004. Socialisticheskij realizm v lingvistike (k istorii funkcional'noj stilistiki v SSSR) [Socialist realism in linguistics: toward a history of fucntional stylistics in USSR].

In L. A. Verbickaja (ed.), *Teoreticheskie problemy jazykoznanija.* St. Petersburg: Filfak SPbGU, 607-20.

Du Bois, John W., S. Schuetze-Coburn, S. Cumming & D. Paolino. 1992. *Discourse transcription.* Santa Barbara Papers in Linguistics, 4. Santa Barbara: UCSB.

Edwards, J. A. 1995. Principles and alternative systems in the transcription, coding and mark-up of spoken discourse. In G. Leech, G. Myers & J. Thomas (eds.) *Spoken English on computer: Transcription, mark-up and application.* London: Longman, 19-34.

Edwards, J. A. 2001. The transcription of discourse. In D. Schiffrin et al. (eds.), 321-348.

Evans, Vyvyan, & Melanie Green. 2006. *Cognitive linguistics: An introduction.* Mahwah, NJ: Lawrence Erlbaum.

Fedorova, Olga V., Ekaterina A.Delikishkina, Svetlana A. Malyutina, Anna M. Uspenskaya & Anastasija A. Fejn. 2010. Èksperimental'nyj podxod k issledovaniju referencii v diskurse: interpretacija anaforičeskogo mestoimenija v zavisimosti ot ritoričeskogo rasstojanija do ego antecedenta [Experimental approach to reference in discourse: Effects of rhetorical structure on pronoun intepretation]. Aleksandr E. Kibrik (ed.). Computational Linguistics and Intellectual Technologies. *Papers from the Annual International Conference "Dialogue" (2010).* Bekasovo, Moscow region. Moscow: Izdatel'stvo RGGU, 525-530.

Ford, M., & V. M. Holmes. 1978. Planning units in sentence production. *Cognition* 6. 35-53.

Goldberg, Adele E. (ed.). 1996. *Conceptual structure, discourse and language.* Stanford: CSLI.

Goldman-Eisler, F. 1972. Pauses, clauses, sentences. *Language and Speech* 15. 103-113.

Herring, Susan C. (ed.). 1996. *Computer-mediated communication: Linguistic, social and cross-cultural perspectives.* Amsterdam: Benjamins.

Iwasaki, Shoichi. 1993. The structure of the intonation unit in Japanese. In Soonja Choi (ed.), *Japanese and Korean Linguistics* 3. Stanford, CA: CSLI, 39-53.

Iwasaki, Sh. 1996. The syntactic and functional structures of intonation unit in Thai. *Pan-Asiatic Linguistics.* Salaya (Thailand): Institute of language and culture for rural development, 750-761.

Janko, Tat'jana E. 2001. *Kommunikativnye strategii russkoj rechi [Communicative strategies of Russian speech].* Moscow: Jazyki slavjanskix kul'tur.

Janko, Tat'jana E. 2006. Intonacija svjaznogo teksta [Intonation of coherent text]. In: *Komp'juternaja lingvistika i intellektual'nye texnologii. Trudy mezhdunarodnoj konferencii Dialog-2006*, 591-596.

Janko, Tat'jana E. 2008. *Intonacionnye strategii russkoj rechi v sopostavitel'nom aspekte [Intonational strategies of Russian speech from a comparative point of view].* Moscow: Jazyki slavjanskix kul'tur.

Kacnel'son, Solomon D. 1972. *Tipologija jazyka i rechevoe myshlenie [The typology of language and linguistic cognition].* Leningrad: Nauka.

Kafkafi, N., C. L. Mayo, D. Drai, D. Golani & G.I. Elmer. 2001. Natural segmentation of the locomotor behavior of drug-induced rats in a photobeam cage. *Journal of Neuroscience Methods* 109. 111-121.

Kibrik, Aleksandr E. 1983. Lingvisticheskie postulaty [Linguistic postulates]. In: *Mexanizmy vyvoda i pererabotki znanij v sistemax ponimanija jazyka. Trudy po iskusstvennomu intellektu.* Uchenye zapiski Tartuskogo universiteta, vyp. 621, 24-39.

Kibrik, Andrej A. 1999. Cognitive inferences from discourse observations: Reference and working memory. In Karen van Hoek et al. (eds.), 29-52.

Kibrik, Andrej A. 2000. *Cognitive discourse analysis: Some results.* Paper presented at the 7th International Pragmatics Conference. Budapest, July 2000.

Kibrik, Andrej A. 2001. Cognitive discourse analysis: Some results. In T. Enikő Németh (ed.), 164-80.

Kibrik, Andrej A. 2003. *Analiz diskursa v kognitivnoj perspektive [Discourse analysis in a cognitive perspective]*. Doctor of Science (habilitation) Dissertation. Moscow: Institute of Linguistics, RAN.

Kibrik, Andrej A. 2006. Multi-factorial choices in speaking. In Boris M. Velichkovsky et al. (eds.), *The Second Biennial Conference on Cognitive Science*, vol. 1. St. Petersburg: SPb State University, Philological Faculty, 86-87.

Kibrik, Andrej A. 2008a. Est' li predlozhenie v ustnoj rechi? [Is there sentence in spoken language?] In A.V.Arkhipov et al. (eds.) *Fonetika i nefonetika. K 70-letiju Sandro V. Kodzasova.* Moscow: Jazyki slavjanskoj kul'tury, 104-115.

Kibrik, Andrej A. 2008b. Prosodicheskij portret govorjashchego kak instrument transkribirovanija ustnogo diskursa [Speaker's prosodic portrait as a tool for transcribing spoken discourse]. In *Komp'juternaja lingvistika i intellektual'nye texnologii. Trudy mezhdunarodnoj konferencii Dialog-2008*, 195-191.

Kibrik, Andrej A. 2009. Modus, zhanr i drugie parametry klassifikacii diskursov [Mode, genre, and other parameters of discourse classification]. *Voprosy Jazykoznanija* 2009(2). 3–21.

Kibrik, Andrej A. In press. *Reference in discourse*. Oxford: Oxford University Press.

Kibrik, Andrej A. & Vera I. Podlesskaya. 2006. Problema segmentacii ustnogo diskursa i kognitivnaja sistema govorjashchego [The problem of spoken discourse segmentation and the speaker's cognitive system]. In Valery D. Solovyev (ed.), *Kognitivnye issledovanija* 1. Moscow: Institut psixologii RAN, 138-58.

Kibrik, Andrej A. & Vera I. Podlesskaya (eds.). 2009. *Rasskazy o snovidenijax: Korpusnoe issledovanie ustnogo russkogo diskursa [Night Dream Stories: A corpus study of spoken Russian discourse].* Moscow: Jazyki slavjanskix kul'tur.

Kibrik, Andrej A. & Vera I. Podlesskaya. In press. Vstavochnye konstrukcii v ustnom diskurse [Inset constructions in spoken discourse]. In V.Z.Dem'jankov, V.Ja. Porxomovskij. *V prostranstve jazyka i kul'tury: zvuk, znak, smysl*. Moscow: Jazyku slavjanskoj kul'tury.

Kodzasov, Sandro V. 1996a. Kombinatornaja model' frazovoj prosodii [A combinatory model of phrasal prosody]. In: *Prosodicheskij stroj russkoj rechi*. Moscow: IrjA RAN, 85-123.

Kodzasov, Sandro V. 1996b. Foneticheskaja baza dannyx IRJa RAN kak istochnik prosodicheskix svedenij [The phonetic data base of IRJa RAN as a source of prosodic evidence]. In: *Prosodicheskij stroj russkoj rechi*. Moscow: IRJa RAN, 124-141.

Kodzasov, Sandro V. 1996c. Zakony frazovoj akcentuacii [The laws of phrasal accentuation]. In: *Prosodicheskij stroj russkoj rechi*. Moscow: IrjA RAN, 181-204.

Kodzasov, Sandro V. 2002. Fazovaja simvolika tona [Phasal symbolism of tone]. In: N.D.Arutjunova (ed.) *Logicheskij analiz jazyka. Semantika nachala i konca*. Moscow: Indrik.

Korotaev, Nikolay A. & Andrej A. Kibrik. 2008. Illokucija soobshchenija v ustnyx rasskazax [The declarative illocution in spoken stories]. In *Trudy mezhdunarodnoj konferencii "Korpusnaja lingvistika – 2008"*. St. Petersburg: SPbGU, 214-220.

Kovtunova, Irina I. 1976. *Sovremennyj russkij jazyk: porjadok slov i aktual'noe chlenenie [Modern Russian: word order and the theme-rheme articulation]*. Moscow: Prosveshchenie.

Krivnova, Olga F. 1989. Intonacionnoe chlenenie kak sredstvo upravlenija proceduroj smyslovogo raspoznavanija [Intonational segmentation as a tool of managing semantic recognition]. In *Eksperimental'naja fonetika*, 112-124.
Krivnova, Olga F. 2007. *Ritmizacija i intonacionnoe chlenenie teksta (opyt eksperimental'nogo issledovanija). [Rythmization and the intonational segmentation of text: An experimental study]*. Doctor of Science (habilitation) dissertation. Moscow: Moscow State University.
Lakoff, George. 1987. *Women, fire, and dangerous things*. Chicago: University of Chicago Press.
Lakoff, George. 1990. The invariance hypothesis: Is abstract reason based on image-schemas? *Cognitive Linguistics* 1 (1). 39-74.
Langacker, Ronald. 1987/1991. *Foundations of cognitive grammar*. Vol. 1, 2; Stanford: Stanford University Press.
Lapteva, Olga A. 1976. *Russkij razgovornyj sintaksis [Russian colloquial syntax]*. Moscow: Nauka.
Levelt, Willem J. M. 1989. *Speaking. From intention to articulation*. Cambridge, MA: MIT Press.
Levinson, Stephen C. 1983. *Pragmatics*. Cambridge: Cambridge University Press.
Linell, P. 1982. *The written language bias in linguistics*. Linköping, Sweden: University of Linköping.
Litvinenko, Alla O. 2000. *Teorija ritoricheskoj struktury kak universal'nyj instrument opisanija diskursa [Rhetorical Structure Theory as a universal tool of discourse description]*. M.A. thesis. Moscow: Moscow State University.
Longacre, Robert. 1983. *The grammar of discourse*. New York: Plenum Press.
Makarov, Mixail L. 2003. *Osnovy teorii diskursa [Fundamentals of discourse theory]*. Moscow: Gnozis.
Mann, William, & Sandra A. Thompson. 1988. Rhetorical structure theory: toward a functional theory of text organization. *Text* 8 (3). 243-81.
McNeill, David. 1992. *Hand and mind: what gestures reveal about thought*. Chicago: University of Chicago Press.
Miller, Jim, & Regina Weinert. 1998. *Spontaneous spoken language*. Oxford: Clarendon Press.
Mukhina, T. V., A. B. Cherepov & K. V. Anokhin. 2003. *Analysis of animal path trajectories in terms of individual behavioral acts*. Department of Systemogenesis, P. K. Anokhin Institute of Normal Physiology, Russian Academy of Medical Sciences. (Unpublished manuscript)
Németh, T. Enikő (ed.). 2001. *Cognition in language use*. Antwerp: International Pragmatics Association.
Nikolaeva, Tat'jana M. 1977. *Frazovaja intonacija slavjanskix jazykov [Phrasal intonation in Slavic languages]*. Moscow: Nauka.
Nikolaeva, Tat'jana M. 1982. *Semantika akcentnogo vydelenija [The semantics of accentual prominence]*. Moscow: Nauka.
Nuyts, Jan. 2007. Cognitive linguistics and functional linguistics. In Dirk Geeraerts & Hubert Cuyckens (eds.), *Handbook of cognitive linguistics*. Oxford: Oxford University Press, 543-65.
Ochs, Elinor. 1979. Transcription as theory. In E. Ochs, B. B. Schieffelin (eds.) *Developmental pragmatics*. N. Y.: Academic Press. 43-72.

Odé, C. 2008. *Transcription of Russian intonation: A free interactive research tool and learning module*. http://www.fon.hum.uva.nl/tori/.

Pawley, A., & F. Syder. 2000. The one clause at a time hypothesis. In H. Riggenbach (ed.) *Perspectives on fluency*. Ann Arbor: University of Michigan Press, 163-191.

Podlesskaya, Vera I., & Andrej A. Kibrik. 2006. Korrekcija v usnoj monologicheskoj rechi po dannym korpusnogo issledovanija [Repair in oral monologic speech: A corpus study. *Russkij jazyk v nauchnom osveshchenii* 12-2. 7-55.

Renkema, Jan. 2004. *Introduction to discourse studies*. Amsterdam: Benjamins.

Sanders, Ted, & Wilbert Spooren. 2007. Discourse and text structure. In Dirk Geeraerts & Hubert Cuyckens (eds.), *Handbook of cognitive linguistics*. Oxford: Oxford University Press, 916-42.

Savel'eva-Trofimova O.A. 2008. *Neklauzal'nye diskursivnye edinicy (na materiale korpusa "Rasskazy o snovidenijax") [Nonclausal discourse units: Evidence from the Night Dream Stories corpus]*. M.A.Thesis. Moscow: Moscow State University.

Schegloff, Emanuel, & Harvey Sacks. 1973. Opening up closings. *Semiotica* 7 (4). 289-327.

Schiffrin, Deborah, Deborah Tannen & Heidi E. Hamilton (eds.). 2001. *The handbook of discourse analysis*. Malden, MA: Blackwell.

Scollon, Ron & S. B. K. Scollon. 2001. *Intercultural communication: A discourse approach*. Malden, MA: Blackwell.

Shcherba, Lev V. 1955. *Fonetika francuzskogo jazyka [French phonetics]*. Moscow: Izdatel'stvo literatury na inostrannyx jazykax.

Sirotinina, Olga B. 1974. *Sovremennaja razgovornaja rech' i ee osobennosti [Modern oral colloquial speech and its peculiarities]*. Moscow: Prosveshchenie.

Sirotinina, Olga B. (ed.) 2003. *Razgovornaja rech' v sisteme funkcional'nyx stilej sovremennogo russkogo literaturnogo jazyka [Colloquial speech in the system of functional styles of modern literary Russian]*. Moscow: URSS.

Smith, Carlota. 2003. *Modes of discourse: The local structure of texts*. Cambridge: Cambridge University Press.

Solganik, Grigorij Ja. 2003. *Stilistika teksta [Text stylistics]*. Moscow: Nauka.

Stelma, J. H. & L. J. Cameron. 2007. Intonation units in spoken interaction: Developing transcription skills. *Text and Talk* 27.3. 361-393.

Svetozarova, N.D., N.B. Vol'skaja, A.V. Pavlova & L.F. Shitova. 1988. Prosodicheskaja organizacija russkoj spontannoj rechi [Prosodic organization of spontaneous Russian speech]. In N.D. Svetozarova (ed.) *Fonetika spontannoj rechi*. Leningrad: Izd-vo Leningradskogo universiteta, 141-182.

Swales, John. 1990. *Genre analysis: English in academic and research settings*. Cambridge: Cambridge University Press.

Taboada, M. & W. C. Mann. 2006. Rhetorical Structure Theory: Looking back and moving ahead. *Discourse Studies* 8. 423-59.

Tao, H.-Y. 1996. *Units in Mandarin conversation*. Amsterdam: John Benjamins.

Thompson, Sandra A., & E. Couper-Kuhlen. 2005. The clause as a locus of grammar and interaction. *Discourse Studies* 7. 481-506.

Tomlin, Russell S. 1995. Focal attention, voice and word order: An experimental cross-linguistic study. In Pamela Downing & Michael Noonan (eds.), *Word order in discourse*. Amsterdam: Benjamins, 517-54.

van Dijk, Teun A., & Walter Kintsch. 1983. *Strategies of discourse comprehension*. New York: Academic Press.

van Dijk, Teun A. (ed.) 1997. *Discourse studies : a multidisciplinary introduction.* Vol. 1, 2; London: Sage Publications.

van Hoek, Karen, Andrej A. Kibrik & Leo Noordman (eds.). 1999. *Discourse studies in cognitive linguistics.* Amsterdam: Benjamins.

Vygotsky, Lev S. 1934/1994. *Thought and language.* Cambridge, MA: MIT Press.

Wouk, F. 2008. The syntax of intonation units in Sasak. *Studies in Language* 32-1. 137-162.

Xitina, M.V. 2004. *Delimitativnye priznaki ustno-rechevogo diskursa [Delimitative features of spoken discourse].* Moscow: MGLU.

Zacks, Jeffry M., & Barbara Tversky. 2001. Event structure in perception and conception. *Psychological bulletin* 127(1). 3-21.

Zemskaja, Elena A. (ed.) 1973. *Russkaja razgovornaja rech'. [Russian colloquial speech.]* Moscow: Nauka.

Zemskaja, Elena A. (ed.) 1983. *Russkaja razgovornaja rech'. Fonetika. Morfologija. Lexika. Zhest. [Russian colloquial speech. Phonetics. Morphology. Lexicon. Gesture.]* Moscow: Nauka.

Zemskaja, Elena A., M.V. Kitajgorodskaja & E.N. Shirjaev. 1981. *Russkaja razgovornaja rech'. Obshchie voprosy. Slovoobrazovanie. Sintaksis [Russian colloquial speech. General issues. Word formation. Syntax].* Moscow: Nauka.

Zemskaja, Elena A., & L. A. Kapanadze (eds.) *Russkaja razgovornaja rech'. Teksty [Russian colloquial speech. Texts.]* Moscow: Nauka, 1978.

Zlatoustova, Ljubov' V. 1981. *Foneticheskie edinicy russkoj rechi [Phonetic units of Russian speech].* Moscow: Izdatel'stvo Moskovskogo universiteta.

Zvegincev, V.A. 1996. *Mysli o lingvistike [Thoughts about linguistics].* Moscow: Izdatel'stvo MGU.

Metaphorical expressions in legal language: evidence from Polish[1]

Piotr Twardzisz, University of Warsaw

In recent years research in Slavic languages has increasingly been concerned with an area known as Language for Specific Purposes (LSP), which is contrasted with Language for General Purposes (LGP). With an impressive amount of research and a growing number of publications focusing on terminological issues, comparatively little attention has been paid to the figurative side of LSP. One of the flagships of LSP, owing to its hermetic lexicon and frustratingly obscure syntax, is legal language or the language of law. The two terms are not necessarily equivalent. While the former typically covers language used to talk and write about all kinds of legal matters, by professionals and non-professionals alike, the latter is applied to the kind of language usually seen in various documents used by the legal profession: contracts, deeds, court documents, normative acts, statutes, etc. For the sake of simplicity, in this paper the two terms, legal language and the language of law, will be used without much distinction. The terminology pervading legal documents as well as specific fixed expressions, archaic phrases and vocabulary, peculiar word order, multiply-complex sentences, to name but a few, are used as incontrovertible evidence for the LSP-LGP dichotomy by its proponents. Semantic considerations within LSP have focused on the prevailing monosemy of terms as opposed to polysemy, which is pervasive in LGP. The aim of the present paper is to compensate for the shortage of semantic research with an analysis of the metaphoric character of an instance of LSP, namely Polish legal language.

1. Legal language as an example of LSP

One may doubt whether what is referred to as legal language can in fact be identified as (a) *language*, in any sense of the term. In his book titled *Legal language*, Peter Tiersma (1999:142) approaches the issue of what legal language is by asking the question: "what is legal language exactly? A separate language? A dialect? Jargon? Argot? A style?". The author points out that none of the above can be viewed as fully satisfactory, taking each of these separately. In this

---

1 I would like to thank the participants of the 2008 Metaphor Festival in Stockholm for their questions and discussion of an earlier version of this paper. Special thanks to Dr. Michael Hornsby and the anonymous reviewer for reading this paper and suggesting improvements and corrections. All remaining mistakes and inconsistencies are my own responsibility.

paper I assume that legal language constitutes an instance of LSP, without attempting to establish its precise status, that is whether it is a language, dialect, style, etc. Logically then, as an instance of LSP, legal language will count here as a language on a priori grounds.

LSP's distinct character, as compared with LGP, has been observed in a number of studies, notably Lukszyn (2007), Lukszyn and Zmarzer (2001). Among Polish LSP scholars, for example in Grucza (2006), the prevailing opinion is that specialist languages ought to be studied as autonomous languages. According to Grucza (2006:34), specialist languages are not complete languages, in the linguistic sense of the word *language*, and are closely related to some 'general' or 'basic' language. Although the phonetics and grammars of specialist languages overlap with those of their respective general languages fairly completely, their lexicons and textual patterns overlap only partially. Moreover, Grucza maintains that specialist languages are not variants of general languages, because one cannot use either specialist language or general language with reference to the same scope of reality. Therefore, he concludes, specialist languages are specific human languages created by specialists for purposes of professional communication within respective professional communities.

Most of the research in LSP carried out by Polish linguists concentrates on terminology, for instance Lukszyn (2007). LSP scholars have remained rather unaware of the role of figurative language, and metaphor in particular, within LSP. By definition, language used by professionals, for instance, IT specialists, civil engineers, medical doctors or lawyers, is meant to be unambiguous and as precise as possible, which is viewed as the opposite of what constitutes the norm in everyday speech or literary language.

Legal language, as an instance of LSP, has been analyzed from three different points of view: (1) as a phenomenon, (2) translation and (3) terminology.

As a phenomenon, Polish legal language has been analyzed in Pieńkos (2003), Malinowski (2006) and others. Most authors agree on certain distinctive features characteristic of this kind of language, namely a lack of emotional expressions, rich terminology, lexical borrowings, frequent nominalizations, use of connectors for text cohesion, to name but a few. However, there is no unanimous agreement on the monosemy-polysemy divide.

Malinowski (2006:121) points out that despite the existence of many potential meanings the problem of the polysemy of a deontic verb does not pose serious difficulties and is not a hindrance to the understanding of an utterance in legal language, thanks to the contextual identification of meaning. However, one needs to approach differently the meaning of verbs used in the construction of primary utterances. Those are mostly verbs taken from general contemporary Polish, for example *działać* 'act', *określać* 'determine', *podlegać* 'be subject to', *wynagradzać* 'award', *porozumiewać* 'communicate', *wykonywać* 'carry out, implement', *wydawać* 'issue (with)', *wymieniać* 'mention'. When determining the meaning of each of these verbs, one has to rely both on context in which the verb is used and refer to the semantic and syntactic characteristics of the contemporary Polish language. As far as nominal polysemy is concerned, Malinowski (2006:144) brings up the example of the lexeme *głowa* 'head' as seen in (1) the anatomical sense of body part, (2) the tax sense *głowa medium*, (3) the statistical sense *głowa mieszkańca* 'resident's head' and (4) the international sense *głowa państwa* 'head of state'. Furthermore, Malinowski adds that the practical lack of polysemic expressions is one of the distinctive features of legal language which distinguishes it from general language.

In the area of translation, Pieńkos (2003:289) claims that from among all LSPs, legal language is characterised by the largest degree of polysemy which results from the variety of legal systems available all over the world. Some research devoted to translation issues in the area of legal language has been carried out in Jopek-Bosiacka (2006), Kielar (2007), Kierzkowska (2002) and others.

In Polish LSP research, Lukszyn and Zmarzer (2001:52), who specialize in terminological studies, believe that legal texts (similarly to texts concerning politics, art and sport) are characterized by a high frequency of use of the so-called hypo-terms which function as the basic conventional language signs. A lesser degree of frequency is shown by terms. The lowest degree of frequency of use is exhibited by quasi-terms. Hypo-terms can be exemplified by the use of specific verbs, nouns and adjectives which in legal discourse may evoke different connotations to those characteristic of everyday use. As for *term*, Lukszyn and Zmarzer (2001:21) define it as a word, simple or complex, with a precisely defined conceptual structure, virtually unambiguous and not undergoing any interpretation of an emotional character. While terms are theoretical notions, or signs coding conceptual units, nomens are classified as tools or forms of searching for terms (Lukszyn and Zmarzer 2001:32–40), in a way similar to quasi-terms, which also constitute forms of searching for terms. A quasi-term, or a nomen, constitutes an intermediate stage in the transition process, largely consisting in semantic transposition, from a general lexical item towards a term. The final stage in this process, a term, is a stable and entrenched unit or a kind of semantic constant.

Summing up this introductory section, let us note that cognitive linguistics remains underrepresented in the research conducted in Polish legal language. Conceptual-semantic properties of legal quasi-terms are the target of research conducted by Bączkowska (2009) who analyzes legal language through cognitive semantics and corpus linguistics. Also, another notable exception is Kosecki (2003, 2005), who analyzes metaphor and metonymy in legal texts relying, among others, on the theory of metaphor proposed by Lakoff and

Johnson (1980). Regrettably, the above-mentioned publications are difficult to come by for a wider audience.

2. Metaphor in English legal language

Similarly to Polish, there are important publications in English which also address legal language in general or for translation purposes (e.g. Tiersma 1999; Mellinkoff 1963). Groot (1998) argues for legal language being very much a system-bound language while Kocbek (2006) claims no direct correlation between legal language and legal systems. Generally, cognitive linguists have avoided serious involvement with research in legal English. However, there has been some interest in the study of metaphor in legal English shown mainly by legal scholars. The two seminal books in the field of language and law, namely Mellinkoff (1963) and Tiersma (1999), though exhaustively and thoroughly covering the interface between law and language, do not seem to attach any importance to the phenomenon of metaphor in legal language.

A contemporary shift from visually-oriented to aurally-oriented figures of speech in American legal language has been observed by Bernard J. Hibbitts, professor of law at the University of Pittsburgh School of Law, in his paper titled 'Making sense of metaphors. Visuality, aurality, and the reconfiguration of American legal discourse'. Expressions such as *we observe the law*, *in the eyes of the law*, *high courts review the decisions of inferior tribunals*, etc., construe law as a matter of looking. Also, we often talk of law as if it were something to look at, for instance, a text, a body or a structure. Hibbitts further observes that there is a tendency, among feminist, African American, Hispanic and Jewish legal scholars, to employ aural metaphors in their writings. For some of those scholars, law is a matter of *voice*, *speaking* and even, occasionally, of *singing*.

In his paper 'The myth of corporate veil protection: Are your assets at risk?', Graham describes the notion of the *corporate veil* as a legal metaphor which stands for the liability shield that a limited liability company or a corporation is expected to create around its owners and their personal assets. The idea is that whenever a court *pierces the corporate veil*, it disregards the existence of a higher form of business organization as, for instance, the owners of a corporation failed to comply with some corporate requirements and formalities. Given that, the business is no longer seen as a corporation but as a sole proprietorship or general partnership and the owners' personal assets are no longer immune to litigation.

Metaphor in the context of legal language has been given serious treatment by Schane (2006). Apart from pointing out expressions taken from legal discourse which originated as metaphors, for instance, *a meeting of the minds*, *a ripening of obligations*, *a binding agreement*, *a broken contract*, etc., Schane acknowledges the need for another kind of metaphor — the 'legal fiction', in the sense of Fuller (1967). An example of that may be the well-known 'the corporation as a person' fiction, upon which we will focus below. As companies or corporations can, in the legal sense, be involved in many activities believed to be reserved for humans, such as owning property or entering into agreements, it comes as no surprise that many legal systems often deal with corporations as if they were indeed persons.

3. Metaphors in legal contracts in the Polish language

I subscribe to the view expressed by Lakoff and Johnson (1980:5) that: "The essence of metaphor is understanding and experiencing one kind of thing in terms of another." Legal language is full of terms which designate abstract notions, most of which are not directly accessible, and is very poor in the so-called candidates for concepts that are understood directly. It will be proposed here that legal language has to compensate for this shortcoming with metaphorical expressions that provide some means of understanding these inaccessible notions. It is much easier to conceive of a company as a (legal) person than of something as elusive as *company*. How else, if not by means of a metaphorical expression, can one code the relationship between a shareholder and an intangible object such as a share? One way of doing that is to project a share onto a domain where it is viewed as a tangible object that can be held or possessed. For example:[2]

(1) *Każdy Wspólnik może posiadać więcej niż jeden udział.*
'Each Shareholder can hold [lit. possess] more than one share.'

In what follows we will take the view going in the opposite direction to the one stating that metaphor is mere semantic decoration or rhetorical device. What will be proposed below is the view that even hermetic texts such as legal contracts are founded on metaphor which is responsible for the way we construct and understand such texts. In legal discourse most concepts appear to be understood in terms of other concepts, an observation which appears to overlap with that made by Lakoff and Johnson (1980:56).

The fact that people carry out various duties *przed* 'in front of' institutions or offices probably passes as an insignificant and inconspicuous element of larger legal discourse. However, one should note that there is no principled account of this spatial relation obtaining in a context like:

2 The following edited examples are taken from Bogudziński et al. (2000).

(2) *Przed wszystkimi organami, instytucjami oraz urzędami [...]*
'In front of all bodies, institutions and offices [...]'

unless one assumes, as Lakoff and Johnson (1999:34) propose, that "[w]e project fronts and backs onto objects. [...] The concepts *front* and *back* are body-based. They make sense only for beings with fronts and backs." In other words, what we experience here is a mapping of elements between the concept of a human organism, with its front and back, and the concept of an institution, which receives the metaphoric sense of a living being, presumably a human.

3.1. *A COMPANY IS A PERSON*

One of the most common forms of business activity in Poland is *spółka z ograniczoną odpowiedzialnością* 'limited liability company (LLC)', which is frequently conceived of as a person. One way of looking at an LLC is to assume the legal perspective, according to which, an LLC satisfies a numbers of characteristics of a (legal) person, for instance, formation, bearing liability, carrying out various duties, termination, etc. Also, one can view an LLC from the conceptual point of view, which we, as human beings, owing to our embodied minds, are particularly prone to do. As stated by Lakoff and Johnson (1999:3–4), "[t]he mind is inherently embodied. Thought is mostly unconscious. Abstract concepts are largely metaphorical" and "[r]eason [...] arises from the nature of our brains, bodies, and bodily experience." Granted that our minds are embodied, entities such as companies, with which humans interact on a daily basis, are themselves conceived of as embodiments. The conception of interaction between people and companies will be significantly facilitated by conceiving companies as human beings.

Similarly to people, especially productive ones, companies can carry out different activities such as *działać* 'operate', *tworzyć* 'create' and *likwidować* 'close' (entities), *wyznaczać* 'appoint' (people), and so on, as in:

(3) a. *Spółka może działać na obszarze Rzeczypospolitej Polskiej.*
'The Company can operate in the Republic of Poland.'
b. *Spółka może tworzyć i likwidować oddziały.*
'The Company can create and close branches.'
c. *Spółka wyznacza następujące osoby do [...].*
'The Company appoints the following persons for [...].'

People are prototypical owners and possessors of goods and objects. Given the metaphor in question, companies, conceived of as humans, should be viewed as possessors or owners:

(4) *Spółka nie posiada zaległych zobowiązań finansowych.*
'The Company does not have any arrears in its financial liabilities.'

Companies, like humans, show interest:

(5) *Spółka oświadcza, że jest zainteresowana nabyciem [...].*
'The Company states that it is interested in purchasing [...].'

Communication is one of the most fundamental abilities and skills that human beings are endowed with. And so are companies, conceived of as people, when they demand, express consent, appeal, etc., as shown in (5) above and in (6) below:

(6) a. *Spółka ma prawo żądać [...].*
'The Company is entitled to demand [...].'
b. *Spółka wyraża zgodę, aby [...].*
'The Company expresses consent for [...].'
c. *Spółka może się zwracać bezpośrednio do [...].*
'The Company can appeal directly to [...].'
d. *Doradca prawny jest zobowiązany poinformować Spółkę o [...].*
'The Legal Adviser is obliged to inform the Company of [...].'

As observed by Lakoff and Johnson (1980:33), physical objects are commonly specified as persons. In the above examples, also legal entities, that is companies, which are non-physical objects, exhibit numerous traits commonly thought of as human-like. However, one observation has to be made concerning an intrinsic asymmetry between the two participants of this metaphor, namely a company and a person. While a company can be conceived of as a person, the opposite relation does not show with any regularity.[3] A company is an element of the target domain, which is more abstract, and is reached via mapping from the source domain with its more tangible element, namely a person.

3.2. *A COMPANY IS AN OBJECT*

Legal documents often focus on two vital points in a company's life. These are the points in time when a company is set up and ultimately closed down. Conventionally, in Polish companies are portrayed as objects that can be tied up with ropes or strings or untied, respectively, as in:

3 An apparent exception to this principle can be found in a sentence like *Kowalski to solidna firma* 'lit. Kowalski (is) a solid firm', meaning that one can rely on Kowalski in any dealings with him (business or otherwise).

(7) a. *Wspólnicy zawiązują niniejszym spółkę cywilną [...].*
'The Partners hereby establish [lit. tie up] a civil partnership [...].'
b. *Rozwiązanie Spółki jest poprzedzone przeprowadzeniem likwidacji.*
'Dissolution [lit. untying] of the Company is preceded by its liquidation.'

In the above examples, a company is mapped onto an object that is tied up with a rope or string or itself is rope-like. Moreover, one cannot avoid the allusion to the potential lifespan of a company, with its two vital points, birth and death. For this kind of conceptualization to obtain, a company needs to be mapped onto a temporal entity, which is proposed in 3.3 below.

3.3. *A COMPANY IS A TEMPORAL ENTITY*

Somewhat similarly to the COMPANY IS A PERSON metaphor, a company can be conceived of as any entity, not necessarily a human being, whose lifespan is limited. Viewed as such, companies are created at some stage:

(8) a. *Czas trwania Spółki jest nieograniczony.*
'The [lit. time of] duration of the Company is unlimited.'
b. *Wspólnicy postanawiają utworzyć spółkę z ograniczoną odpowiedzialnością.*
'The Partners have decided to establish a limited liability company.

Although there does not seem to be any rigid borderline between the examples under 3.2 and those under 3.3, as both involve some initial creation and final dissolution, there is a significant distinction between them. While the ones under 3.2 do not involve any reference to existence, those under 3.3 do.

3.4. *AN OBJECT IS A PERSON*

In the previous section we have seen that in written legal discourse companies frequently exhibit traits characteristic of animate beings. Now, let us examine the metaphor in which objects or things are conceptually related to people.

Different kinds of documents can be conceived of as consisting of body-parts, for example arms, which can be used to cover something. Similarly to humans, documents such as *pełnomocnictwo* 'power of attorney', *faktura* 'invoice', etc. have arms:

(9) a. *Pełnomocnictwo do sprzedaży Rzeczy, które obejmuje [...].*
'A power of attorney to sell the Item, which covers [lit. embraces] [...].'
b. *Faktura obejmująca wynagrodzenie za sporządzenie analizy [...].*
'The invoice covering [lit. embracing] the remuneration for preparing the analysis [...].'

Towary 'goods', like patients, can undergo examination:

(10) *Badanie towarów i pobieranie ich próbek [...].*
'Testing goods and collecting samples of them [...].'

Regulaminy 'regulations' can commence life in a manner observed in animate beings:

(11) *Niniejszy Regulamin wchodzi w życie z dniem jego uchwalenia.*
'These Regulations come into effect [lit. come to life] on the date they are passed.'

The above metaphorical mapping, also known as personification in more traditional accounts, common in non-specialist contexts, appears to be frequently employed in legal contexts.

As an aside, one may wonder whether the same concept, say PERSON, can on one occasion be vaguer than, say, OBJECT, and on some other occasion, more concrete. If it is true, as Lakoff and Johnson (1980:112) claim, that "[w]e tend to structure the less concrete and inherently vaguer concepts [...] in terms of more concrete concepts, which are more clearly delineated in our experience," then one would expect some stability of directionality. So, if PERSON were less concrete and inherently vaguer than OBJECT, then the PERSON IS AN OBJECT metaphor would be what one would expect. But then, we have established the OBJECT IS A PERSON metaphor, which would entail the opposite viewing, namely OBJECT being less concrete and inherently vaguer than PERSON.

This apparent contradiction should not be seen as such for the simple reason that it is quite normal to view a given entity on a particular occasion as vague and on a different occasion as concrete. The actual choice of which of the two entities involved in a given metaphorical arrangement is vaguer and which is more concrete is left to the conceptualizer who decides how to establish directionality between the two domains. Also, the mutual relations between the concepts of people and the concepts of objects have established themselves well in our conceptual systems. We tend to smoothly conceive of humans as objects (e.g. *Ty szmato!* 'you scumbag/slut!; lit. you rag!') as well as of objects (e.g. cars, bikes, yachts, etc.) as humans. Therefore the above reservation does not seem to pose any serious threat to the entire mechanism of metaphorization.

3.5. *AN INTANGIBLE ENTITY IS AN OBJECT*

It is presumably our natural tendency to construe a less concrete concept in terms of a tangible object with a clear shape that we are familiar with through our experience. That experience with concrete objects constitutes the basis for viewing events, activities and ideas as more substantial beings (cf. Lakoff and Johnson 1980:25, 112). Intangibles such as *nieruchomość* 'property', *przedmiot dzierżawy* 'leased property', *rachunek bankowy* 'bank account', *prawo* 'right', *obowiązek* 'duty', etc. can be conceived of as objects that one can carry:

(12) a. *Przeniesienie własności Nieruchomości na rzecz Spółki [...].*
'The transfer of ownership to the Property to the Company [...].'
b. *Dzierżawca jest zobowiązany przekazać Wydzierżawiającemu Przedmiot dzierżawy w stanie nie pogorszonym.*
'The Lessee shall be obliged to transfer the Leased Property to the Lessor in no worse condition.'

Bank accounts, rights, duties, legal services, economic activity, etc. are all carriable objects:

(13) a. *Rachunek bankowy nr: [...] prowadzony przez bank [...].*
'Bank account no.: [...] kept [lit. led/conducted] by [...].'
b. *Żaden ze Wspólników nie może przenieść praw i obowiązków wynikających z niniejszego Porozumienia na rzecz osób trzecich.*
'Neither Partner can transfer his rights and duties resulting from this Memorandum of Understanding to third parties.'
c. *Doradca prawny będzie prowadził obsługę prawną Spółki.*
'The Legal Adviser shall provide [lit. conduct] legal services for the Company.'
d. *Oświadczam, że prowadzę działalność gospodarczą [...].*
'I declare that I am conducting economic activity [...].'

Intangibles such as costs are burdens or heavy objects:

(14) *Koszty sporządzenia niniejszej Umowy ponoszą Wspólnicy w częściach równych.*
'The costs of preparing these Articles of Association shall be borne by the Partners in equal proportion.'

Duties are things that can be put on, and presumably worn or carried around:

(15) *Nałożenie na Wspólników obowiązku [...].*
'The imposition of a duty on Partners [...].'

Intangibles can be held and/or possessed:

(16) *Każdy Wspólnik może posiadać więcej niż jeden udział.*
'Each Shareholder can hold [lit. possess] more than one share.'

The above metaphorical relation has been effectively established in general language. The following expressions probably appear in a number of languages: *to toss an idea about in one's mind*, *to handle problems*, *dotknąć zagadnienia* 'to touch an issue', *przerzucić obowiązek na kogoś innego* 'to impose [lit. cast] a duty on someone else', etc. Interestingly, the activities performed with the abstract entities, conceptualized as objects, can be limited to a few distinct processes such as moving around, carrying, lifting, holding, etc. In other words, a more abstract concept retrieved from the processes observed is the one designating a change in the location of an object. The metaphorical object needs to be conceived of as the right size for a person to carry out a change of location. To sum up, legal concepts in this domain are seen as concrete objects that people are capable of handling.

3.6. *AN ENTITY IS A CONTAINER*

The idea of viewing people and our visual field as containers probably goes back to Lakoff and Johnson (1980:29–30), who claim that "[e]ach of us is a container, with a bounding surface and an in-out orientation." Also, legal texts provide us with numerous examples in which abstract concepts are viewed as containers with bounding surfaces and in-out orientations.

The following are examples of abstract notions such as *dobra wiara* 'good faith', *sprawa* 'issue' and *głosowanie* 'voting' which are all viewed as container-like objects, as in:

(17) a. *Strony w dobrej wierze uzgodnią pomiędzy sobą pozostałe warunki.*
'The parties shall agree in good faith the remaining conditions.'
b. *W sprawach nie uregulowanych w niniejszym Porozumieniu [...].*
'In issues not governed by this Memorandum of Understanding [...].'
c. *Zarząd podejmuje uchwały w głosowaniu jawnym.*
'The Management Board passes resolutions in open voting.'

Less abstract entities such as different kinds of documents often serve as metaphorical containers. Indeed, documents do contain either the graphic or electronic representation of various messages or bodies of text. Language for general purposes is full of instances explicitly evoking the notion of a container and the contained, for example *list of contents*, *in this book*, *in the second paragraph*, *in the third line from the top*, etc. From there, it is not too far to the conception of documents as known in legal language such as *protokół* 'minutes', *kodeks handlowy* 'commercial code' and *statut* 'statute':

(18) a. *Protokół winien zawierać miejsce i datę [...].*
'The minutes should contain the place and date [...].'
b. *Czynności przewidziane w Kodeksie Handlowym / w Statucie [...].*
'The activities foreseen in the Commercial Code / in the Statute [...].'

Furthermore, one can trace fairly distant metaphorical extensions designating container-like entities in such elusive concepts as *imię* 'name', *forma* 'form' or *tajemnica* 'confidentiality':

(19) a. *Występowanie w moim imieniu przed wszystkimi organami [...].*
'Appearing on my behalf [lit. in my name] in front of all bodies [...].'
b. *Zleceniodawca oświadcza, że działa w formie spółki z ograniczoną odpowiedzialnością.*
'The Client states that he operates in the form of a limited liability company.'
c. *Utrzymanie w tajemnicy wszelkich informacji [...].*
'To maintain [lit. in] the confidentiality of all the information […].'

It is reasonable to assume that most of the metaphorical language involving the idea of containment is used around us largely unnoticed. With a multitude of prepositional phrases headed by the preposition *w* 'in' or expressions involving the verb *zawierać* 'contain', our conception of containers, apart from rather obvious cases, in the above examples and many others, remains subconscious.

3.7. *Residual cases*

Undoubtedly, the metaphorical patterns listed in 3.1-3.6 do not constitute an exhaustive list of all possible metaphorical templates for Polish legal language. Still, through their frequent occurrence in legal documents, they may be taken as highly representative among all metaphoric expressions within Polish legal language.

Apart from what has been established above, at least a few other metaphorical patterns can be proposed owing to their relatively noticeable frequency of use. Take, for instance, *pieniądze* 'money' or *zysk* 'profit' which can be mapped onto the domain of liquids owing to their liquid-like qualities. The notion of pouring is explicitly voiced in the example below:

(20) *Do kapitału zapasowego będzie przelewane przynajmniej 8% czystego zysku rocznego Spółki.*
'At least 8% of annual net profit shall be transferred [lit. poured] to supplementary capital.'

In conceptual opposition to the above categorization, *mandaty* 'mandates' and *kadencje* 'terms in office' are related to flames that get extinguished:

(21) *Mandaty członków Zarządu wygasają z dniem [...].*
'The mandates for members of the Management Board expire [lit. are extinguished] on the date […].'

An interesting metaphorical relation takes place when one refers to the categories *kapitał zakładowy* 'share capital' or *wartość* 'value'. The domain that instantly enters in a relation with *wartość* and such like is a vertical axis against which all measurements and adjustments are made. And so we have:

(22) a. *Wysokość kapitału zakładowego: [...] PLN.*
'[lit. the height of] Share capital of [...] PLN.'
b. *Podwyższenie wartości dotychczasowych udziałów [...].*
'Increasing [lit. raising] the nominal value of the existing shares [...].'

The above conceptualizations are very well established in the language of economics where most abstract concepts, such as inflation, unemployment, sales, etc., are related to vertical axes. Phenomena characteristic of financial markets are described by way of vertically-oriented movements, for instance *spadek* 'drop, slump' or *wzrost* 'increase'. Legal texts which concern financial or economic matters import those economic metaphors.

4. Conclusions

The above list of expressions does not leave much doubt over whether Polish legal language is metaphorical. Polish legal contracts appear to be full of figurative expressions. Six clearly delineated metaphors have been established which are believed to be in frequent use. Those are: A COMPANY IS A PERSON, A COMPANY IS AN OBJECT, A COMPANY IS A TEMPORAL ENTITY, AN OBJECT IS A PERSON, AN INTANGIBLE ENTITY IS AN OBJECT, AN ENTITY IS A CONTAINER. Besides, there are also several residual cases found, which, on the one hand, are not easy to categorize, and on the other, do not cluster many relevant instantiations. They are symptomatic, though, of the open-ended quality of the entire conceptual area which hosts the different metaphorical categories enumerated thus far. So, the above list of metaphorical templates established for Polish legal language for contracts cannot be taken as exhaustive and exclusive. Nevertheless it can be presumed as dominating in this type of discourse.

An interesting question to ask here is whether metaphor is a mechanism whose validity within legal language can be compared to the role of metaphor in some other types of specialist discourse, for instance, political, journalistic, economic, etc. So far there has not been much recognition of the role of metaphor in legal language voiced in the linguistic literature. This apparent shortage of research in the area of metaphorical legal language may be dictated by the still prevailing, and too readily accepted, view that legal language is a hermetic language known for its semantic precision which stands in opposition to a phenomenon as elusive as metaphor. This deeply-rooted assumption about metaphor-free legal language seems to have rubbed off on many researchers, both on the applied linguistic front as well as the legal one.

Another essential question may be whether metaphor is an omnipresent phenomenon characteristic of legal language per se or only certain instances of legal language. This short article addresses only selected cases of Polish written legal discourse, further narrowed down to several types of legal contracts. To posit a metaphorical nature of legal language as such, a far more detailed and widespread analysis would have to be undertaken. A somewhat similar attempt has been made in Twardzisz (forthcoming) based on a sample of a few dozen legal contracts in English, with similar findings.

References

Bączkowska, Anna. 2009. Językoznawstwo kognitywne i korpusowe w badaniach nad przekładem języka prawnego na przykładzie *Kodeksu Spółek Handlowych* i *Kodeksu Pracy*. [Cognitive linguistics and corpus linguistics in research on the translation of legal language: the case of *Polish Commercial Companies Code* and *Labour Code*]. http://abscience.org/mojeartykuly/documents/translat_Torun.doc (accessed 30 March 2009).

Bogudziński, Jacek, Konrad Buczkowski & Andrzej Kaznowski. 2000. *Compendium of legal templates*. Warsaw: C. H. Beck.

Fuller, Lon L. 1967. *Legal fictions*. Stanford, CA: Stanford University Press.

Graham, Stanford A. 2008. The myth of corporate veil protection: Are your assets at risk? http://bulletproofveil.com/pdf/whitepaper.pdf (accessed 23 July 2008).

Groot, de Gérard-René. 1998. Language and law. In E. H. Hondius (ed.), *Netherlands Reports to the Fifteenth International Congress of Comparative Law*, 21–32. Antwerp and Groningen: Intersentia Rechtswetenschappen.

Grucza, Sambor. 2006. Idiolekt specjalistyczny – idiokultura specjalistyczna – interkulturowość specjalistyczna [Specialist idiolekt – specialist idioculture – specialist interculturalness]. In Jan Lewandowski, Małgorzata Kornacka & Waldemar Woźniakowski (eds.), *Teksty specjalistyczne w kontekstach międzykulturowych i tłumaczeniach* (Series: Języki Specjalistyczne 6), 30-49. Warsaw: Department of Specialist Languages, University of Warsaw.

Hibbitts, Bernard J. 1994. Making sense of metaphors. Visuality, aurality, and the reconfiguration of American legal discourse. 16 *Cardozo Law Review* 229. http://faculty.law.pitt.edu/hibbitts/meta_int.htm (accessed 5 July 2008).

Jopek-Bosiacka, Anna. 2006. *Przekład prawny i sądowy* [Legal and court translation]. Warsaw: Państwowe Wydawnictwo Naukowe.

Kielar, Barbara. 2007. Wiedza specjalistyczna tłumacza – na przykładzie tekstów prawnych [The translator's specialist knowledge – legal texts in focus]. In Małgorzata Kornacka (ed.), *Teksty specjalistyczne jako nośniki wiedzy fachowej* (Series: Języki Specjalistyczne 7), 19-33. Warsaw: Department of Specialist Languages, University of Warsaw.

Kierzkowska, Danuta. 2002. *Tłumaczenie prawnicze* [Legal translation]. Warsaw: Wydawnictwo Polskiego Towarzystwa Tłumaczy Ekonomicznych, Prawniczych i Sądowych (TEPIS).

Kocbek, Alenka. 2006. Language and culture in international legal communication. *Managing Global Transitions* 4 (3). 231–247.

Kosecki, Krzysztof. 2003. Metafora pojęciowa *Czas to Pieniądz* a struktura umów o pracę w prawie polskim i europejskim [Conceptual metaphor *Time Is Money* and the structure of employment contracts in Polish and European law]. *Rozprawy Humanistyczne* I. 73–78. Włocławek: Higher Vocational School in Włocławek Press.

Kosecki, Krzysztof. 2005. On metonymies in the names of selected legal contracts. *Rozprawy Humanistyczne* IV. 189–96. Włocławek: Higher Vocational School in Włocławek Press.

Lakoff, George & Mark Johnson. 1980. *Metaphors we live by*. Chicago: The University of Chicago Press.

Lakoff, George & Mark Johnson. 1999. *Philosophy in the flesh. The embodied mind and its challenge to western thought*. New York: Basic Books.

Lukszyn, Jerzy. 2007. Tekst specjalistyczny pod lingwistyczną lupą [Specialist text through a linguistic magnifying glass"]. In Małgorzata Kornacka (ed.), *Teksty specjalistyczne jako nośniki wiedzy fachowej* (Series: Języki Specjalistyczne 7), 51-70. Warsaw: Department of Specialist Languages, University of Warsaw.

Lukszyn, Jerzy & Wanda Zmarzer. 2001. *Teoretyczne podstawy terminologii* [Theoretical foundations of terminology]. Warsaw: Department of Specialist Languages, University of Warsaw.

Malinowski, Andrzej. 2006. *Polski język prawny. Wybrane zagadnienia* [Polish legal language. Selected issues]. Warsaw: LexisNexis.

Mellinkoff, David. 1963. *The language of the law*. Boston: Little, Brown & Co.

Pieńkos, Jerzy. 2003. *Podstawy przekładoznawstwa. Od teorii do praktyki* [Foundations of translatology. From theory to practice]. Cracow: Kantor Wydawniczy Zakamycze.

Schane, Sanford. 2006. *Language and the law*. London and New York: Continuum.

Tiersma, Peter. 1999. *Legal language*. Chicago: The University of Chicago Press.

Twardzisz, Piotr. (forthcoming). Metaphors in commercial contracts. In Nils-Lennart Johannesson, David C. Minugh & Christina Alm-Arvius (eds.), *Selected Papers from the Stockholm 2008 Metaphor Festival*. Stockholm: Department of English, Stockholm University.

Metaphor validation in polysemous structures: a case study of Serbo-Croatian bilingual dictionaries

Danko Šipka, Arizona State University

1. Introduction

The present paper reports empirical research on the materials from a Serbo-Croatian – English dictionary of non-standard lexicon (Šipka 2000) aimed at applying and empirically testing the theoretical apparatus presented in Steen (2007). The materials stemming from Šipka (2000) were selected for this case study owing to quantitative and the qualitative abundance of metaphoric extensions in its entries.

The presentation in this paper will take the following course. Theoretical considerations will be addressed at the very outset with equal attention devoted to Steen's account of the models of metaphor and metalexicographic principles of dictionary compilation. This will be followed by the analysis of empirical materials from the aforementioned dictionary. The final segment of this paper will provide concluding remarks about the usefulness of particular models in practical lexicographic work.

2. Theoretical Considerations

Steen (2007:3-26) establishes three major delineations: grammar versus usage, language versus thought and symbol versus behavior and employs them as follows: "Combining these contrasts leads to a 2*2*2 configuration, producing eight distinct areas for locating different types of language research" (Steen 2007:4). Each of the eight fields in the grid generates its pertinent research question.

While other research questions advanced by Steen may be of interest in lexicography, the most pertinent one, at least from the standpoint of the dictionary production process, is the following:

"When does a conceptual structure related to a conventionalized linguistic form count as metaphorical?" (Steen 2007:14).

The answer to this question translates into concrete dictionary compilation decisions, such as sense identification and grouping. The aforementioned question and ensuing lexicographic policies are rooted in the model of metaphor selected for the project.

Steen (2007:48-57) identifies the following four models of metaphor:

a. The two-domain approach, which "...models metaphor as a mapping from a conceptual source domain to a conceptual target domain. When we speak of *defending* a position in an argument, Conceptual Metaphor Theory says we conceptualize ARGUMENT AS WAR, with war functioning as a conceptual source domain which informs and structures our view of the conceptual target domain of argumentation" (Steen 2007:49).

b. The many-space approach, which "...replaces the two conceptual domains with at least four conceptual space. Two of the four spaces correspond to the source and target domains in the classic model, and two other spaces capture the common ground between the two spaces (the generic space) as well as the emergent structure deriving from the mapping in context (the blended space)" (Steen 2007:51).

c. The class-inclusion approach, which "... is based on the interplay between three conceptual categories. In the expression *My job is a jail*, the class-inclusion approach holds that there is a conceptual target category for the topic term job, a conceptual source category for the vehicle term jail and a conceptual superordinate category 'of things or situations that the metaphor vehicle exemplifies' (e.g., situations that are confining, oppressive, etc.)" (Steen 2007:52).

d. The career of metaphor approach, which allows different models for different kinds of metaphors: "...for novel metaphor, only two conceptual structures are required, corresponding to source (or 'base') and target in the two-domain approach; but for conventional metaphor, three concepts are required, corresponding to the topic concept and the subordinate and superordinate vehicle concepts in class-inclusion theory" (Steen 2007:53).

The question at hand here is if the two-domain approach, being the simplest one, suffices to solve all required lexicographic tasks, or does one, alternatively, need to deploy other models. Lexicography, being an applied linguistic discipline, prefers solving problems with the simplest possible solution, as it involves least time and effort.

The purpose of a bilingual dictionary is expressed succinctly in Zgusta's seminal Manual of Lexicography: "The basic purpose of a bilingual dictionary is to coordinate with the lexical units of one language those lexical units of another language which are equivalent in their lexical meaning" Zgusta (1971:294) and repeated in Zgusta (2006:230): "The task of the bilingual lexicographer is to find such lexical units in the target language as are equivalent to the lexical units of the source language, and to coordinate them."

In order to establish equivalence, the lexicographer needs to identify senses of L1 entries.

The difficulty of this task is emphasized in theoretical works and empirical studies alike. Yong and Peng (2007:143) note: "Meaning discrimination is one of the most intricate tasks involved in bilingual lexicographic communication" and Kilgarriff (1998) in his survey of lexicographers finds "Splitting: identifying the senses" to be the second most difficult segment of lexicographic work (with "Finding right wording" to be the single most difficult task).

The dictionary under discussion here is intended for reception only, which means that the problem of differentiating senses being different from finding the right equivalent, rightfully advanced by Ривелис (2007:164-167) does not bear reference here.

The aforementioned dictionary project, Šipka (2000), followed recommendations elaborated in Meer (1999:205), summarized as follows: "In my view dictionary-makers will be much better placed to alert learners to the subtle shades of meaning of many non-literal uses."

3. Empirical Data

The dictionary selected for empirical research, Šipka (2000), features extraordinary abundance of polysemous structures, where over 40% of all senses can be ascribed to metaphorical extensions.

One of the most prominent features of nonstandard lexemes is the abundance of semantic extensions, most notably the metaphoric ones. This, coupled with extreme volatility of nonstandard lexemes and their meanings necessitate the need to clearly mark metaphorical extensions and thus "train" the user to recognize any future extensions not attested in the present dictionary.

The distribution of different factors in the formation of the nonstandard lexicon in question is provided in Table 1. Degrees of semantic extension refer to the depth of extension in the dictionary entry microstructure. Thus, in *ribica* 'small fish' > 'vagina' > 'female', 'vagina' is 1° semantic extension and 'female' is 2° semantic extension.

Table 1. The origins of nonstandard Serbo-Croatian

Factor	Count	Percent
1° Semantic Extension	8861	37.79
2° Semantic Extension	1191	5.08
3° Semantic Extension	215	.92
4° Semantic Extension	45	.19
5° Semantic Extension	12	.05
6° Semantic Extension	6	.026
Modification	1923	8.20
Modification, 1° Semantic Extension	848	3.62
Modification, 2° Semantic Extension	80	.34
Modification, 3° Semantic Extension	11	.05
Modification, 4° Semantic Extension	5	.02
Modification, 5° Semantic Extension	3	.01
Modification, 6° Semantic Extension	2	.01

Factor	Count	Percent
Modification, Word-Formation	30	.13
Modification, Word-Formation,1° Semantic Extension	36	.15
Word-formation	1295	5.52
Word-Formation, 1° Semantic Extension	3662	15.62
Word-Formation, 2° Semantic Extension	375	1.60
Word-Formation, 3° Semantic Extension	77	.33
Word-Formation, 4° Semantic Extension	12	.05
Word-Formation, 5° Semantic Extension	3	.01
Word-Formation, 6° Semantic Extension	1	.01
Word-Formation, 7° Semantic Extension	1	.01
Lack of any modifications/extensions	4753	20.27

The semantic extensions identified in the dictionary point to the following three relevant distinctions explained in turn:

1. substandard-internal	standard→substandard
2. semantic-internal	phonetic→semantic
3. culture-bound	more general

The first category above points to the fact that a substandard semantic extension can have its source domain in substandard or in standard lexicon. The first type of extension can be seen in *munđaroš* 'cheater, swindler' with its destination domain 'smuggler' (both meanings are substandard) or *eki* 'pragmatic operator used to express surprise' and 'question tag' (both meanings are substandard). The second situation can be illustrated by: *devojka* 'young female' (standard) and 'prostitute' (substandard) or standard *taster* 'key (e.g. on a keyboard)' and substandard 'informer'.

The first value within the second category above has been thoroughly elaborated in numerous studies. We have, for example, the metaphor *Udaljeno mesto je kao rupa* 'A remote place is like a hole' in an internal semantic extension *rupa* 'hole' - 'remote place, hellhole' or *Riđi ljudi izgledaju kao mrkva* 'Redheads look like carrots' which is behind the extension *mrkva* 'carrot' and 'redhead, carrot top'. The second value is seldom mentioned in the literature. This concept covers the situations in which the partial similarity of forms triggers semantic extension. Thus, substandard Serbo-Croatian has a series of lexemes to denote a homosexual with the following source domains: *pedala* 'pedal', *pedikir* 'pedicure', *pešak* 'pedestrian', *peškir* 'towel', *Peđa* 'proper name', *Pežo* 'Peugeot, make of car'. The only link between these meanings and 'male homosexual' is that they are phonologically similar to *peder*, the most common substandard word for a male homosexual.

Finally, the extension can be culture-bound or more general, which can be seen from the examples of culture-bound metaphoric extensions presented in Table 2:

Table 2. Culture-bound metaphorical extensions

lexeme	source domain	link	target domain
1. krmak	'boar' (the animal)	a non-Muslim is like a boar	'non-Muslim'
2. leš	'corpse'	a senior citizen is already a corpse	'senior citizen'
3. pop	'priest'	a homosexual is like a priest (no wife)	'homosexual'
4. dijete	'child'	a female child is just a child, i.e. not the (male) child one wishes to have	'female child'

Example 1. is determined by one's religious/ethnic affiliation (Bosnian Muslim) and certain religious practices. Example 2. is determined by one's membership in an age group, Example 3. by the Catholic religion (celibacy is compulsory for Catholic priests), finally, sexism is the basis for Example 4.

On the other hand: *rasturiti* 'destroy' which extends into 'win' or *plakati* 'cry', which extends into 'complain' has much wider cross-cultural distribution for the simple fact that they are based on certain (almost) universal cognitive and affective faculties of the human species.

The (sub)cultural webs which determine any particular semantic extension can be very intricate and non-transparent for the individuals outside that given (sub)cultural milieu. The lexeme *fataljist* 'person who cleans parks', for example, is determined by the following linguistic facts: a. the existence of a substandard verb *fatati* 'to catch', b. the existence of the noun *list* 'leaf', c. the existence of a sterotype that an Albanian accent turns *l* into *lj*, which changes *list* into *ljist*, d. the existence of the noun *fatalist* 'fatalist'. On the other hand, there is a common perception that Albanians (therefore b. above) performed such jobs in the former Yugoslavia. The lexeme *fatalist* is changed into 'Albanian' *fataljist*, re-segmented into: *fata-ljist* 'catches leaves' but still remains present (a person with the socialeconomic status of a garbage man is perceived as a fatalist). The dictionary corpus contains numerous examples of such intricate networks.

The most frequent instances links for semantic extensions include:

All females are desirous of sex
All females are prostitutes
Elderly people are already dead
A female is equal to her vagina
A handicapped person is stupid
Inhabitants of the countryside are mentally deficient
Male homosexuals are ridiculous and non-collegial
Outgroup members are lower beings
Sexual intercourse is an act of violence with a male in the active role

For example, the verbs meaning 'chop', 'drill', 'gnaw', 'hit', 'kick', 'nail', 'poke', 'prick', 'stab', 'slaughter', 'tear apart', etc. are all used to designate sexual intercourse.

Again, the nature and frequency of these extensions offers an insight into the conceptions rooted in different Serbo-Croatian cultures. The most salient features mirrored in the language are the lack of tolerance for anything beyond traditional patriarchal patterns of behavior and beliefs, as well as strong male chauvinism.

The hitherto account of metaphoric extensions relied on the two-domain approach and indeed a vast majority of cases can easily be processed using the two-domain model. Let us now turn to more intricate cases to see if the use of other models would be warranted.

There exists a class of metaphoric extensions exemplified by *anđeo* 'angel' in standard > 'policeman' in slang, *kustos* 'custodian' in standard > 'waiter' in slang, *zid* 'wall' in standard > 'neighbor' in slang, or *celivati* 'to kiss' in standard > 'to hit, to punch' in slang. The extension here entails irony, which would suggest the need for a blended space and thus the many-spaces approach.

Similarly, one can identify cases where the class-inclusion approach is warranted, e.g. *robija* 'strict prison' in standard > 'school', 'hard work' in slang or *zatvor* 'prison' in slang > 'school' in slang.

However, such situations where other models may be appropriate are extremely rare (around 1% of all semantic extensions) and being unfamiliar with the appropriate model even in such cases is not likely to lead into failure to identify semantic extensions.

4. Conclusions

This particular case study shows that, purely theoretically, one can find examples which would call for all four models of metaphor, given that the fourth model incorporates the three others. However, for all practical reasons, lexicographers and their users can function perfectly fine using, even intuitively, the two-domain model alone. What is much more important is the fact that metaphors can be embedded into quite intricate culture-bound networks and that they can interact with other linguistic factors, as shown in Table 2 and Table 1 respectively. Lexicographers should thus be trained and users should be educated about this. A clear grasp of the two-domain model along with an awareness of the complex interaction of phonology and semantics, language and culture are most likely to be translated into sound lexicographic practice.

References

Kilgarriff, Alan. 1998. The hard parts of lexicography. In *International Journal of Lexicography*; 11: 51 – 54.

Meer van der, Geart. 1999. Metaphors and Dictionaries: The Morass of Meaning or How to Get Two Ideas for One, *International Journal of Lexicography*, Vol. 12, No. 3, 195-208

Новиков, Л.А. 1982. *Семантика русского языка*. Москва: Высшая школа.

Ривелис, Евгений. 2007. *Как возможен двуязычный словарь*, Stockholm: Acta Universitatis Stockholmiensis.

Steen, Gerard J. 2007. *Finding Metaphor in Grammar and Usage*. Amsterdam: John Benjamins.

Šipka, Danko. 2000. *SerboCroatian – English Colloquial Dictionary*. Washington, DC: Dunwoody Press.

Yong, Heming and Jing Peng. 2007. *Bilingual Lexicography from a Communicative Perspective*. Amsterdam: John Benjamins.

Zgusta, Ladislav. 2006. *Lexicography Then and Now. Selected Essays*. Tübingen: Niemeyer.

Zgusta, Ladislav. 1971. *Manual of Lexicography*. The Hague: Mouton.

Gerhild Zybatow / Philip Dudchuk / Serge Minor / Ekaterina Pshehotskaya (eds.)

Formal Studies in Slavic Linguistics

Proceedings of Formal Description of Slavic Languages 7.5

Frankfurt am Main, Berlin, Bern, Bruxelles, New York, Oxford, Wien, 2010.
374 pp., num. tab. and graph.
Linguistik International. Editor in charge: Lew Zybatow. Vol. 25
ISBN 978-3-631-58680-8 · hardback € 62,80*

This book contains a collection of papers presented at the *European Conference on Formal Description of Slavic Languages 7.5* which took place in Moscow in December 2008. It brings together formal research from various areas of theoretical linguistics based on the data from Slavic languages. The papers in this book are organized into three parts. Part One includes nine papers in formal syntax and morphology of Slavic languages. Part Two is devoted to formal semantics and pragmatics and contains five papers. Finally, another five works dealing with interface issues constitute Part Three of this volume.

Content: Syntax · Binding · Agreement · Case · Gapping · Clitic Doubling · Correlatives · Morphology · Gender · Semantics · Indefinites · Reciprocals · Aspect · Argument Structure · Scrambling · Information Structure · Pragmatics · Slavic Languages · Formal Description of Slavic Languages

Frankfurt am Main · Berlin · Bern · Bruxelles · New York · Oxford · Wien
Distribution: Verlag Peter Lang AG
Moosstr. 1, CH-2542 Pieterlen
Telefax 00 41 (0) 32 / 376 17 27

*The €-price includes German tax rate
Prices are subject to change without notice

Homepage http://www.peterlang.de